CERAMIC MATERIALS – PROGRESS IN MODERN CERAMICS

Edited by **Feng Shi**

Ceramic Materials – Progress in Modern Ceramics
Edited by Feng Shi

Published by InTech

Janeza Trdine 9, 51000 Rijeka, Croatia

Publishing Process Manager Ivona Lovric

Technical Editor Teodora Smiljanic

Cover Designer InTech Design Team

Additional hard copies can be obtained from orders@intechopen.com

Ceramic Materials – Progress in Modern Ceramics, Edited by Feng Shi
p. cm.
ISBN 978-953-51-0476-6

For EU product safety concerns:
IN TECH d.o.o., Prolaz Marije Krucifikse Kozulić 3, 51000 Rijeka, Croatia, info@intechopen.com or visit our website at intechopen.com.

Meet the editor

Dr Feng Shi is an associate professor in the College of Physics and Electronics, Shandong Normal University, P.R.China, whose research addresses novel functional materials with an eye toward practical applications such as microwave dielectric ceramics, bulk ceramics, as well as ceramic thin films. In the past five years, more than 40 papers in ceramics have been published in Applied Spectroscopy Reviews, Dalton Transactions, the Journal of Alloys and Compounds, Applied Surface Science, Journal of Materials Science - Materials in Electronics, and the International Journal of Materials Research. Presently, Feng Shi is the reviewer of many journals like Crystal Growth & Design; Journal of the American Ceramic Society; Journal of Colloid and Interface Science; Journal of Nanoparticle Research; Journal of Raman Spectroscopy; Current Nanoscience; Material Science and Engineering B; and Materials Science in Semiconductor Processing.

Contents

Preface

As we all know, ceramic materials are inorganic, non-metallic, solid, and inert materials. Things are made from them by the action of heat and subsequent cooling, which may be crystalline or partly crystalline. The definition of ceramic is often restricted to inorganic crystalline materials, as opposed to the noncrystalline glasses, which involve several steps of the ceramic process, and their mechanical properties behave similarly to ceramic materials.

The history of ceramics is as old as civilization, and our use of ceramics is a measure of the technological progress of a civilization. Ceramics have important effects on human history and human civilization. Earlier transitional ceramics, several thousand years ago, were made by clay minerals such as kaolinite. Modern ceramics are classified as advanced and fine ceramics. Both include three distinct material categories: oxides such as alumina and zirconia, nonoxides such as carbide, boride, nitride, and silicide, as well as composite materials such as particulate reinforced and fiber reinforced combinations of oxides and nonoxides. These advanced ceramics, made by modern chemical compounds, can be used in the fields of mechanics, metallurgy, chemistry, medicine, optical, thermal, magnetic, electrical and electronics industries, because of the suitable chemical and physical properties. In particular, photoelectron and microelectronics devices, which are the basis of the modern information era, are fabricated by diferent kinds of optical and electronic ceramics. In other words, optical and electronic ceramics are the base materials of the modern information era.

Bulk ceramics are made into the desired shape by reaction in situ, or by "forming" powders into the desired shape, and then sintering to form a solid body. However, ceramic thin films can be made by chemical or physical deposition. Grains, secondary phases, grain boundaries, pores, micro-cracks, structural defects, and hardness microindentions consist of the microstructure of the ceramics, which are generally indicated by the fabrication method and process conditions.

To explain more about advanced ceramics, this book (organized by InTech – Open Access Publisher) has been written by different authors, who focus on modern ceramics. A feature of this text is that we keep in mind that many of today's high-tech ceramic materials and processing routes have their origin in the potter's craft, microstructures, and properties. Throughout the text we make connections to these related fields. The text covers ceramic materials, from the fundamentals to industrial applications, including a consideration of safety and their impact on the modern technologies, including nano-ceramic, ceramic matrix composites, nanostructured ceramic membranes, porous ceramics, and sintering theory models of modern ceramics.

We thank all the authors and all the editors who contributed greatly to this book, and we hope that readers find it interesting.

Dr. Feng Shi
Shandong Normal University
China

Part 1

Electronic Ceramics

1

Characterization of PLZT Ceramics for Optical Sensor and Actuator Devices

Ribal Georges Sabat
Royal Military College of Canada
Canada

1. Introduction

Perovskite Lead Lanthanum Zirconate Titanate (PLZT) ceramics have the following chemical formula $Pb_{1-x}\ La_x\ (Zr_y\ ,\ Ti_{1-y})_{1-0.25x}\ V^B{}_{0.25x}\ O_3$ and are typically known as PLZT (100x/100y/100(1-y)). Compositional changes within this quaternary ferroelectric system, especially along the morphotropic phase boundaries, can significantly alter the material's properties and behaviour under applied electric fields or temperature variations. This allows such a system to be tailored to a variety of transducer applications. For instance, PLZT ceramics have been suggested for use in optical devices (Glebov et al. 2007; Liberts, Bulanovs, and Ivanovs 2006; Wei et al. 2011; Ye et al. 2007; Zhang et al. 2009) because of their good transparency from the visible to the near-infrared, and their high refractive index ($n \approx 2.5$), which is advantageous in light wave guiding applications (Kawaguchi et al. 1984; Thapliya, Okano, and Nakamura 2003). PLZT compositions near the tetragonal and rhombohedral ferroelectric phases and anti-ferroelectric/cubic phases, typically with compositions (a/65/35) with $7<a<12$, are known as relaxor ferroelectrics, since they exhibit a frequency-dependent diffuse ferroelectric-paraelectric phase transition in their complex dielectric permittivity. Relaxor ferroelectrics are particularly attractive in transducer applications because they can be electrically or thermally induced into a ferroelectric phase possessing a large dipole moment accompanied by a large mechanical strain, and revert back to a non-ferroelectric state upon the removal of the field or temperature. They also exhibit a slim hysteretic behaviour in the transition region, upon the application of an electric field, making them ideal for precise control actuator applications.

In this chapter, I will conduct a review on some of the fundamental material properties of relaxor ferroelectric PLZT ceramics, which include the dielectric, ferroelectric, electromechanical, electro-optical and thermo-optical behaviours. Further details on each section can be found in the references (Lévesque and Sabat 2011; Sabat, Rochon, and Mukherjee 2008; Sabat and Rochon 2009b; Sabat and Rochon 2009c; Sabat and Rochon 2009a).

2. Dielectric properties

The temperature and frequency dependence of the dielectric properties of any ferroelectric material are essential features to study, since they provide insight to possible transducer characteristics, such as the electrostrictive and electro-optic effects, which are both consequences of dipole moments arising from ion displacements.

Transparent PLZT (9.5/65/35) ceramics, having a thickness of 0.64 mm, were cut in 10 mm squares. Forty-nanometer layers of gold were sputtered on opposing faces and conducting wires were glued to each surface to act as electrodes. The complex dielectric permittivity was measured using an impedance analyzer at a frequency range from 0.12 to 5000 kHz. The temperature at which the measurements were taken could be varied since the samples were placed inside a thermal chamber. The probing ac electric field of the impedance analyzer was set at amplitude of 1 V and the heating rate was approximately 1 °C/min, starting at -60°C up to 100°C. The relative permittivity and loss tangent can be respectively calculated from the real and imaginary parts of the dielectric permittivity.

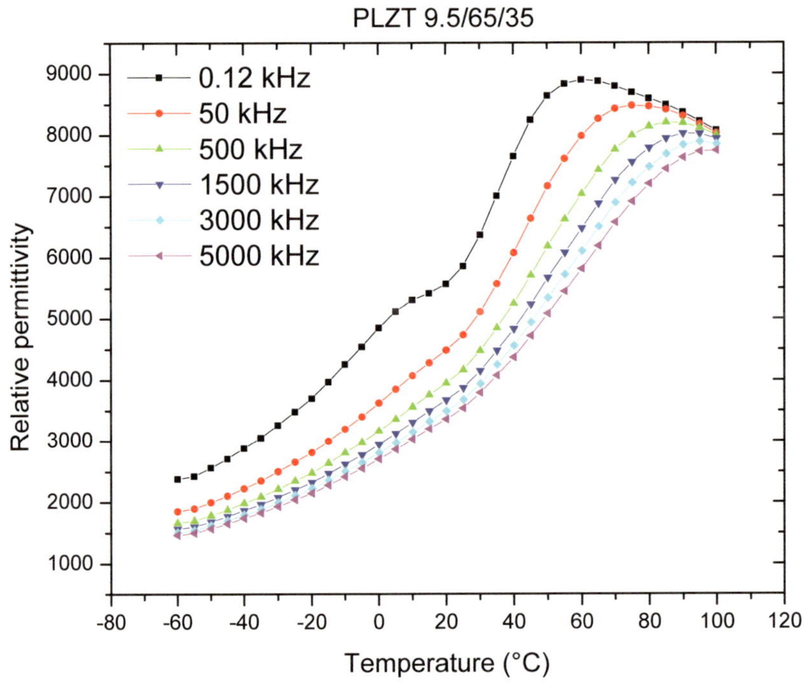

Fig. 1. Real relative permittivity of PLZT (9.5/65/35) as a function of temperature and frequency.

Figures 1 and 2 respectively show the temperature dependence of the relative permittivity and loss tangent of relaxor ferroelectric PLZT (9.5/65/35). As the temperature increases from -60°C to 100°C, the relative permittivity generally increased due to the unfreezing of domains. Between 0°C and 10°C, a broad peak can be seen in the lower frequency curves. This peak corresponds to the diffuse phase transition in this relaxor ceramic from the ferroelectric to the paraelectric state (also called the relaxor phase). Further heating continued to increase the relative dielectric permittivity until a maximum was achieved, at which point, the crystal's structure became cubic. This maximum in the permittivity, which is frequency dependent, occurs at the Curie temperature. Evidence of these phase transitions can also be seen in the loss tangent graph in figure 2.

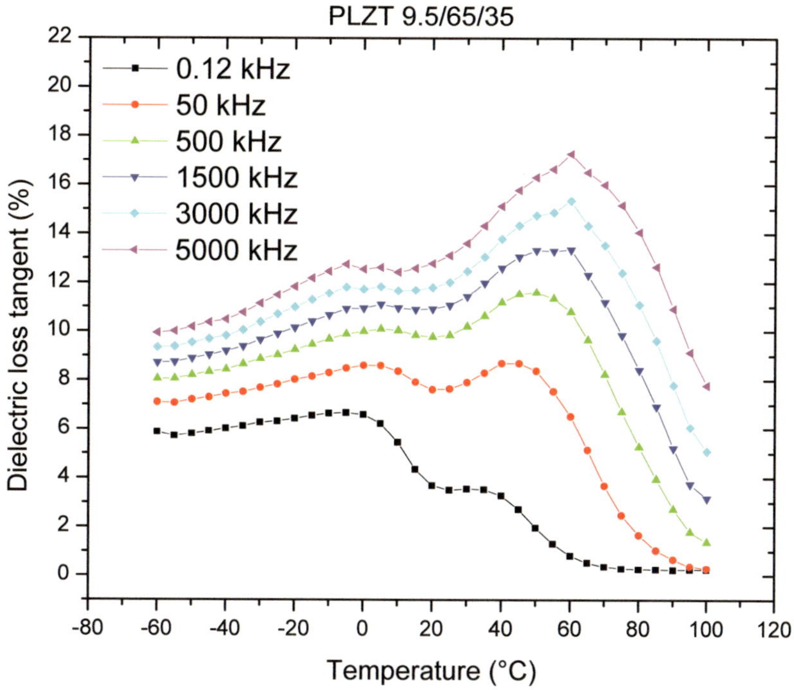

Fig. 2. Loss tangent of PLZT as a function of temperature and frequency.

The relative permittivity in figure 1 seems to decrease with increasing frequency, while the loss tangent in figure 2 increases at higher frequencies. These two observations go hand-in-hand since the frequency response of the complex permittivity is highly affected by the ability of ferroelectric domains and dipoles to rotate with the applied electric field. At higher frequencies, the ceramic material is no longer able to store as much electric energy in the dipoles and the relative permittivity decreases. As a consequence, a larger portion of the input energy is transferred to heating the ceramic and the loss tangent increases.

3. Ferroelectric properties

A Sawyer-Tower circuit (Sawyer and Tower 1930), with a $9.8\mu F$ series capacitance, was used to measure the ferroelectric hysteresis at room temperature. Figures 3 and 4 respectively show the electric displacement of PLZT (9.5/65/35) and PLZT (9.0/65/35) ceramics as a function of a dc bias electric field. The field was first increased from zero to +1.7 MV.m^{-1}, back down to -1.7 MV.m^{-1}, and finally up to zero. This cycle lasted 50 seconds and was repeated 3 consecutive times. Typical relaxor ferroelectric hysteretic curves were observed for these two compositions.

Figures 3 and 4 clearly illustrate how such a small change in the chemical composition of the PLZT can strongly affect the material's properties: PLZT (9.5/65/35) samples appear to possess a higher electric displacement compared to PLZT (9.0/65/35) at identical field values, but the hysteresis is slightly slimmer for the (9.0/65/35) composition samples. From Haertling's room temperature phase diagram of PLZT (Haertling 1987), it can be seen that the relaxor ferroelectric compositions studied here are located near the intersection of several other crystal phases including the ferroelectric-tetragonal, ferroelectric-rhombohedral and the antiferroelectric phase. Remnants of antiferroelectric hystereses can be found in both figures, but it's more evident for PLZT (9.5/65/35).

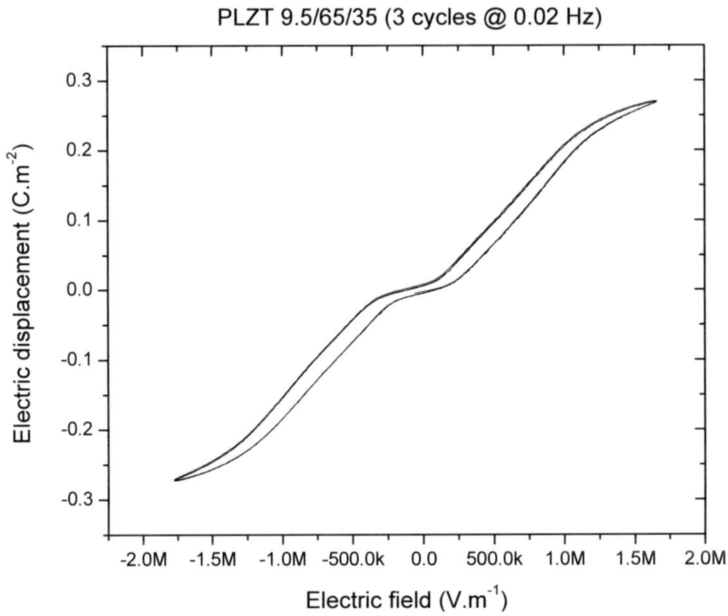

Fig. 3. Electric displacement of PLZT (9.5/65/35) as a function of dc electric fields.

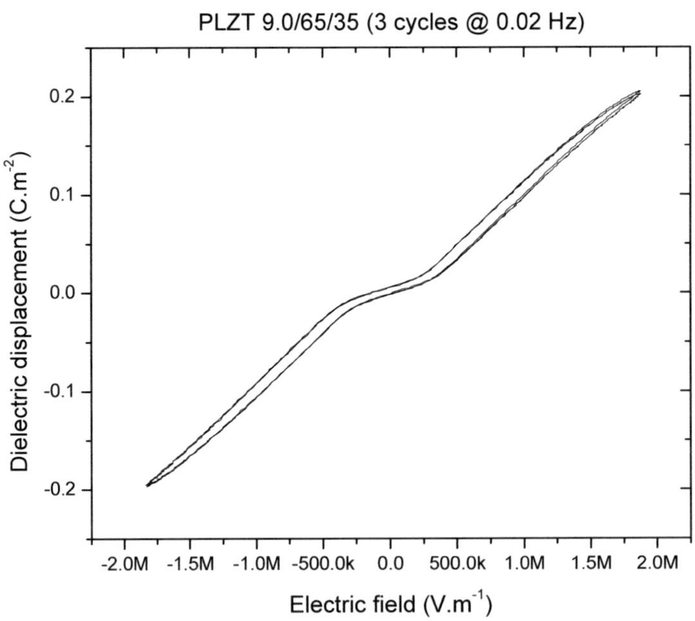

Fig. 4. Electric displacement of PLZT (9.0/65/35) as a function of dc electric fields.

According to the dielectric results in the previous section, decreasing the temperature at which these measurements were taken should increase the hysteresis gap (ferroelectric behaviour), and increasing the temperature should further decrease the hysteresis. This was observed by Carl et al. (Carl and Geisen 1973) for PLZT (9.0/65/35). The slim ferroelectric behaviour of these PLZT compositions makes them ideal for use as precision sensors and actuators, since they have almost no remnant polarization when the field is removed. Hence, the risk of depoling is eliminated in this case.

4. Electrostrictive and piezoelectric properties

4.1 Theory

Electrostriction refers to the elastic deformation of all dielectric materials upon the application of an electric field. Unlike piezoelectricity, the electrostrictive strain is quadratic to the electric field and reversal of the field doesn't reverse the strain direction. The basic phenomenology of electrostriction in materials is discussed in detail in many texts (Lines and Glass 1977; Jona and Shirane 1962; Mason 1958; Mason 1950). For a dielectric material under isothermal, adiabatic and stress-free conditions, upon the application of an electric field E_k, the strain tensor S_{ij} can be written as:

$$S_{ij} = \sum_{1}^{3} d_{ijk}E_k + \sum_{1}^{3} \gamma_{ijkl}E_kE_l + \ldots \tag{1}$$

Where d_{ijk} is the piezoelectric coefficient and γ_{ijkl} is the electrostriction coefficient. For the case where the applied electric field is in the 3-direction, which is taken to be the direction perpendicular to a sample's electrodes, equation (1) becomes:

$$S_3 = d_{33}E_3 + \gamma_{333}E_3^2 + \ldots \tag{2}$$

If all the coefficients of equation (2) are known, one can accurately predict the longitudinal strain under a varying electric field for a given piezoelectric or electrostrictive material, and even for a material exhibiting both piezoelectric and electrostrictive effects, such as irreversible electrostrictive materials. For ideal reversible electrostrictive materials, which possess no remnant polarization at zero electric field, the odd power term of the electric field in equation (2) vanishes. However, we will consider the relaxor PLZT ceramics studied in this chapter as irreversible electrostrictives, to account for any ferroelectric behaviour under dc bias fields, and we will therefore include both terms of the electric field in equation (2).

If a sinusoidal electric field with a dc bias component $E_3 = E_{DC} + E_0\cos(\omega t)$ is applied to an irreversible electrostrictive material, such as relaxor PLZT, equation (2) becomes:

$$S_3 = d_{33}\left(E_{DC} + E_0\cos(\omega t)\right) + \gamma_{333}\left(E_{DC} + E_0\cos(\omega t)\right)^2 + \ldots \tag{3}$$

Which can be re-arranged as follows:

$$S_3 = \ldots + \left[d_{33}E_0 + 2\gamma_{333}E_{DC}E_0 + \ldots\right]\cos(\omega t) + \left[\tfrac{1}{2}\gamma_{333}E_0^2 + \ldots\right]\cos(2\omega t) + \ldots \tag{4}$$

Hence, first, second and other harmonics should be present in the strain response of the PLZT ceramic. Experimentally, peaks should appear in the Fourier-transformed strain measurements at one and two times the driving frequency. By fitting the frequency-domain experimental strain peaks to the corresponding term in equation (4), the strain-electric-field longitudinal material coefficients (d_{33} and γ_{333}) can be obtained.

4.2 Experiment

To measure the strain developed by the PLZT ceramics upon the application of an electric field, a Doppler laser vibrometer can be used. The vibrometer uses the heterodyne technique to acquire the characteristic motion of the vibrating samples, as opposed to the homodyne method which uses the electronic equivalent of the fringe counting principle to measure strain. The vibrometer's helium-neon laser beam was pointed at one of the gold-coated surfaces of a vibrating PLZT sample and scattered back from it; the opposing sample surface was glued onto a fixed copper plate. The velocity and amplitude of the vibrating sample surface creates a phase or frequency modulation of the scattered laser light due to the Doppler effect. Using suitable decoders, the sample velocity information can be recovered from the frequency modulation of the Doppler signal, while the displacement information can be reconstructed from the phase modulation available at the same time.

If the velocity decoder on the vibrometer is used, the displacement information can still be obtained by integrating the velocity signal as a function of time. Fourier analysis can be performed to transform the displacement-amplitude-versus-time signal into the frequency domain. Even though the laser vibrometer can achieve picometer resolution, it can't detect strain by a dc voltage alone. Only alternating compressive and expansive strain can be measured with this method. Fortunately, this includes combined ac and dc field excitations.

Since the relaxor PLZT compositions used in this study have a cubic crystal structure at room temperature, the expectation is that the strain generated should be mostly quadratic, especially if no dc bias is applied. Figure 5 shows the Fourier-transformed displacement amplitude as a function of frequency for a PLZT (9.5/65/35) ceramic under a combination of both ac and dc bias electric fields. As expected, several frequency peaks are seen; this indicates that the material is vibrating at harmonic frequencies to the applied field.

Figure 6 shows the ac strain amplitude versus ac field amplitude, measured up to the fourth harmonic, of PLZT (9.5/65/35) without any dc bias. The second harmonic electrostrictive strain is pre-dominant. It rapidly increased with the ac field and reached a maximum of 0.00026 $m.m^{-1}$ at 0.5 $MV.m^{-1}$. The theoretical curve seen in this figure is the result of fitting the data collected at 240 Hz to the second harmonic term in equation (4).

Figure 7 shows the ac strain amplitude versus dc bias fields, measured up to the fourth harmonic, with a driving 0.37 $MV.m^{-1}$ peak-to-peak ac field at 120 Hz for PLZT (9.5/65/35). In this case, the first harmonic piezoelectric strain is dominant and seems to increase with the dc bias field until a maximum is reached at 1.2 $MV.m^{-1}$ dc. The theoretical curve seen in this figure is the result of fitting the data collected at 120 Hz to the first harmonic term in equation (4), while fixing the ac field value. This general behaviour of relaxor ferroelectrics has been previously observed for PMN electrostrictive ceramics (Masys et al. 2003).

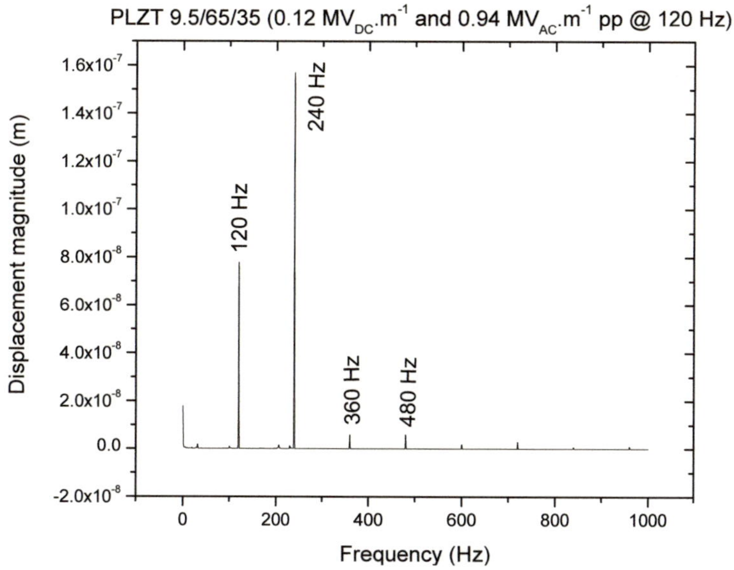

Fig. 5. Displacement magnitude of PLZT (9.5/65/35) as a function of frequency.

The increased strain with increasing dc bias in figure 7 can be explained by previous dielectric measurements of PLZT (9.0/65/35) ceramics as a function of both temperature and dc bias (Bobnar et al. 1999): They have observed a sharp increase in the dielectric permittivity with increasing dc bias fields at temperatures close to the ferroelectric-relaxor phase transition, indicating that the dc bias is inducing the creation of electric dipoles at this transition, and hence increasing the overall piezoelectric response.

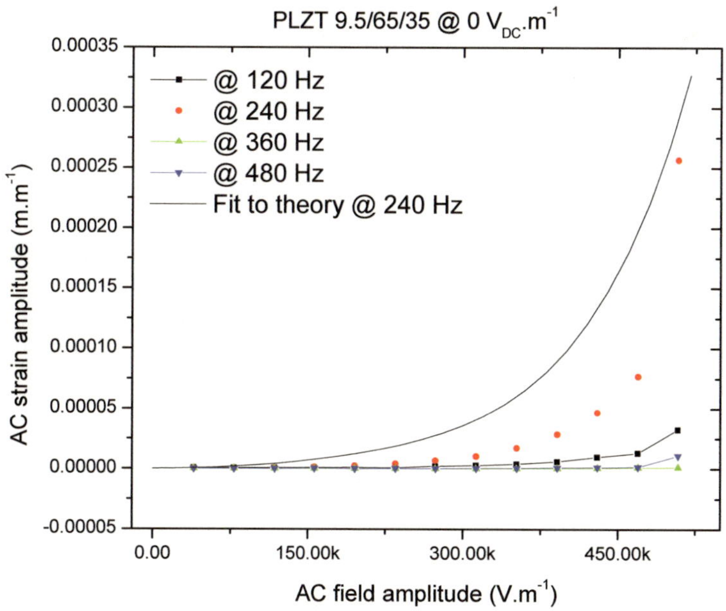

Fig. 6. Strain amplitude of PLZT (9.5/65/35) as a function of ac electric field.

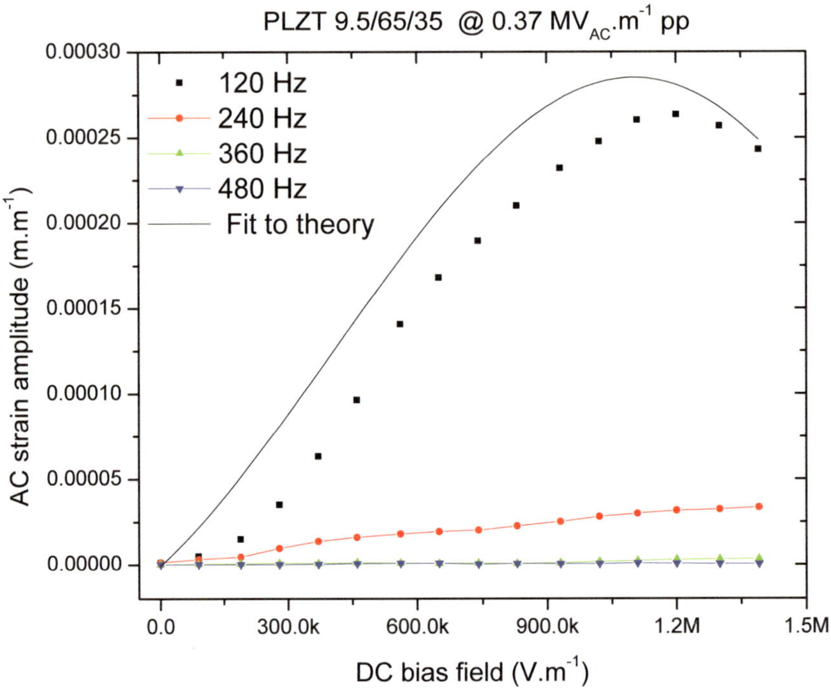

Fig. 7. Strain amplitude of PLZT (9.5/65/35) as a function of dc bias electric field.

This behaviour could also be explained as a consequence of the dc bias field induced reorientation of the polar nano-regions, favouring their alignment in the direction of the field (Tagantsev and Glazounov 1999). It is equally possible that, since the sample's composition is located near the ferroelectric tetragonal/rhombohedral boundary on the PLZT phase diagram, a dc bias induced phase transition from paraelectric to ferroelectric would increase the number of available polarization states, thus, maximizing the strain.

Next, the ac strain amplitude was plotted as a function of dc bias fields for various driving ac fields, as seen in figure 8. The first-harmonic piezoelectric strain increased with both the ac and dc fields until a maximum of approximately 0.8 x 10^{-3} m.m^{-1} occurred at 1.1 MV.m^{-1} dc and 1.09 MV.m^{-1} ac peak-to-peak. These results once again confirm studies of the dielectric behaviour of PLZT (9.0/65/35) (Bobnar et al. 1999), in which above a critical field, called E_c, a phase transition from relaxor to ferroelectric occurs in the PLZT structure. It is perhaps the presence of both phases simultaneously that give rise to the piezoelectric strain; further increasing of the fields would just render the samples more and more ferroelectric, therefore decreasing the strain.

The theoretical fitting curves seen in figures 6 to 8 were all fitted simultaneously to the first and second harmonic terms in equation (4), and the longitudinal piezoelectric and electrostrictive material coefficients were determined for PLZT (9.5/65/35) and PLZT (9.0/65/35), as seen in table 1. It can be noted that the experimental results and the theory are generally in good agreement. The small discrepancy between theory and experiment can be associated with random experimental uncertainties.

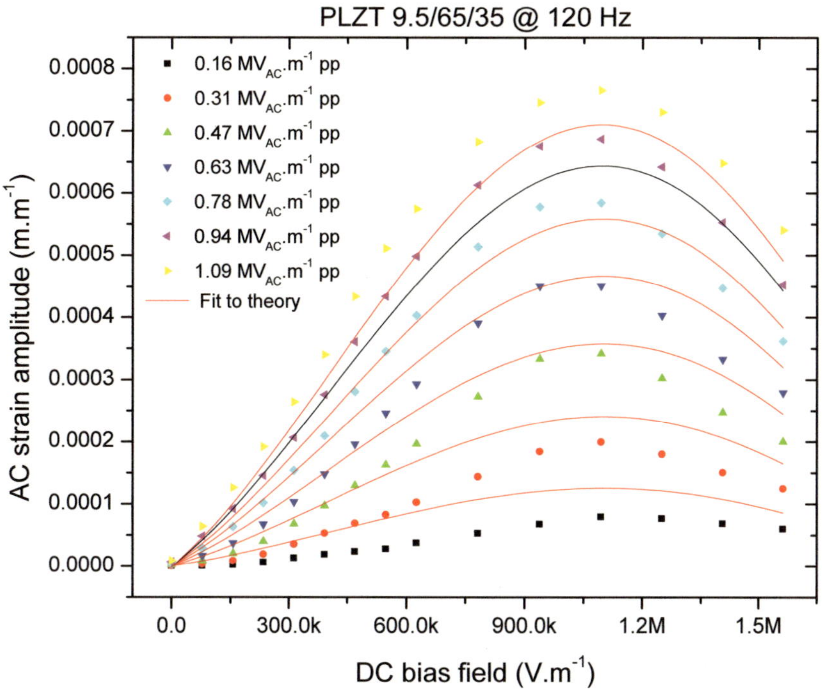

Fig. 8. Strain amplitude of PLZT (9.5/65/35) as a function of dc bias electric field at various ac field amplitudes.

Material coefficient	Units	(9.5/65/35) PLZT		(9.0/65/35) PLZT	
		Value	Error	Value	Error
d_{33}	$(m.V^{-1})$ x 10^{-11}	-0.34	2.44	5.65	4.93
γ_{333}	$(m^2.V^{-2})$ x 10^{-16}	6.14	0.64	3.18	0.83

Table 1. Strain material coefficients of PLZT.

5. Electro-optic and thermo-optical properties

5.1 Fabry-Pérot method

Many previously published papers have reported on the dc field electro-optic properties of PLZT ceramics (Haertling and Land 1971; Haertling 1971; Fogel, BarChaim, and Seidman 1980; Goldring et al. 2003). In this section, the electro-optic effects of large driving ac fields and the superposition of ac and dc electric fields are studied on relaxor PLZT ceramics using an interferometric technique. Two-hundred-nanometers thick gold electrodes were sputtered on both sides of each PLZT sample while leaving ~ 3 mm gap in the centre of each face, as illustrated in figure 9.

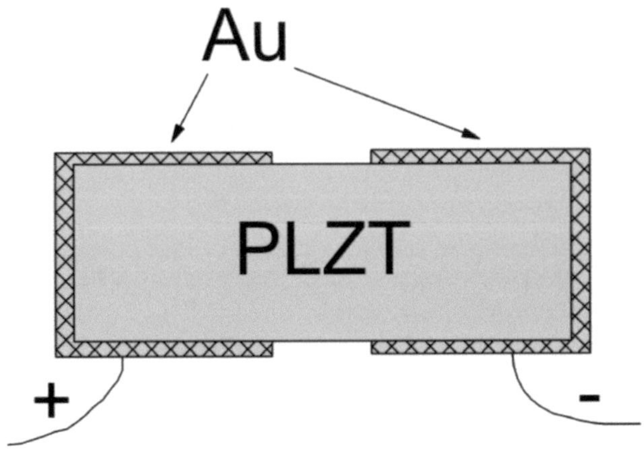

Fig. 9. Top view representation of the gold-sputtered electrodes on a PLZT sample.

The transmission of a laser light beam through the sample will exhibit a Fabry-Pérot interference pattern as the light incidence angle changes (Hecht 1987). Upon the application of any electric field, the angular interference fringes will shift due to a change in the optical path length within the sample. Depending on the input light's polarization, the individual refractive index variations, denoted Δn_1 and Δn_3, can be measured individually. The 3-subscript indicates the direction of the applied field, while the 1-subscript is the direction orthogonal to it. Figure 10 shows an example of a Fabry-Pérot interference pattern.

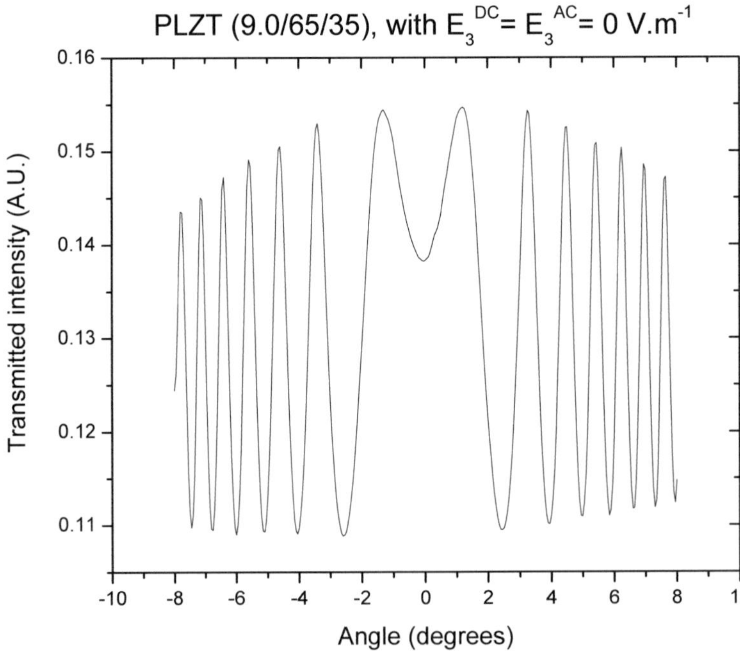

Fig. 10. Fabry-Pérot interference on a PLZT (9.0/65/35) sample with no applied fields.

The condition for the intensity maxima could be written as:

$$(nd)^2 - d^2 \sin^2 \theta = \left(\frac{m\lambda}{2}\right)^2, \qquad m = 0,1,2... \tag{5}$$

where n is the refractive index of the PLZT, d its thickness, θ is the angle of incidence of the laser and λ is the wavelength of the laser. Assuming a change in both the refractive index and the thickness of the PLZT under an electric field, one would obtain from equation (5) the change in the optical path length between a zero field and an applied field condition:

$$\Delta(nd) = \frac{d}{2n}\left(\sin^2 \theta_2 - \sin^2 \theta_1\right) \tag{6}$$

Where θ_1 is the incidence angle corresponding to an intensity maximum at a zero field condition and θ_2 is the same intensity maximum angle under an arbitrary applied field E_3. But, since $\Delta(nd) = n\Delta d + d\Delta n$, further development would lead to:

$$\Delta n = \frac{\left(\sin^2 \theta_2 - \sin^2 \theta_1\right)}{2n} - S_3 n \tag{7}$$

Where S_3 is the longitudinal strain due to the arbitrary transverse electric field. It was previously found that the ratio of $\Delta n / S$ for unclamped relaxor PLZT ceramics was approximately 16 (Carl and Geisen 1973), therefore, the second term in equation (7) should equate to approximately $0.16\Delta n$, since relaxor PLZT compositions have an approximate refractive index of 2.5 at room temperature. Under mechanical clamping conditions, as is the case illustrated in figure 9, the sample would have restricted movement and hence the second term on the right hand side of equation (7) would become even smaller.

Assuming an increase in the temperature of the tested samples under applied ac fields possibly due to the rapid rotation of domains, thermal expansion of the PLZT ceramic can be contributing to the strain. Therefore, equation (7) can be modified as follows:

$$\Delta n = \frac{\left(\sin^2 \theta_2 - \sin^2 \theta_1\right)}{2n} - S_3 n - n\ \alpha\ \Delta T \tag{8}$$

Where α is the linear thermal expansion coefficient and ΔT is the temperature variation. Haertling calculated that the average linear thermal expansion coefficient of various relaxor PLZT ceramics was around 3.9×10^{-6} °C^{-1} (Haertling 1971). Therefore, the third term on the right hand side of equation (8) would contribute approximately $10^{-5}\Delta T$ to the refractive index change Δn due to thermal expansion. Since there is an experimental error of approximately $\pm 5 \times 10^{-5}$ attributed to the refractive index change found using this method, a hypothetical increase in the temperature of the sample by just a few degrees would have a negligible effect on Δn. Therefore, the bulk of the change in the optical path length due to an applied electric field and possible thermal expansion should mostly be related to the refractive index change alone. Equation (8) can be approximated as:

$$\Delta n \approx \frac{\left(\sin^2\theta_2 - \sin^2\theta_1\right)}{2n} \tag{9}$$

5.2 Room temperature results

The first set of measurements was conducted on the effect of a dc electric field E_3^{DC} on the change in the refractive indices Δn_3 (same direction as the applied electric field) and Δn_1 (perpendicular to the applied electric field and the plane of incidence). In order to measure Δn_3, two polarizers, placed before and after the sample, were set to allow only the Transverse Magnetic (TM) component of the laser light to propagate; the TM component of light have their $\vec{E}$-vector parallel to the applied electric field on the sample. On the other hand, Δn_1 was measured by having both polarizers allow only the Transverse Electric (TE) component of the laser light to pass through to the light detector; the TE component have their $\vec{E}$-vector perpendicular to the applied electric field.

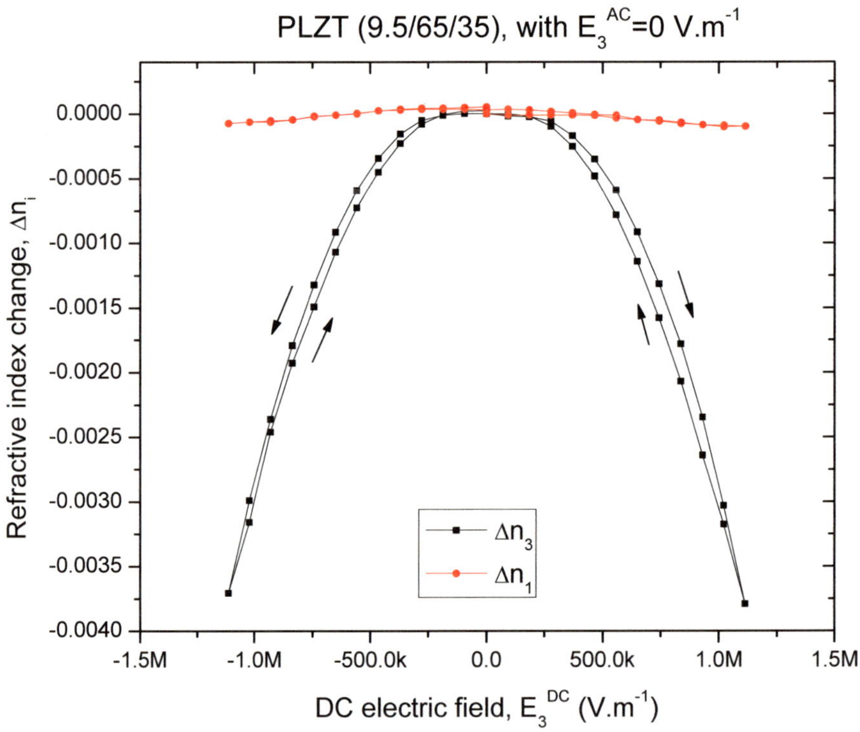

Fig. 11. Refractive index change Δn_i as a function of dc electric fields E_3^{DC} for PLZT (9.5/65/35).

Figure 11 shows the change in the refractive indices Δn_1 and Δn_3 as a function of an applied dc electric field E_3^{DC} for a PLZT (9.5/65/35) ceramic. The dc field was cycled from 0 to +1.1 MV.m^{-1} down to -1.1 MV.m^{-1}, and back up to 0. Δn_1 remained near zero while Δn_3 decreased quadratically with the dc field, due to the electro-optic Kerr effect, until a minimum of -

0.0037 was reached. It was expected that Δn_3 be much larger than Δn_1 since the dc field favours the creation and alignment of domains along its direction, and hence causing a greater change in the refractive index along its direction. Only a minor hysteresis is observed in the electro-optic behaviour; this mirrors the dielectric results discussed previously.

The temperature of the PLZT ceramic was monitored as an ac electric field, with 1 kHz frequency, was applied. A significant increase of up to 25°C was measured for PLZT (9.5/65/35) at $E_3^{AC} = 620\ kV.m^{-1}$. This elevation in the samples' temperature can be attributed to the increased movement of domains trying to align with the alternating field direction and would cause a thermal expansion of the sample. The maximum thermal strain generated should decrease the refractive index change Δn_i of a tested sample by a maximum of 2.5 x 10^{-4}. According to the dielectric measurements, such a temperature increase would also affect the crystal structure of the PLZT by moving it away from the paraelectric-ferroelectric transition region, around room temperature, towards the paraelectric or relaxor phase.

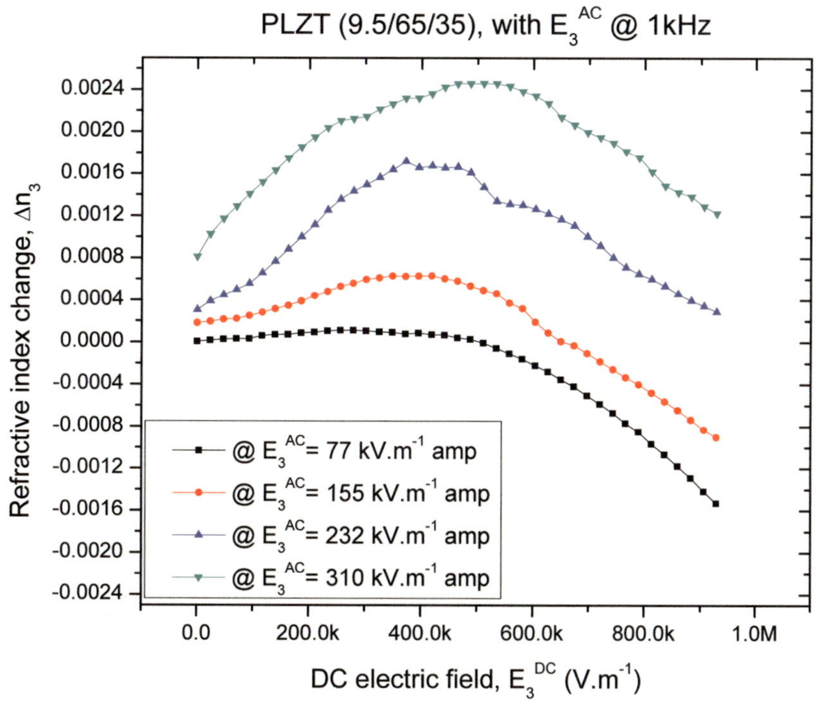

Fig. 12. Refractive index change Δn_3 as a function of dc electric fields E_3^{DC} at various ac field amplitudes E_3^{AC} @ 1 kHz frequency for PLZT (9.5/65/35).

Figure 12 shows the refractive index change Δn_3 as a function of a dc electric field E_3^{DC} in superposition with various values of an ac field E_3^{AC} at 1 kHz frequency. The effect of increasing ac fields were found to be opposite to increasing dc fields for PLZT (9.5/65/35) ceramics; Δn_3 decreased with the dc field, but increased with the ac field. A broad peak can be seen in this figure around $E_3^{DC} = 0.5\ MV.m^{-1}$ and this peak becomes more pronounced as the ac field amplitude increased. These results hint to the presence of a critical threshold dc field E_c, below which, increasing ac fields cause the destruction, misalignment and rapid rotation of domains, exhibited by an increase in the overall temperature of the sample and the subsequent transition to the relaxor phase, and above which, increasing dc bias fields favour the stabilization of the ferroelectric phase and promotes the presence of electric dipoles and their alignment along the dc field direction.

5.3 Temperature effects on the electro-optic properties

The same experimental set-up, as described above, was used, except the samples were now placed in a temperature-controlled chamber. As seen in figure 13, the temperature of a PLZT (9.5/65/35) was varied from 20°C down to -30°C, up to 60°C and down again to 20°C. Only a slight hysteresis was found for these results. The temperature dependence of the refractive index change Δn of PLZT (9.5/65/35), without any applied field, was found to be linear in the above range, with a positive slope of $\Delta n / \Delta T = (8.57 \pm 0.25) \times 10^{-5}\ °C^{-1}$. The third term in equation (8), which corresponds to the linear thermal expansion, contributed approximately 11% to the calculated slope.

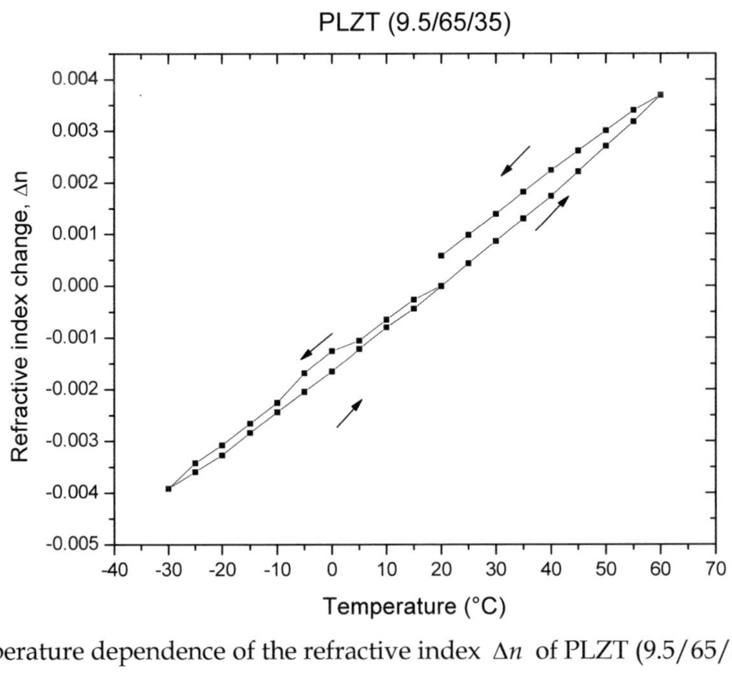

Fig. 13. Temperature dependence of the refractive index Δn of PLZT (9.5/65/35).

The temperature coefficient of the refractive index of this tested PLZT composition is in good agreement with the value of $(7.1 \pm 0.3) \times 10^{-5}\ °C^{-1}$ previously obtained using the thermal-lens method (Falcão et al. 2006).

An increase in a ceramic's temperature would increase its specific volume, thus decreasing its density, and consequently, its refractive index. But, according to Prod'homme

(Prod'homme 1960), it also increases the electronic polarizability of the sample due to a decrease in size of the atomic groupings responsible for the polarization, as the structure tends to a more dissociated state, which in turn causes the refractive index to increase gradually. These two contradicting effects can be explained by the results of figure 13, and equation (8): the linear thermal expansion decreased Δn as the temperature increased, but the shift in the Fabry-Pérot pattern clearly indicates a much stronger increase of Δn with increasing temperature. Therefore, the electronic polarizability had a greater impact on the temperature dependence of the refractive index of the PLZT ceramic compared to the thermal expansion.

We have seen that relaxor ferroelectric PLZT ceramics, with compositions (a/65/35) with 7<a<12, undergo a thermally or electrically induced diffuse and frequency-dependent paraelectric (or relaxor) to long-range ferroelectric phase transition. PLZT (9.5/65/35) ceramics undergo this phase change around 5°C. However, no evidence of this phase change is seen in figure 13. This is likely because the specific refractivity of materials R, which is a measure of the electronic polarization, is unaffected by this particular phase transition. This refractivity constant is defined by the Lorentz-Lorenz relation:

$$R = \frac{n^2 - 1}{n^2 + 2} \cdot V \tag{10}$$

Where n is the refractive index and V is the specific volume. Differentiating equation (10) with respect to temperature yields (Prod'homme 1960):

$$\frac{\Delta n}{\Delta T} = \frac{(n^2 - 1)(n^2 + 2)}{6n}(\varphi - \beta) \tag{11}$$

Since the first term of the right hand side of equation (11) can be considered a constant, the temperature coefficient of the refractive index $\Delta n / \Delta T$ is a measure of the difference between the electronic polarization coefficient ϕ and the volumetric thermal expansion coefficient β, where ϕ and β are defined as:

$$\varphi = \frac{1}{R}\frac{dR}{dT} \tag{12}$$

$$\beta = \frac{1}{V}\frac{dV}{dT} \cong 3 \cdot \alpha \tag{13}$$

Therefore, with the results of figure 13, the polarization coefficient ϕ for PLZT (9.5/65/35) is (4.14 ± 0.09) x 10^{-5} °C^{-1}, by using an approximate value of $\beta \cong 3 \cdot \alpha = 1.2$ x 10^{-5} °C^{-1}.

Figures 14 and 15 show the temperature dependence of Δn_3 and Δn_1 under a dc bias field $E_3^{DC} = 774$ kV.m^{-1} for PLZT (9.5/65/35). The application of a dc bias, at a given temperature, decreased Δn_3 while Δn_1 remained unchanged. It was found that $\Delta n_3 / \Delta T$ increased over $\Delta n / \Delta T$, but $\Delta n_1 / \Delta T$ remained somewhat unchanged. This indicates that the dc bias only increased the electronic polarizability of the test sample along the 3-axis, the direction of the applied field. All the temperature coefficients of the refractive indices for PLZT (9.5/65/35) and PLZT (9.0/65/35) were obtained and are summarized in table 2. Since the signs of all the calculated $\Delta n / \Delta T$ are positive, this means that the polarization coefficient ϕ of these materials is increasing at a larger rate than the volumetric thermal expansion β with increasing temperature.

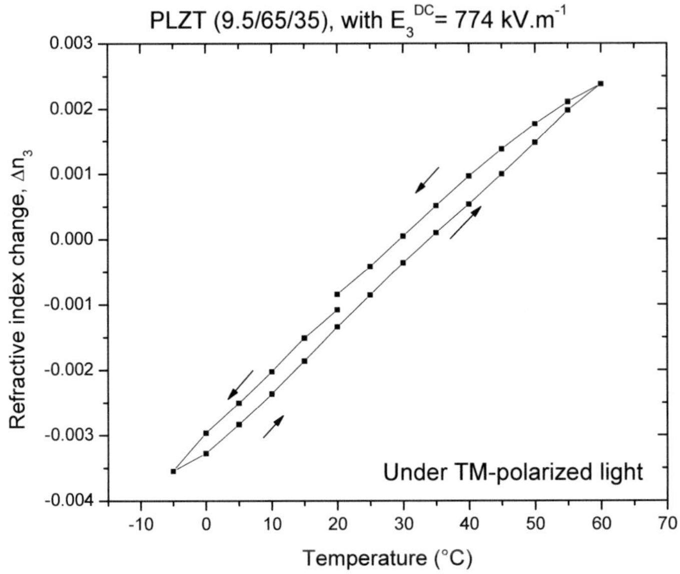

Fig. 14. Temperature dependence of the refractive index Δn_3 of PLZT (9.5/65/35) under dc bias.

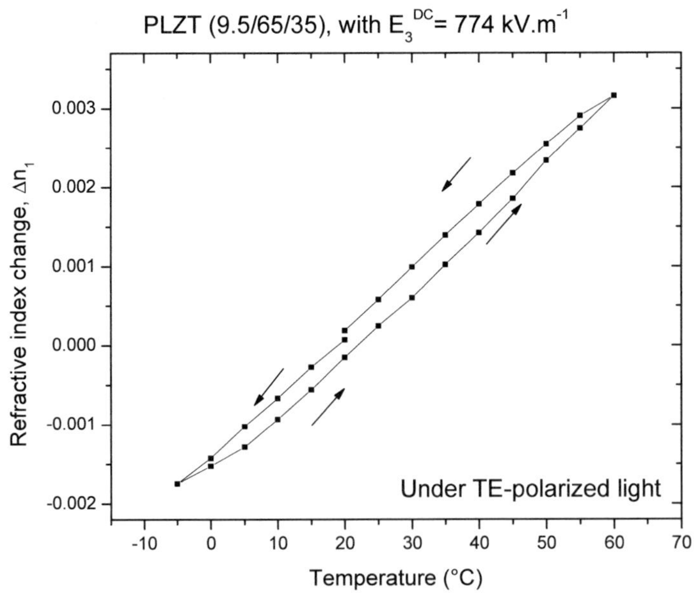

Fig. 15. Temperature dependence of the refractive index Δn_1 of PLZT (9.5/65/35) under dc bias.

Parameter (x 10^{-5} °C^{-1})	E_3^{DC} (kV.m^{-1})	(9.5/65/35) PLZT		(9.0/65/35) PLZT	
		Value	Error	Value	Error
$\frac{\Delta n}{\Delta T}$	0	8.57	0.25	7.28	0.16
$\frac{\Delta n_3}{\Delta T}$	774	9.41	0.42	8.11	0.30
$\frac{\Delta n_1}{\Delta T}$	774	7.85	0.32	7.47	0.26

Table 2. The measured temperature coefficients of the refractive index of relaxor PLZT ceramics.

Figure 16 shows the refractive index change Δn_3 for PLZT (9.5/65/35) as a function of dc bias fields up to 1 MV.m^{-1} and at temperatures of 0, 20 and 40°C.

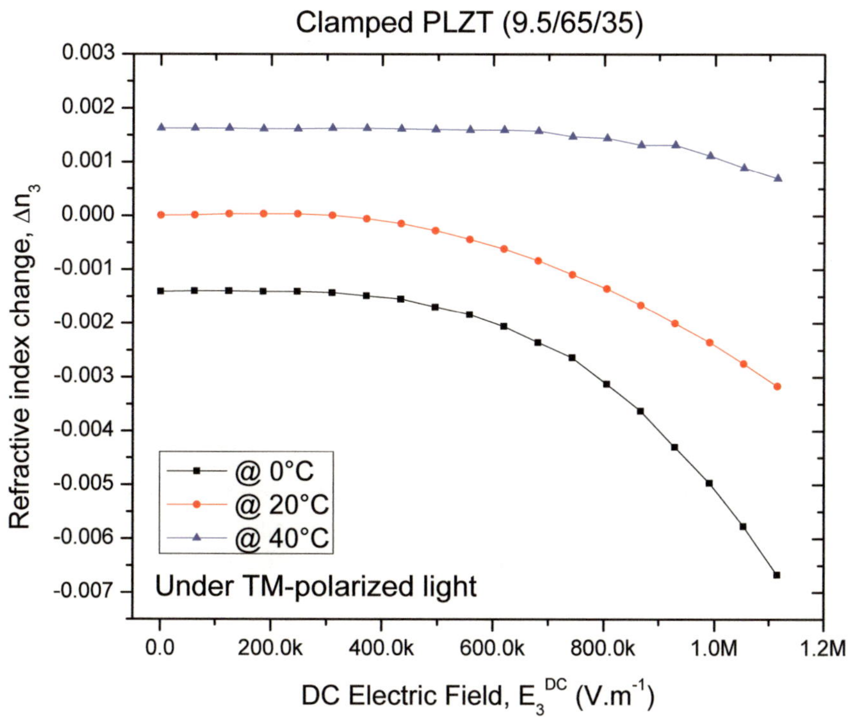

Fig. 16. Dc bias electric field dependence of the refractive index Δn_3 of PLZT (9.5/65/35), at 0, 20 and 40°C.

The Kerr quadratic electro-optic coefficient K_{33} in the presence of a dc field E_3^{DC} can be obtained from (Narasimhamurty 1981):

$$\Delta n_3 = -\frac{1}{2} n^3 K_{33} (E_3^{DC})^2 \quad (14)$$

The various K_{33} coefficients were calculated using equation (14) and reported in table 3. It was found that K_{33} seems to decrease with increasing temperature for both PLZT (9.5/65/35) and (9/65/35). This is due to the ferroelectric to relaxor phase change occurring around the tested temperature range.

	K_{33} x 10^{-16} ($m^2.V^{-2}$)			
Temperature	**(9.5/65/35) PLZT**		**(9.0/65/35) PLZT**	
	Value	Error	Value	Error
0°C	5.12	0.04	2.18	0.02
20°C	3.33	0.02	1.54	0.02
40°C	0.79	0.02	0.88	0.02

Table 3. The quadratic electro-optic coefficient K_{33} of PLZT at various temperatures.

5.4 Thermal-lensing effects

Since PLZT absorbs strongly in the far-infrared, a CO_2 laser beam ($\lambda = 10.6\ \mu m$) can be used to create a Gaussian temperature distribution at the surface of the ceramic. This temperature distribution will consequently alter the localized refractive index of the PLZT, emulating an optical lens. Hence, a low power visible He-Ne laser, travelling along the same path as the CO_2 laser, will get focalized after passing through the thermal lens. The focal distance of the created lens is adjustable and will depend on the input power of the CO_2 laser. Figure 17 gives a graphical representation of this phenomenon.

A PLZT (9/65/35) ceramic was positioned on glass slide substrate, which was moved by means of a translation stage. It was brought very near the CO_2 focal point while a CCD camera was positioned underneath the glass slide at a distance of 13.5 cm. Figure 18 shows what was observed. As the CO_2 laser power was increased, a focusing of the red light occurs.

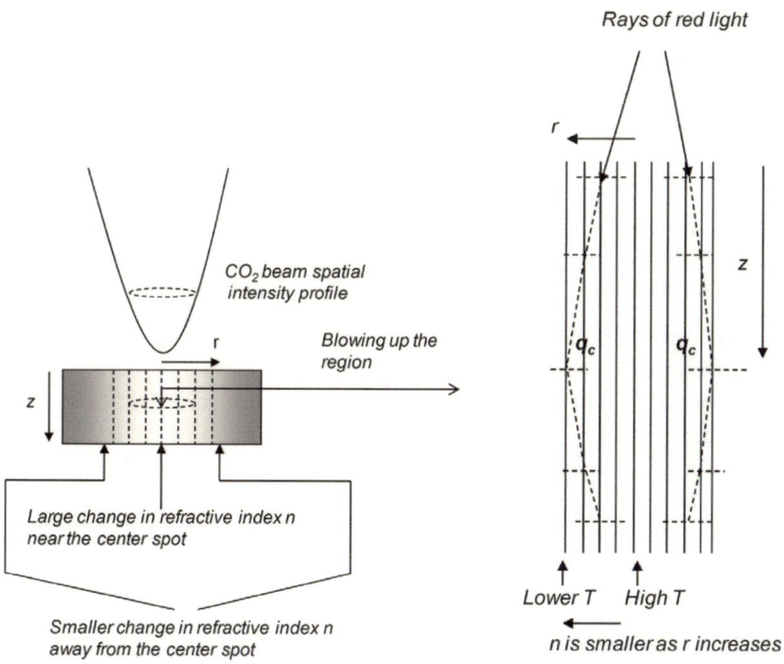

Fig. 17. CO_2 beam spatial intensity profile and refractive index gradient near the thick PLZT ceramic surface and self-focusing of a ray of red light resulting from a refractive index gradient near the surface.

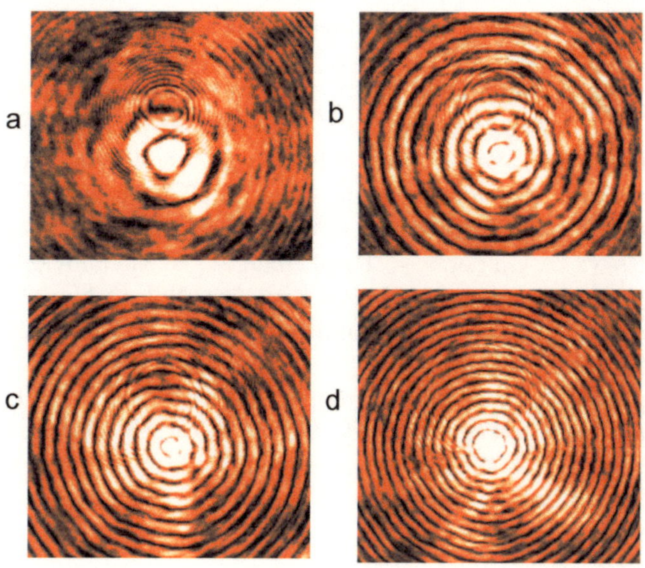

Fig. 18. Thermal lensing for on PLZT (9/65/35). A lens of 35 mm was used in front of the camera to produce a better view. CO^2 laser power a) 2.5 W b) 2.8W c) 3.2W d) 3.5 W.

6. Conclusion

Ferroelectrics will also play a major role in the next generation of optical devices. They have the potential to be the backbone of wireless laser communication systems or ground-based fibre optics, capable of carrying large amounts of information at ultrafast speeds. Notwithstanding their industrial and technological uses, ferroelectric materials will also continue to provide consumers with accessories and gadgets that will pleasure and simplify their everyday life. With the recent push on miniaturization of practical devices and the advancement of nanotechnology, ferroelectric materials may be fabricated to exhibit transducer effects at a scale small enough to be useful at the nanometre resolution.

In order to invent and develop future practical applications, we must have a greater fundamental understanding of the properties of these versatile ceramics. This can be accomplished by measuring and characterizing their behaviour under a variety of possible operating conditions. In this chapter, the transducer properties of Lead Lanthanum Zirconate Titanate (PLZT) ceramics were presented. Their dielectric, electromechanical and optical properties have been systematically studied and could provide valuable information for engineers seeking the development or enhancement of practical applications.

7. References

Bobnar, V., Kutnjak, Z., Pirc, R., & Levstik, A. (1999). Electric-field-temperature phase diagram of the relaxor ferroelectric lanthanum-modified lead zirconate titanate. *Phys. Rev. B,* Vol. 60, No. 9, pp. 6420-6427.

Carl, K., & Geisen, K. (1973). Dielectric and optical properties of a quasi-ferroelectric PLZT ceramic. *Proc. IEEE,* Vol. 61, No. 7, pp. 967-974.

Falcão, E. A., Pereira, J. R. D., Santos, I. A., Nunes, A. R., Medina, A. N., Bento, A. C., Baesso, M. L., Garcia, D., & Eiras, J. A. (2006). Thermo optical properties of transparent PLZT 10/65/35 ceramics. *Ferroelectrics,* Vol. 336, No. 1, pp. 191.

Fogel, Y., BarChaim, N., & Seidman, A. (1980). Longitudinal electrooptic effects in slim-loop and linear PLZT ceramics. *Appl. Opt.,* Vol. 19, No. 10, pp. 1609-1617.

Glebov, A. L., Smirnov, V. I., Lee, M. G., Glebov, L. B., Sugama, A., Aoki, S., & Rotar, V. (2007). Angle selective enhancement of beam deflection in high-speed electrooptic switches. *IEEE Photonics Technology Letters,* Vol. 19, No. 9, pp. 701.

Goldring, D., Zalevsky, Z., Goldenberg, E., Shemer, A., & Mendlovic, D. (2003). Optical characteristics of the compound PLZT. *Appl. Opt.,* Vol. 42, No. 32, pp. 6536-6543.

Haertling, G. H. (1987). PLZT electrooptic materials and applications—a review. *Ferroelectrics,* Vol. 75, pp. 25-55.

——— (1971). Improved hot-pressed electrooptic ceramics in the (pb,la)(zr,ti)O_3 system. *Journal of the American Ceramic Society,* Vol. 54, No. 6, pp. 303-309.

Haertling, G. H., & Land, C. E. (1971). Hot-pressed (pb,la)(zr,ti)O_3 ferroelectric ceramics for electrooptic applications. *Journal of the American Ceramic Society,* Vol. 54, No. 1, pp. 1-10.

Hecht, E. (1987). *Optics* (2nd), Addison-Wesley, .

Jona, F., & Shirane, G. (1962). *Ferroelectric crystals*Macmillan, , New York.

Kawaguchi, T., Adachi, H., Setsune, K., Yamazaki, O., & Wasa, K. (1984). PLZT thin-film waveguides. *Appl.Opt.,* Vol. 23, No. 13, pp. 2187-2191.

Lévesque, L., & Sabat, R. G. (2011). Thermal lensing investigation on bulk ceramics and thin-film PLZT using visible and far-infrared laser beams. *Optical Materials,* Vol. 33, No. 3, pp. 460.

Liberts, G., Bulanovs, A., & Ivanovs, G. (2006). PLZT ceramics based EFISH cell for nonlinear heterodyne interferometry. *Ferroelectrics,* Vol. 333, No. 1, pp. 81.

Lines, M. E., & Glass, A. M. (1977). *Theory and applications of ferroelectric materials*Oxford University Press, , Oxford.

Mason, W. P. (1958). *Physical acoustics and the properties of solids*Van Nostrand, , New York.

———. (1950). *Piezoelectric crystals and their applications to ultrasonics*Van Nostrand, , New York.

Masys, A. J., Ren, W., Yang, G., & Mukherjee, B. K. (2003). Piezoelectric strain in lead zirconate titanate ceramics as a function of electric field, frequency, and dc bias. *J. Appl. Phys.,* Vol. 94, No. 2, pp. 1155-1162.

Narasimhamurty, T. S. (1981). *Kerr quadratic electro-optic effect: Pockels' phenomenological theory, in:* Photoelastic and Electro-Optic Properties of CrystalsAnonymous , pp. 359-362, Plenum Press, , New York.

Prod'homme, L. (1960). Thermal change in the refractive index of glasses. *Phys. Chem. Glass.,* Vol. 1, No. 4, pp. 119-122.

Sabat, R. G., & Rochon, P. (2009a). The dependence of the refractive index change of clamped relaxor ferroelectric lead lanthanum zirconate titanate (PLZT) ceramics on AC and DC electric fields measured using an interferometric method. *Ferroelectrics,* Vol. 386, No. 1, pp. 105.

———. (2009b). Interferometric determination of the temperature dependence of the refractive index of relaxor PLZT ceramics under DC bias. *Optical Materials,* Vol. 31, No. 10, pp. 1454.

———. (2009c). Investigation of relaxor PLZT thin films as resonant optical waveguides and the temperature dependence of their refractive index. *Appl.Opt.,* Vol. 48, No. 14, pp. 2649-2654.

Sabat, R. G., Rochon, P., & Mukherjee, B. K. (2008). Quasistatic dielectric and strain characterization of transparent relaxor ferroelectric lead lanthanum zirconate titanate ceramics. *J. Appl. Phys.,* Vol. 104, No. 5, pp. 054115.

Sawyer, C. B., & Tower, C. H. (1930). Rochelle salt as a dielectric. *Phys.Rev.,* Vol. 35, No. 3, pp. 269-273.

Tagantsev, A. K., & Glazounov, A. E. (1999). Does freezing in $PbMg_{1/3}Nb_{2/3}O_3$ relaxor manifest itself in nonlinear dielectric susceptibility? *Appl. Phys. Lett.,* Vol. 74, No. 13, pp. 1910-1912.

Thapliya, R., Okano, Y., & Nakamura, S. (2003). Electrooptic characteristics of thin-film PLZT waveguide using ridge-type mach-zehnder modulator. *J.Lightwave Technol.,* Vol. 21, No. 8, pp. 1820.

Wei, F., Sun, Y., Chen, D., Xin, G., Ye, Q., Cai, H., & Qu, R. (2011). Tunable external cavity diode laser with a PLZT electrooptic ceramic deflector. *IEEE Photonics Technology Letters,* Vol. 23, No. 5, pp. 296.

Ye, Q., Dong, Z., Fang, Z., & Qu, R. (2007). Experimental investigation of optical beam deflection based on PLZT electro-optic ceramic. *Opt. Express,* Vol. 15, No. 25, pp. 16933-16944.

Zhang, Jingwen W., Yingyin K. Zou, Kewen K. Li, Qiushui Chen, Hua Jiang, Xuesheng Chen, and Piling Huang. 2009. Laser action with Nd3 doped electro-optic lead lanthanum zirconate titanate ceramics. Paper presented at Conference on Lasers and Electro-Optics/International Quantum Electronics Conference.

2

Electrode Size and Dimensional Ratio Effect on the Resonant Characteristics of Piezoelectric Ceramic Disk

Lang Wu[1], Ming-Cheng Chure[1], Yeong-Chin Chen[2], King-Kung Wu[1] and Bing-Huei Chen[3]
[1]*Department of Electronics Engineering, Far-East University,*
[2]*Department of Computer Science & Information Engineering, Asia University,*
[3]*Department of Electrical Engineering, Nan Jeon Institute of Technology,*
Taiwan

1. Introduction

After discovered at 1950, lead zirconate titanate [Pb(Zr,Ti)O_3; PZT] ceramics have intensively been studied because of their excellent piezoelectric properties [Jaffe et al., 1971; Randeraat & Setterington, 1974; Moulson & Herbert, 1997; Newnham & Ruschau, 1991; Hertling, 1999]. The PZT piezoelectric ceramics are widely used as resonator, frequency control devices, filters, transducer, sensor and etc. In practical applications, the piezoelectric ceramics are usually circular, so the vibration characteristics of piezoelectric ceramic disks are important in devices design and application. The vibration characteristics of piezoelectric ceramics disk had been study intensively by many of the researchers [Shaw, 1956; Guo et al., 1992; Ivina, 1990a, 2001b; Kunkel et al., 1990; Masaki et al., 2008]. Shaw [Shaw, 1956] measured vibrational modes in thick barium titanate disks having diameter-to-thickness ratios between 1.0 and 6.6. He used an optical interference technique to map the surface displacement at each resonance frequency, and used a measurement of the resonance and antiresonance frequency to calculate an electromechanical coupling for each mode. Guo et al., [Guo et al., 1992] presented the results for PZT-5A piezoelectric disks with diameter-to-thickness ratios of 20 and 10. There were five types of modes being classified according to the mode shape characteristics, and the physical interpretation was well clarified. Ivina [Ivina, 1990] studied the symmetric modes of vibration for circular piezoelectric plates to determine the resonant and anti-resonant frequencies, radial mode configurations, and the optimum geometrical dimensions to maximize the dynamic electromechanical coupling coefficient. Kunkel et al., [Kunkel et al., 1990] studied the vibration modes of PZT-5H ceramics disks concerning the diameter-to-thickness ratio ranging from 0.2 to 10. Both the resonant frequencies and effective electromechanical coupling coefficients were calculated for the optimal transducer design. Masaki et al., [Masaki et al., 2008] used an iterative automated procedure for determining the complex materials constants from conductance and susceptance spectra of a ceramic disk in the radial vibration mode.

The phenomenon of partial-electroded piezoelectric ceramic disks also study by some researchers. Ivina [Ivina, 2001] analyzed the thickness symmetric vibrations of piezoelectric disks with partial axisymmetric electrodes by using the finite element method. According to the spectrum and value of the dynamic electromechanical coupling coefficient of quasi-thickness vibrations, the piezoelectric ceramics can be divided into two groups. Only for the first group can the DCC be increased by means of the partial electrodes, which depends on the vibration modes. Schmidt [Schmidt, 1972] employed the linear piezoelectric equations to investigate the extensional vibrations of a thin, partly electroded piezoelectric plate. The theoretical calculations were applied to the circular piezoelectric ceramic plate with partial concentric electrodes for the fundamental resonant frequency. Huang [Huang, 2005] using the linear two-dimensional electroelastic theory, the vibration characteristics of partially electrode-covered thin piezoelectric ceramic disks with traction-free boundary conditions are investigated by theoretical analysis, numerical calculation, and experimental measurement.

In this study, the vibration characteristics of a thin piezoelectric ceramic disk with different electrode size and dimensional ratio are study by the impedance analysis method.

2. Vibration analysis of the piezoelectric ceramic disk

Figure 1 shows the geometrical configuration of the piezoelectric ceramic disk with radius R and thickness h. The piezoelectric ceramic disk is assumed to be thin (R>>h) and polarized in the thickness direction. If the cylindrical coordinates (r, θ, z) with the origin in the center of the disk are used. The linear piezoelectric constitutive equations of a piezoelectric ceramic with crystal symmetry C_{6mm}, can be expressed as [IEEE, 1987]:

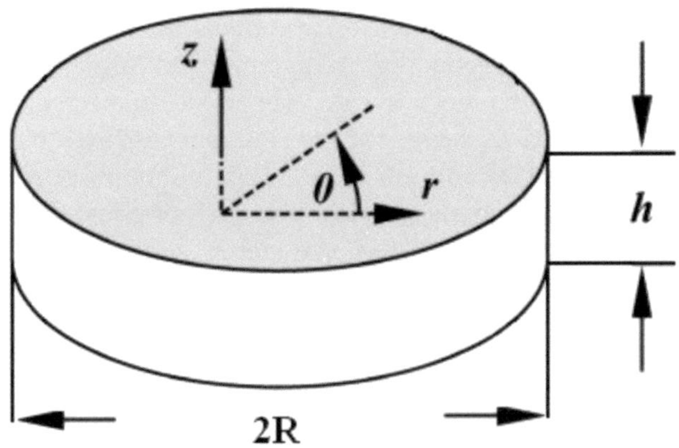

Fig. 1. The geometrical configuration of the piezoelectric ceramic disk.

$$\begin{bmatrix} S_{rr} \\ S_{\theta\theta} \\ S_{zz} \\ S_{\theta z} \\ S_{rz} \\ S_{r\theta} \\ D_r \\ D_\theta \\ D_z \end{bmatrix} = \begin{bmatrix} s_{11}^E & s_{12}^E & s_{13}^E & 0 & 0 & 0 & 0 & 0 & d_{31} \\ s_{12}^E & s_{11}^E & s_{13}^E & 0 & 0 & 0 & 0 & 0 & d_{31} \\ s_{13}^E & s_{13}^E & s_{33}^E & 0 & 0 & 0 & 0 & 0 & d_{33} \\ 0 & 0 & 0 & s_{44}^E & 0 & 0 & 0 & d_{15} & 0 \\ 0 & 0 & 0 & 0 & s_{44}^E & 0 & d_{15} & 0 & 0 \\ 0 & 0 & 0 & 0 & 0 & s_{66}^E & 0 & 0 & 0 \\ 0 & 0 & 0 & 0 & d_{15} & 0 & \varepsilon_{11}^T & 0 & 0 \\ 0 & 0 & 0 & d_{15} & 0 & 0 & 0 & \varepsilon_{11}^T & 0 \\ d_{31} & d_{31} & d_{33} & 0 & 0 & 0 & 0 & 0 & \varepsilon_{33}^E \end{bmatrix} \begin{bmatrix} T_{rr} \\ T_{\theta\theta} \\ T_{zz} \\ T_{\theta z} \\ T_{rz} \\ T_{r\theta} \\ E_r \\ E_\theta \\ E_z \end{bmatrix} \quad (1)$$

where T_{rr}, $T_{\theta\theta}$ and T_{zz} are the longitudinal stress components in the r, θ and z directions; $T_{r\theta}$, $T_{\theta z}$ and T_{rz} are the shear stress components. S_{rr}, $S_{\theta\theta}$, S_{zz}, $S_{\theta z}$, S_{rz} and $S_{r\theta}$ are the strain components. D_r, D_θ and D_z are the electrical displacement components, and E_r, E_θ, and E_z are the electrical fields. $s_{11}^E, s_{12}^E, s_{13}^E, s_{33}^E, s_{44}^E$ and s_{66}^E are the compliance constants at constant electrical field, in which $s_{66}^E = 2(s_{11}^E - s_{12}^E)$. d_{15}, d_{31} and d_{33} are the piezoelectric constants; ε_{11}^T and ε_{33}^T are the dielectric constants.

The electric field vector E_i is derivable from a scalar electric potential V_j:

$$E_r = -\frac{\partial V}{\partial r} \quad (2a)$$

$$E_\theta = -\frac{1}{r}\frac{\partial V}{\partial \theta} \quad (2b)$$

$$E_z = -\frac{\partial V}{\partial z} \quad (2c)$$

The electric displacement vector D_i satisfies the electrostatic equation for an insulator, and shown as:

$$\frac{\partial D_r}{\partial r} + \frac{1}{r}\frac{\partial D_\theta}{\partial \theta} + \frac{1}{r}D_r + \frac{\partial D_z}{\partial z} = 0 \quad (3)$$

Some basic hypotheses are employed for analysis the vibration of thin disk [Rogacheva, N.N., 1994]:

a. Normal stress T_{zz} is very small, so it can be neglected relative to other stresses, hence $T_{zz} = 0$.
b. The rectilinear element normal to the middle surface before deformation remains perpendicular to the strained surface after deformation, and its elongation can be neglected, i.e., $S_{rz}=S_{\theta z} = 0$.
c. Electrical displacement D_z is a constant with respect to the thickness.

In this study, only the radial axisymmetry vibrations of the disk are considered, so $S_{zz}=0$. The electrodes are coated on the z axis, so $E_r=0$ and $E_\theta=0$. The constitutive equations can reduce to:

$$S_{rr} = s_{11}^E T_{rr} + s_{12}^E T_{\theta\theta} + d_{31}E_z \tag{4a}$$

$$S_{\theta\theta} = s_{12}^E T_{rr} + s_{11}^E T_{\theta\theta} + d_{31}E_z \tag{4b}$$

$$S_{r\theta} = s_{66}^E T_{r\theta} = 2(s_{11}^E - s_{12}^E)T_{r\theta} \tag{4c}$$

$$D_z = d_{31}T_{rr} + d_{31}T_{\theta\theta} + \varepsilon_{33}^E E_z \tag{4d}$$

The stresses and the charge density Q on the surface of the disk, can be obtained by inversion of Eq.(4),

$$T_{rr} = \frac{1}{s_{11}^E(1-\sigma^2)}(S_{rr} + \sigma S_{\theta\theta}) - \frac{d_{31}}{s_{11}^E(1-\sigma)}\frac{2V_3}{h} \tag{5a}$$

$$T_{\theta\theta} = \frac{1}{s_{11}^E(1-\sigma^2)}(S_{\theta\theta} + \sigma S_{rr}) - \frac{d_{31}}{s_{11}^E(1-\sigma)}\frac{2V_3}{h} \tag{5b}$$

$$T_{r\theta} = \frac{1}{s_{66}^E}S_{r\theta} = \frac{S_{r\theta}}{2(s_{11}^E - s_{12}^E)} = \frac{S_{r\theta}}{2s_{11}^E(1+\sigma)} \tag{5c}$$

$$D_z = d_{31}(S_{rr} + S_{\theta\theta}) + \varepsilon_{33}^T \frac{2V_3}{h} \tag{5d}$$

where σ is Poisson's ratio and equal to $-(s_{11}^E / s_{12}^E)$, V_3 is the voltage applied in the z-direction.

Assumed the radial extensional displacement of the middle plane as:

$$u_r(r,t)=U(r)e^{i\omega t} \tag{6}$$

where ω is the angular frequency.

The strain-mechanical displacement relations are:

$$S_{rr} = \frac{\partial U}{\partial r} \tag{7a}$$

$$S_{\theta\theta} = \frac{U}{r} \tag{7b}$$

$$S_{r\theta} = \frac{1}{r}\frac{\partial U}{\partial \theta} \tag{7c}$$

Then the stress-displacement relations of the radial vibration are given as:

$$T_{rr} = \frac{1}{s_{11}^E(1-\sigma^2)}\left(\frac{dU}{dr} + \sigma\frac{U}{r}\right) + \frac{d_{31}}{s_{11}^E(1-\sigma)}\frac{2V_3}{h} \tag{8a}$$

$$T_{\theta\theta} = \frac{1}{s_{11}^E(1-\sigma^2)}\left(\frac{U}{r} + \sigma\frac{dU}{dr}\right) + \frac{d_{31}}{s_{11}^E(1-\sigma)}\frac{2V_3}{h} \tag{8b}$$

and the charge density is:

$$Q = -\frac{d_{31}}{s_{11}^E(1-\sigma)}\left(\frac{\partial U}{\partial r} + \frac{U}{r}\right) - \frac{2d_{31}^2}{s_{11}^E(1-\sigma)}\frac{2V_3}{h} + \varepsilon_{33}^T\frac{2V_3}{h} \tag{9}$$

The equation of motion in the radial direction is

$$\frac{\partial T_{rr}}{\partial r} - rT_{\theta\theta} + \frac{1}{r}T_{rr} = \rho\frac{\partial^2 u_r}{\partial t^2} \tag{10}$$

where ρ is the density.

Substitution of Eq.(8) into Eq.(10), find

$$\frac{d^2U}{dr^2} + \frac{1}{r}\frac{dU}{dr} - \frac{1}{r^2}U - \rho\omega^2 s_{11}^E(1-\sigma^2)U = 0 \tag{11}$$

the general solution of Eq.(11) is

$$U(r)=CJ_1(\beta r) \tag{12}$$

where J_1 is the Bessel function of the first kind and first order, C is a constant and

$$\beta^2 = \rho s_{11}^E(1-\sigma^2)\omega^2 \tag{13}$$

For the boundary condition at r = R, it has:

$$\int_{-h/2}^{h/2} T_{rr}dz = 0 \tag{14}$$

So, the constant C is found to be:

$$C = \frac{2Vd_{31}(1+\sigma)}{(1-\sigma)J_1(\beta R) - \beta RJ_0(\beta R)}\frac{R}{h} \tag{15}$$

where J_0 is Bessel function of the first kind and zero order.

When the piezoelectric disk in radial vibration, the current can be developed as[Huang et al, 2004]:

$$I = \frac{\partial}{\partial t}\iint_S D_z ds = i\omega\int_0^{2\pi}\int_0^R \left\{\frac{d_{31}(1+\sigma)}{s_{11}^E(1-\sigma^2)}\left[\frac{dU}{dr}+\frac{U}{r}\right]+\frac{2\varepsilon_{33}^T V}{h}(k_p^2-1)\right\} r dr d\theta$$
$$= i\omega\frac{2\pi R^2 V\varepsilon_{33}^T}{h}\frac{\left[1-\sigma+(1+\sigma^2)\frac{k_p^2}{1-k_p^2}\right]J_1(\beta R)-\beta R J_0(\beta R)}{(1-\sigma)J_1(\beta R)-\beta R J_0(\beta R)} \tag{16}$$

where S is the area of the electrode and

$$k_p = \sqrt{\frac{2d_{31}^2}{\varepsilon_{33}^T s_{11}^E(1-\sigma)}} \tag{17}$$

is the planar effective electromechanical coupling factor.

The frequencies corresponding to the current I approaches infinity is the resonant frequencies. The characteristic equation of resonant frequencies for radial vibrations is given by:

$$\eta J_0(\eta)=(1-\sigma)J_1(\eta) \tag{18}$$

where $\eta=\beta R$.

The antiresonant frequencies for which the current through the piezoelectric ceramic disk equal to zero are determined from the roots of the following equation:

$$\left[1-\sigma+(1+\sigma)\frac{k_p^2}{k_p^2-1}\right]J_1(\eta)=\eta J_0(\eta) \tag{19}$$

From Eqs(12), (18) and (19), under free boundary conditions, the resonant frequency of the piezoelectric ceramic disk with fully coated electrode can be expressed as:

$$f_r = \frac{\eta}{2\pi R}\sqrt{\frac{1}{\rho s_{11}^E(1-\sigma^2)}} \tag{20}$$

3. Experimental process

The piezoceramic disks used in this study were prepared by conventional powder processing technique, starting from high purity raw materials, TiO_2 (Merck, 99%), ZrO_2 (Aldrich, 99%) and PbO (Merck, 99%). The compositions of the ceramics were in the vicinity of the MPB of PZT in the tetragonal range, and doped with minor MnO_2 and Sb_2O_3.

After 2h ball milling with ZrO_2 balls, the mixed powders were calcined in air for 2h at 850°C. The calcined powders were then cold pressed into disk type pellets. The pellets were sintered at 1250°C for 2h with the double crucible arrangement, with $PbZrO_3$ atmosphere powder for PbO compensation.

Two groups of samples were used in this study, one group used for electrode size study, the other group used for dimensional ratio effect. The diameter and thickness of samples used for electrode size study were 15mm and 0.9mm, respectively. The (diameter/thickness) ratio of the samples used for the dimensional ratio effect was from 7.25 to 20.16.

Crystal phase structure was examined by a Rigaku X-ray diffraction (XRD) with CuKα radiation (λ=1.5418Å), scanned from 20° to 80° and scanning speed was 4°/min. Microstructures were analyzed on a polished surface of the specimens by high resolution scanning electron microscopy (SEM). The relative density of piezoelectric ceramics was measured by an Archimedes's method.

The flat surfaces of the samples were polished, cleaned and coated with silver electrode to get a better Ohmic contact for electric properties measurement. There are total five different electrode size used for electrode size study, as list in Table-1. After electrode coating, the samples were poled under 3.0 kV/mm electric field at elevated temperature 100°C for 1h in the silicone oil bath. After polarization, the piezoelectric properties were surveyed with an HP4194A Impedance/Gain-Phase Analyzer and Berlincourt d_{33} meter based on the IEEE standards [IEEE, 1987]. Resonant spectrum of the samples was determined in the frequency range from 100 kHz to 1500 kHz, and the results were recorded by a PC base data acquisition system.

No.	D1	D2	D3	D4	D5
Electrode Size	100%	80%	60%	40%	20%

Table 1. Electrode size used in this study.

4. Results and discussions

Figure 2 shows the SEM micrographs of the sintered ceramics body, and the grain size of the ceramics body was less than 2μm. The XRD pattern of the ceramics before poling was shown in Fig.3. The crystal structure of the ceramics is tetragonal perovskite, with a=4.036Å, c=4.102Å and tetragonality c/a=1.016.

Fig. 2. SEM micrographs of the sintered ceramics body.

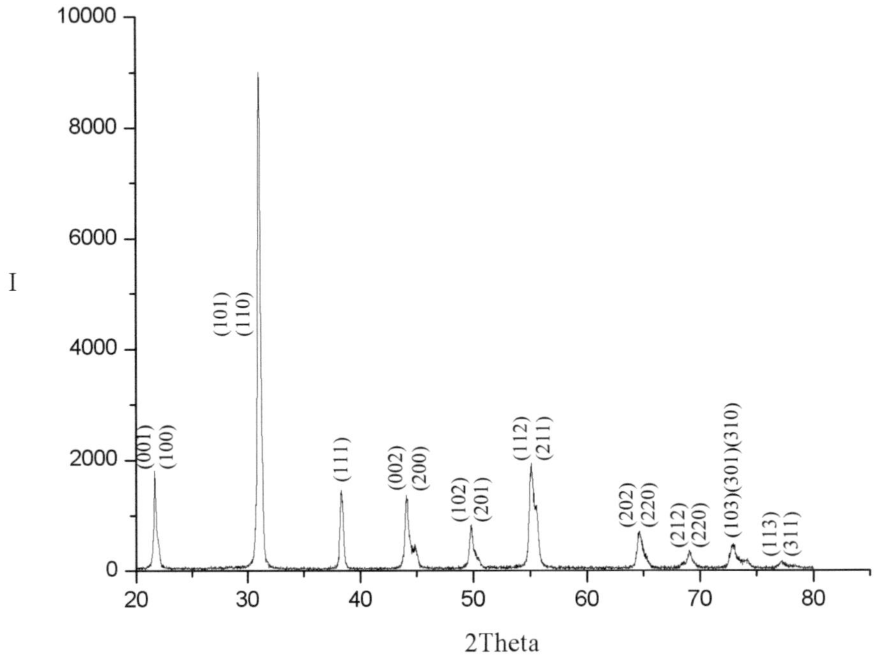

Fig. 3. The XRD pattern of the sintered ceramics body.

In large (diameter/thickness) ratio piezoelectric ceramic disk, the vibration modes are classified into five groups [Ikegami et al., 1974]. They are radial (R) mode, edge (E) mode, thickness shear (TS) mode, thickness extensional (TE) mode and high-frequency radial (A) mode. The resonant frequency of radial (R) mode and high-frequency radial (A) mode are strongly dependent on the diameter of the piezoelectric ceramic disk. And the resonant frequency of edge (E) mode, thickness shear (TS) mode and thickness extensional (TE) mode are strongly dependent on the thickness of the piezoelectric ceramic disk. With the increasing of frequency, the radial (R) mode appears at first, and then are edge (E) mode, thickness shear (TS) mode and thickness extensional (TE) mode, the high-frequency radial (A) mode exist at most higher frequency.

The resonant spectrum ranging from 100 kHz to 1500 kHz of piezoelectric ceramics disks with fully electrode size is show in Fig.4. From the results of Fig.4, it found that when the electrode is fully coating, there are one fundamental radial vibration mode (R), seven radial overtone modes (R1 to R7) and two edge modes (E) were existed in the frequency ranging from 100 kHz to 1500 kHz. The mode shapes of the fundamental radial vibration mode and its overtone are shown in Fig.5. It can see that the wave number increases with the frequency. The mode shape of edge mode is shown in Fig.6, the mode shape of edge mode is similar to that of radial modes except that the axial displacement at the edge is large, whereas it approaches zero in the radial modes [Guo & Cawley, 1992].

In the radial vibration mode, besides the fundamental mode, there are some overtone modes with inharmonic frequency separation. The series resonant frequency for the nth vibration mode of the piezoelectric ceramic thin disk in the radial vibration mode can be expressed as [Randeraat & Setterington, 1974]:

$$f_{(sn)} = \frac{\alpha_n}{2\pi R}\sqrt{\frac{1}{s_{11}^E \rho (1-\sigma^2)}} \tag{21}$$

where s^E_{11} is compliance and $\sigma = -s^E_{12}/s^E_{11}$ was Poisson's ratio for the piezoelectric ceramic used in this study.

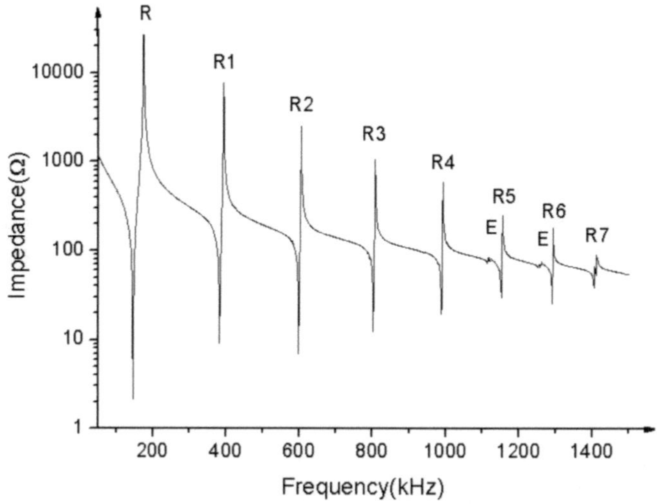

Fig. 4. The resonant spectrums of piezoelectric ceramics disk with fully electrode.

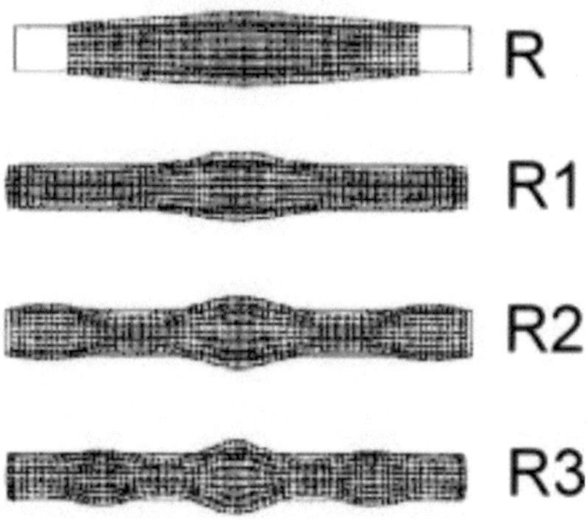

Fig. 5. The mode shapes of the radial vibration modes.

Fig. 6. The mode shape of edge mode.

For the piezoelectric ceramic disk in radial vibration mode, the ratio between the fundamental series resonant frequency and the overtones series resonant frequencies are not integers, but with a values equal to α_n/α_1, in which α_1 is the coefficient corresponding to the fundamental mode and α_n is the coefficient corresponding to the nth order overtone vibration mode as shown in Eq.(21). The relation between overtones mode orders and the coefficient α_n and coefficient ratio α_n/α_1 were shown in Fig.7, and can be fit with second order polynomial:

$$\alpha_n = -1.83482 + 3.96149x - 0.15542x^2 \tag{22}$$

$$\alpha_n/\alpha_1 = -0.89661 + 1.94149x - 0.07613x^2 \tag{23}$$

where x is the mode orders.

The coefficient α_n and coefficient ratio α_n/α_1 increased with the increasing of overtones mode order, but the slope of coefficient ratio α_n/α_1 curve was less than that of coefficient α_n curve. It means that the separation between vibrations modes would decreased with the increasing of overtone mode order.

The planar effective electromechanical coupling factor k_p is decreased with the increasing in overtone mode order also, as shown in Fig.8, it also can be fit with second order polynomial:

$$k_p = 0.70566 - 0.23079x + 0.01959x^2 \tag{24}$$

The decrease of planar effective electromechanical coupling factor k_p with the increase in overtone mode order can be explained by the vibration mode shape. In radial vibration mode, when the electrical voltage is applied in the thickness direction, the disk contracts in the radial direction, and expands in the thickness direction, due to Poisson's ratio effects. The vibration mode shapes for the fundamental radial vibration mode, and the six radial overtone modes are predicted by the finite element method and shown in Fig. 5. From the results of Fig.5, it found that the contraction of the disk in the radial direction is the largest for fundamental radial vibration mode, and it decreases with the increasing of overtone mode order. The meaning of such phenomenon is that the electrical energy transfer to mechanical energy is decreased with the increasing of overtone mode order. As a result, the planar effective electromechanical coupling factor k_p is decreased with the increasing of overtone mode order.

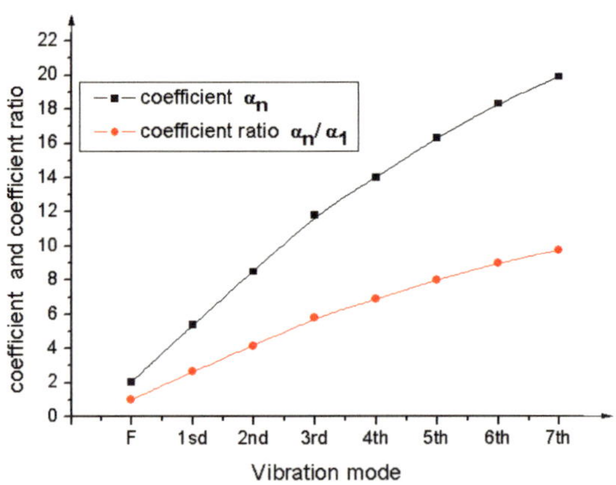

Fig. 7. The variation of coefficient α_n and coefficient ratio α_n/α_1 with mode orders.

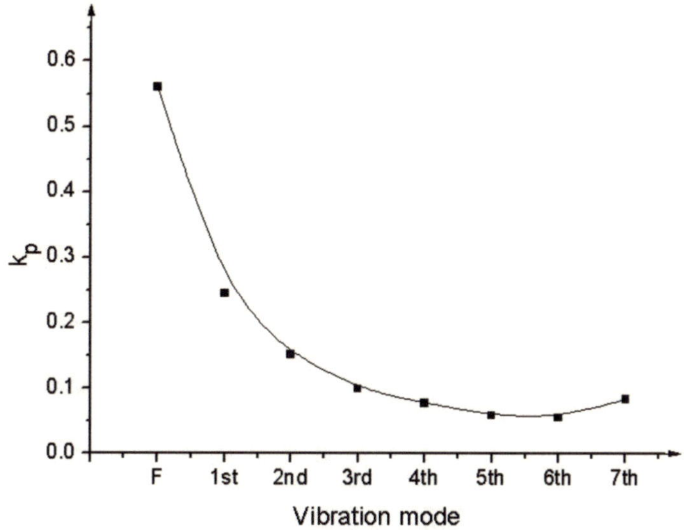

Fig. 8. The variation of electromechanical coupling factor k_p with mode orders.

When the electrode is reduced, the resonant spectrum will shift to higher frequency range, as shown in Fig. 9. The minimum impedance corresponding to series resonant frequency increased with the decreasing of electrode size. The maximum impedance corresponding to parallel resonant frequency also increased with the decreasing of electrode size, but when the electrode size reduced to less than 50%, the maximum impedance is decreased with the decreasing of electrode size.

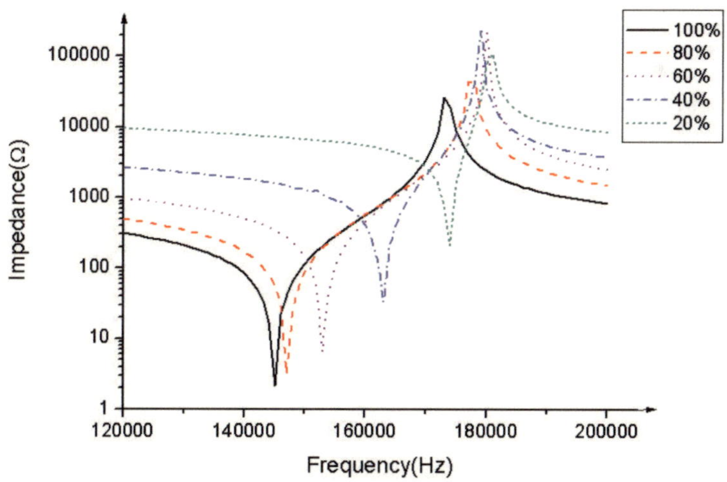

Fig. 9. Resonant spectrum of piezoelectric ceramic disk with different electrode size.

The series resonant and parallel resonant frequencies of fundamental radial vibration modes are increased with the decreasing of electrode size, but the planar effective electromechanical coupling factor is decreased with the decreasing of electrode size, as shown in Fig.10.

The relation between series resonant frequency (f_s) and planar effective electromechanical coupling factor (k_p) with electrode size (D_e) is second order polynomial and can be expressed as:

$$f_s=189-0.80857(D_e)+0.0357(D_e)^2 \quad (25)$$

$$k_p=0.0642+0.1136(D_e)-6.43929\times10^{-5}(D_e)^2 \quad (26)$$

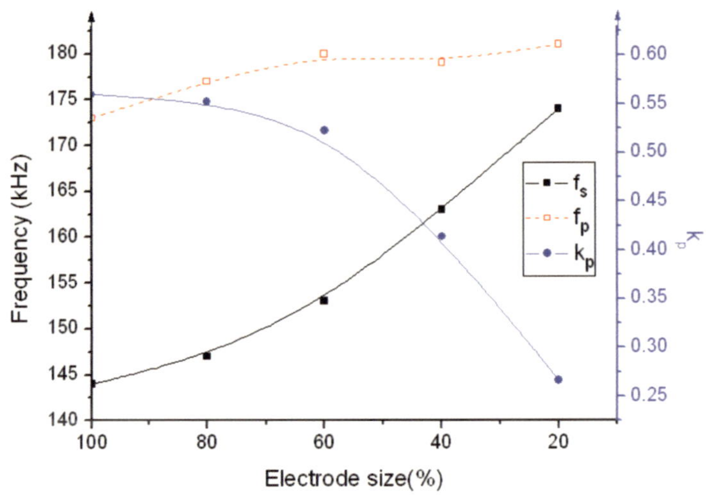

Fig. 10. The relation between electrode size and the electromechanical coupling factor k_p.

The frequency spectrums of different diameter-to-thickness ratio are shown in Fig.11 to Fig.14. From the results of these figures, it found that the numbers of radial vibration overtone mode are increased with the increasing of (D/t) ratio.

Besides the fundamental mode, there are eight radial vibration overtone modes are existed in the measured frequency range for (D/t) = 20.16. When the (D/t) decreased to 16.13, only six radial vibration overtone modes are existed in the measured frequency range. When the (D/t) ratio was larger than 15, only radial vibration mode is existed in the measured frequency range. When the (D/t) was less than 15, the numbers of radial vibration overtone mode reduced, and some other vibration modes will exist in the high frequency range. For (D/t) = 12.08 and 7.25, only five and three radial vibration overtone modes are existed in the measured frequency range, respectively. When the measured frequency higher than 600 kHz, some spurious vibration signals existed in the spectrum, these spurious vibration signals may be caused by other vibration mode; such as twist mode, bending mode or thickness mode. The relation between (overtone mode resonant frequency/fundamental resonant frequency) ratio and (diameter/thickness) ratio was shows in Fig.15. The (overtone mode resonant frequency/fundamental resonant frequency) ratio are increased with the increasing of (diameter/thickness) ratio.

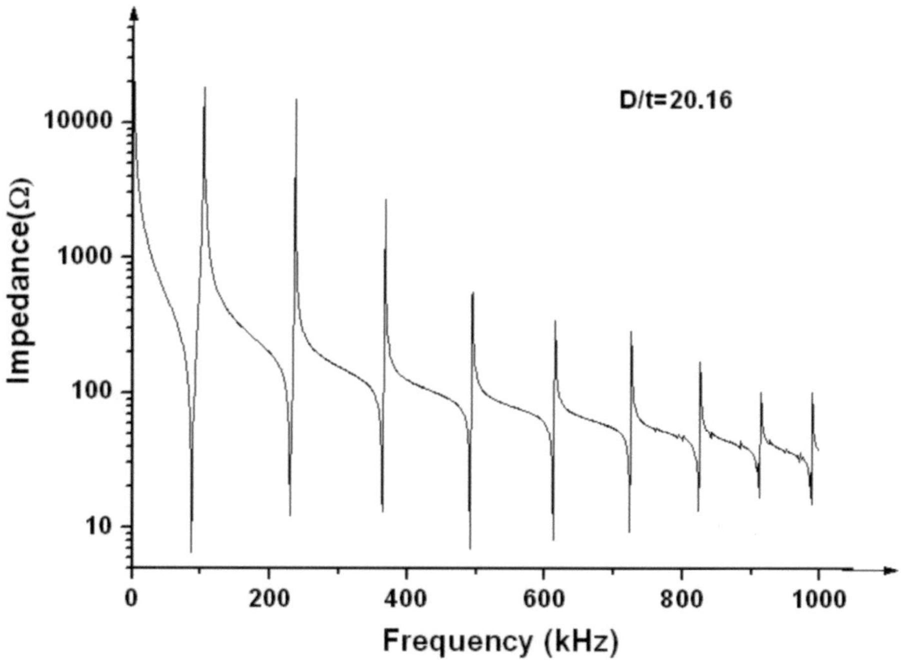

Fig. 11. The frequency spectrums of piezoelectric ceramic disk with D/t=20.16.

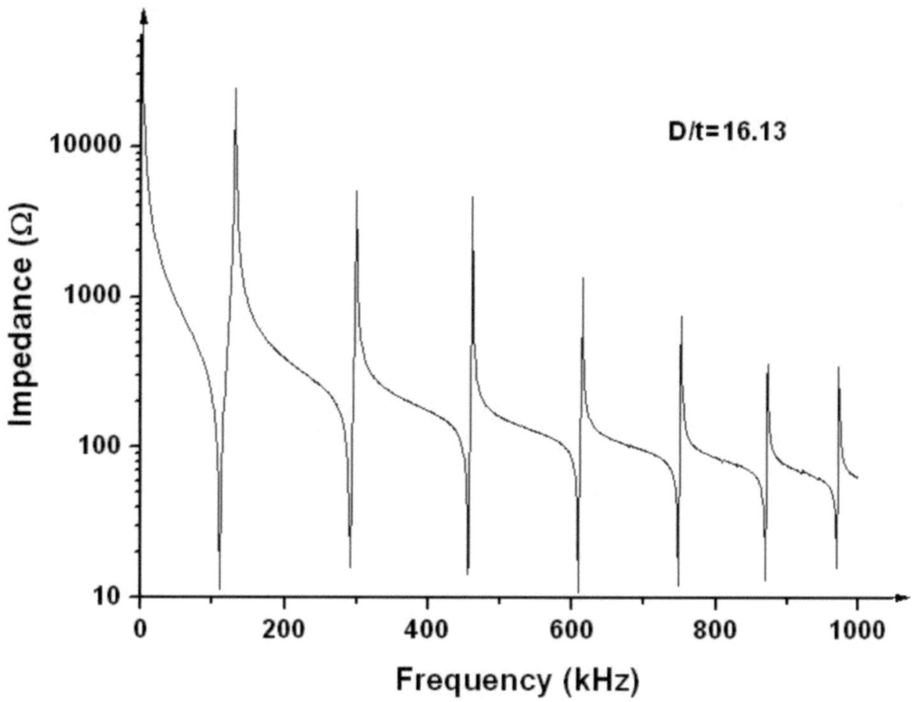

Fig. 12. The frequency spectrums of piezoelectric ceramic disk with D/t=16.13.

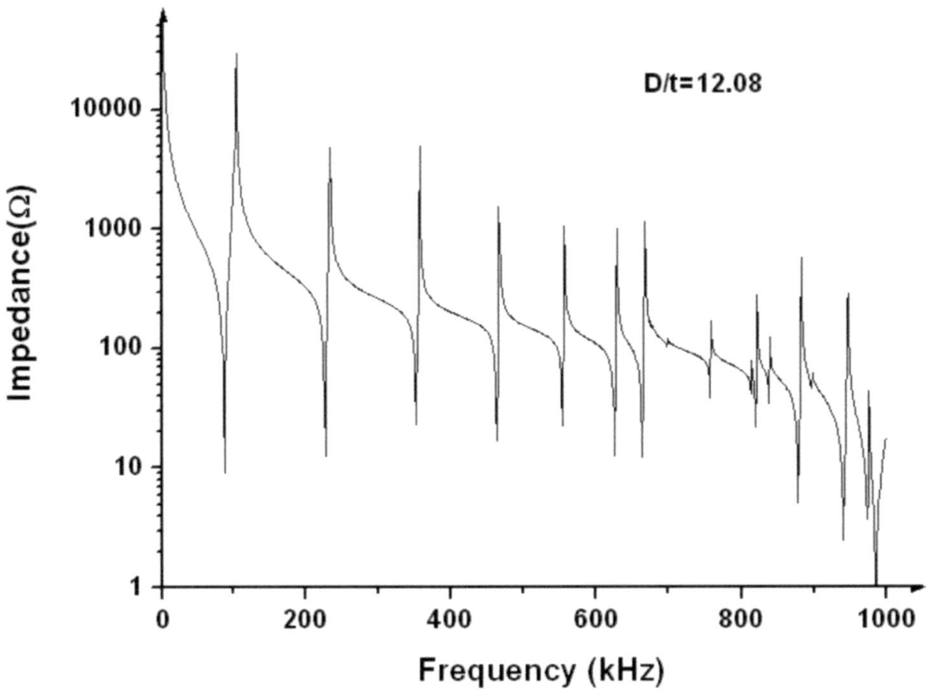

Fig. 13. The frequency spectrums of piezoelectric ceramic disk with D/t=12.08.

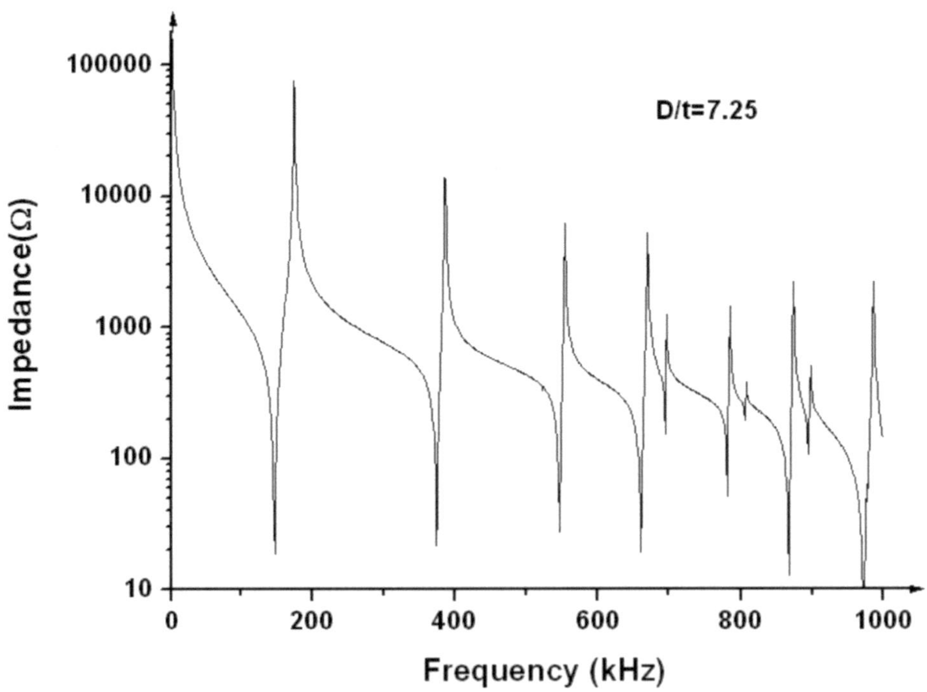

Fig. 14. The frequency spectrums of piezoelectric ceramic disk with D/t=7.25.

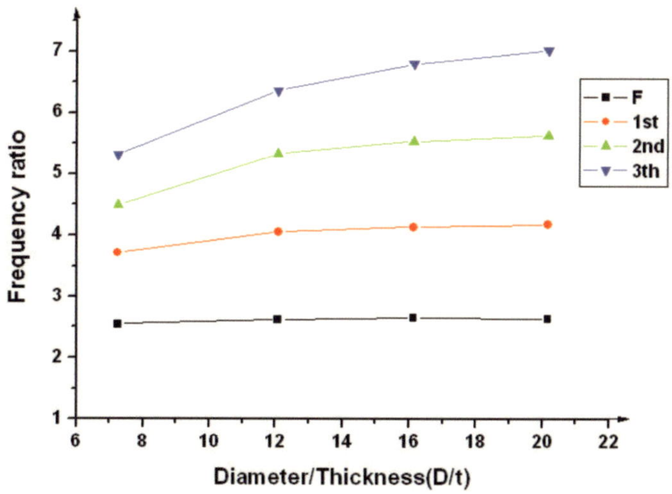

Fig. 15. Relation between (overtone mode resonant frequency/fundamental resonant frequency) ratio and (diameter/thickness) ratio.

5. Conclusion

For piezoelectric ceramic thin disk with larger (diameter/thick) ratio, in the lower frequency range, the major vibration mode is radial vibration mode. In the radial vibration mode, besides the fundamental mode, there are some overtone modes with inharmonic frequency separation. The planar effective electromechanical coupling factor k_p is decreased with the increasing in overtone mode order. The relation between planar effective electromechanical coupling factor and overtone mode order can be fit with second order polynomial.

When the electrode is reduced, the resonant spectrum will shift to higher frequency range, and the planar effective electromechanical coupling factor is decreased with the decreasing of electrode size.

The numbers of overtones mode are increased with the increasing of (D/t) ratio. The (overtone mode resonant frequency/fundamental resonant frequency) ratio are increased with the increasing of (diameter/thickness) ratio.

6. References

GuO, N. & Cawley, P. (1992). Measurement and Prediction of the Frequency Spectrum of Piezoelectric Disks by Modal Analysis, *The Journal of The Acoustical Society of America*, Vol.92, No.6, pp.3379-3388, ISSN 0001-4966

Guo, N.; Cawley, P. & Hitchings, D. (1992). The Finite Element Analysis of the Vibration Characteristics of Piezoelectric Discs, *Journal of Sound and Vibration*, Vol..159, No.1, pp.115- 38, ISSN 0022-460X

Haertling, G.H. (1999). Ferroelectric Ceramics: History and Technology, *Journal of the American Ceramic Society*, Vol.82, Issue.4, pp.797–818, ISSN 1551-2916

Huang, C.H. (2005). Theoretical and Experimental Vibration Analysis for a Piezoceramic Disk Partially Covered with Electrodes, *The Journal of The Acoustical Society of America,* Vol. 118, No.2, pp.751-761, ISSN 0001-4966

Huang, C.H.; Lin, Y.C. &Ma, C.C. (2004). Theoretical Analysis and Experimental Measurement for Resonant Vibration of Piezoceramic Circular Plates, *IEEE Transactions on Ultrasonics Ferroelectrics and Frequency Control,* Vol. 51, No. 1, pp.12–24, ISSN 0885-3010

IEEE Standard on Piezoelectricity, 176-1987, ANSI-IEEE Std. 176, 1987.

Ikegami, S.; Ueda, I. & Kobayashi, S. (1974). Frequency Spectrum of Resonant Vibration in Disk Plates of $PbTiO_3$ Piezoelectric Ceramics, *The Journal of The Acoustical Society of America,* Vol.55, No.2, pp.339-344, ISSN 0001-4966

Ivina, N. F. (1990). Numerical Analysis of the Normal Modes of Circular Piezoelectric Plates of Finite Dimensions, *Soviet Physics – Acoustics,* Vol.35, No.4, pp.385–388, ISSN 0038-562X

Ivina, N. F. (2001). Analysis of the Natural Vibrations of Circular Piezoceramic Plates with Partial Electrodes, *Acoustical Physics,* Vol.47, No.6, pp.714–720, ISSN 1652-6865

Jaffe, B.; Cook, W.R. & Jaffe, H. (1971). *Piezoelectric Ceramics,* Academic Press, ISBN 0-12-379550-8, New York, U.S.A.

Kunkel, H. A.; Locke, S. & Pikeroen, B. (1990). Finite-Element Analysis of Vibrational Modes in Piezoelectric Ceramics Disks, *IEEE Transactions on Ultrasonics Ferroelectrics and Frequency Control,* Vol. 37, No. 4, pp.316–328, ISSN 0885-3010

Masaki, M.; Hashimoto, H.; Masahiko, W. & Suzuki, I. (2008). Measurements of Complex Materials Constants of Piezoelectric Ceramics: Radial Vibrational Mode of a Ceramic Disk, *Journal of the European Ceramic Society,* Vol.28, Issue.1, pp.133–138, ISSN 0955-2219

Moulson, A.J. & Herbert, J.M. (1997). *Electroceramics: Materials, Properties, Applications,* Chapman and Hall, ISBN 0-412-29490-7, London, England

Newnham, R.E. & Ruschau, G.R. (1991). Smart Electroceramics, *Journal of the American Ceramic Society,* Vol.74, Issue.3, pp. 463–480, ISSN 1551-2916

Randeraat, J.V. & Setterington, R.E. (1974). *Piezoelectric Ceramics,* Mullard House, ISBN 0-901232-75-0, London, England

Rogacheva, N.N. (1994). *The Theory of Piezoelectric Shells and Plates,* ISBN-10:084934459X, CRC Press, Boca Raton, Florida, USA

Shaw, E.A.G. (1956). On the Resonant Vibrations of Thick Barium Titanate Disks, *The Journal of the Acoustical Society of America,* Vol.28, No.1, pp.38-50, ISSN 0001-4966

Schmidt, G. H. (1972). Extensional Vibrations of Piezoelectric Plates, *Journal of Engineering Mathematics,* Vol. 6, No. 2, pp.133–142, ISSN 0022-0833

Part 2

Nano-Ceramics

3

Fine Grained Alumina-Based Ceramics Produced Using Magnetic Pulsed Compaction

V. V. Ivanov, A. S. Kaygorodov, V. R. Khrustov* and S. N. Paranin
Institute of Electrophysics UD RAS
Russian Federation

1. Introduction

Despite the numerous studies dedicated to the alumina fabrication and its properties research there is still an interest for this material. The main reason for this is the low cost of alumina and wide range of such unique properties as mechanical, electrophysical, thermal and chemical ones. High alumina values are fully realized on monocrystal leucosapphire only. But monocrystal material (by known reasons) cannot be used widely. Cheaper polycrystalline alumina is characterized by low toughness, brittleness, weak thermal resistance and low wear resistance. The general opinion for the realization of the whole potential of alumina is the minimization of its grain size. This is what the researchers's efforts are directed to. The usage of standard approaches namely, standard two-staged technology including cold uniaxial compaction and thermal sintering appears to be a preferable one. Dopants are used for alumina sintering and grain growth processes control. At that the usage of nanopowders compacted up to high density can lead to a formation of fine structure.

The aim of the work is to fabricate alumina-based ceramics with high wear resistance. It is achieved by the usage of weakly aggregated nanopowders and by the application of magnetic-pulsed compaction as well as by the addition of titania, magnesia and zirconia.

2. Initial powders

In present work the nanopowders, obtained by pulsed methods of electrical explosion of wires (EEW) (Kotov, 2003) and laser evaporation (LE) of the ceramic targets with given content (Kotov et al., 2002) were used. The powders were synthesized in the laboratory of pulsed processes of the Institute of Electrophysics UB RAS. These methods allow to obtain weakly aggregated nanopowders due to pulsed material heating and evaporation regimes and the condensation process managing.

The principal of the EEW powders is the formation of the mixture of boiling metal drops because of heating after the explosion of the conductor under the action of the high dense current (10^4-10^6 A/mm^2). This conductor spreads rapidly into the surrounding media. The flying drops are oxidized in the oxygen containing atmosphere (usually argon and oxygen

* Corresponding Author

mixture). The as-formed oxide particles are carried from the explosive chamber to the devices of caching and separation of the powder (electro filters and cyclones) by the gas flow. Al_2O_3 particle image is given on figure 1 (a).

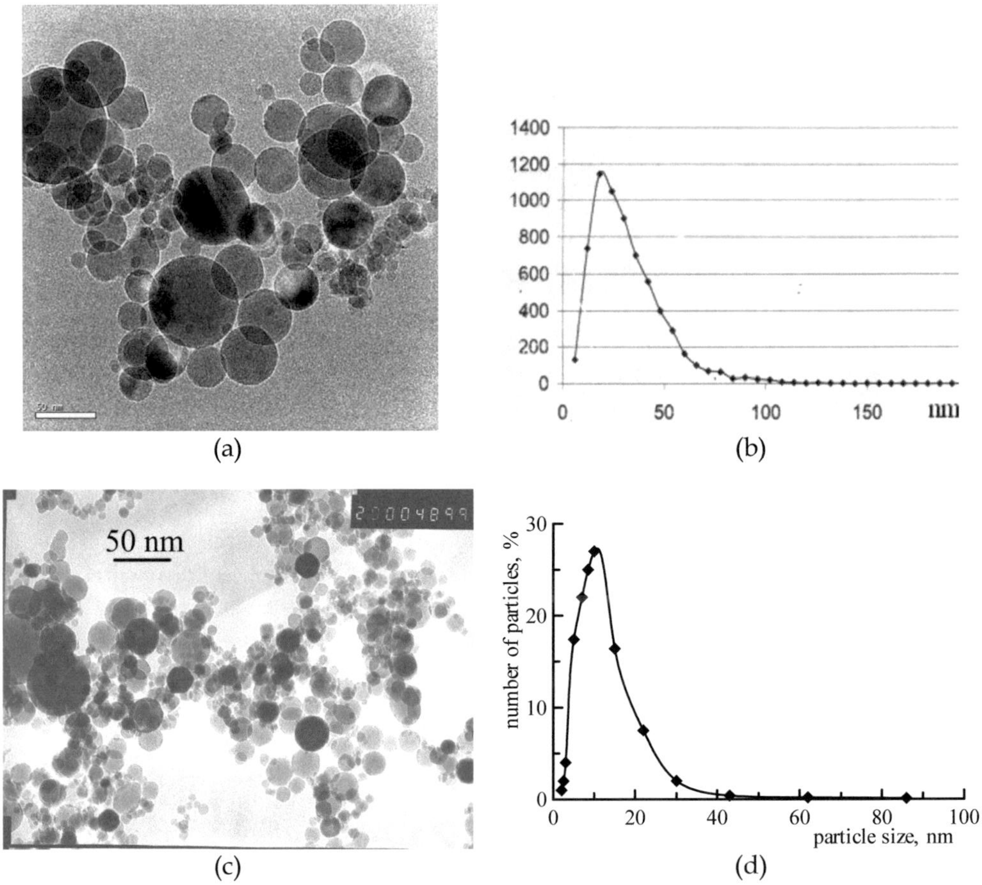

Fig. 1. The appearance of a nanoparticles made by: (a) - EEW process, Al_2O_3 and (c) - LE process, YSZ, with their size distribution (b) and (d), according TEM data.

Laser evaporation method (LE) is based on the usage of pulsed CO_2 laser to evaporate the ceramic target, prepared from micron-sized powders of the given chemical composition. The dispersion of the powders doesn't play any role here. The as-formed oxide particles are carried from the working chamber to the devices of caching and separation of the powder (electro filters and cyclones) by the gas flow (Ar, N_2 and O_2 mixture). For example, figure 1, (c) represents TEM image of the YSZ solid solution particles.

The disperse composition of the nanopowders was characterized by the average particle size of 20 - 30 nm (EEW) and 10 - 15 nm (LE) where the amount of particles that are larger than 0.2 microns was less than 8 wt. %. The nanopowders that were synthesized by pulsed EEW and LE methods are characterized mainly by the spherical-shaped particles and by the size dispersion with the positive asymmetry in the micron region (fig. 1 b and d). The distribution spectrum width of the particle sizes, obtained by LE method is 25% narrower comparing to that of EEW method. Baring in mind that the particles are spherical and that they have relatively smooth surface, the objective characteristics of the powders is the average volume-surface diameter d_{BET} which is determined from the specific surface of the powder, S_{BET} (m^2/g):

$$d_{BET} = 6/\rho\, S_{BET}, \quad (1)$$

where ρ - density of the particle material in g/cm^3.

The peculiarity of this powders is their weak aggregation. The empirical criteria of weak aggregation (in present work) was the formation of the stable suspensions in alcohol (ethanol or isopropanol) under the ultra sonic action of 20 Watts/ml for 3-5 minutes.

In order to eliminate large (>200 nm) particles the powders were separated in alcohol. The mixing of different powders in order to obtain composites was performed here. The ultrasonic mixing during liquid evaporation provided the homogeneous distribution of the particles and prevented the formation of hard aggregates.

The general characteristics (phase composition, specific surface area, average particle size) that are used in present work are presented in tables 1 and 2.

№	Powder type	Phase content, wt.%	S_{BET}, m^2/g	dx, nm
1	Al_2O_3	$0.2\,\gamma + 0.8\,\delta$	72	23
2	AM1	$0.85\,\gamma + 0.15\,\delta$; Mg/Al = 1.5 at.%	69	24

Table 1. The characteristics of the initial powders, obtained by the explosion of the wire (EEW).

Al_2O_3 nanopowder was synthesized by EEW from pure alumina (table 1, p.1) (Kotov, 2003). In order to insert some dopants into alumina the composite compositions were prepared. A part of compositions was made by mixing of EEW and LE nanopowders (table 2, pp. 1 - 4), another part – by EEW alloy of the necessary content (table 1, p.2), and the third part – by the evaporation of the target from powder mixture (table 2 pp. 5, 6). TiO_2 and ZrO_2 doping was realized by the mixing of corresponding nanopowders: AT1 is the mixture of EEW nanopowders Al_2O_3 and TiO_2, AZ10 - EEW nanopowders Al_2O_3 and ZrO_2, A40, A85, - EEW nanopowder of Al_2O_3 and LE powder 2.8YSZ – solid solution of 2,8 mol.% of Y_2O_3 in zirconia (Kotov et al., 2002). Phase and disperse contents of the given composite nanopowders are the combination of the corresponding properties of its components. The homogeneity of particle distribution of both compositions in mixed composite nanopowders was provided by the ultra sonic action with continuous mixing during suspension drying.

The doping by MgO was realized on the stage of explosion of the Al-Mg alloy wire (table 1): AM1 nanopowder.

Two compositions were prepared by the ceramic target evaporation of the known content (table 2, A45 and A93 types) (Kotov, 2002).

The differences in the preparation regime are seen also in phase and granular content of nanopowders. AM1 nanopowder (table 1), obtained by EEC of Al-Mg alloy, on the contrary from EEW of pure alumina, mainly consists from alumina γ-modification. As MgO dissolves in γ and δ-alumina modifications (Smothers & Reynolds, 1954), there were no Mg containing phases found in this powder. The differences in phase content of the EEW alumina and AM1 powders are demonstrated by the diffractograms (a) and (b) on figure 2.

A45 and A93 nanopowders (table 2), obtained by the ceramic target of given content are characterized by the complex phase content and smaller average X-ray size comparing to mixed powders of close content (A40 and A85). The peculiarities of their phase contents are illustrated by the X-ray diffractogramms (fig. 3, a and b).

№	Powder type	Composition,wt.%	Phase content, wt.%	Production method	S_{BET}, m^2/g	dx nm	dx_2 nm
1	AT1	99 (Al_2O_3) + 1 (TiO_2)	Al_2O_3: 0.20 γ + 0.80 δ TiO_2: 0.25 a + 0.75 r	EEW+ suspension mixing	56	22 - δ	48 - r 44 - a
2	A40	40 Al_2O_3 + 60 2.8YSZ	Al_2O_3: 0.20 γ + 0.80 YSZ: t	LE/EEW+ suspension mixing	73	24 - δ	17-t
3	A85	85 Al_2O_3 + 15 2.8YSZ					
4	AZ10	90 Al_2O_3 + 10 ZrO_2	Al_2O_3: 0.20 γ + 0.80 ZrO_2: 0.6 m+0.4 t	EEW+ suspension mixing	72	24 - γ	50 - m 28 - t
5	A45	45 Al_2O_3 + 55 1.6YSZ	18 γ+ 43 t + 39 c	LE of the ceramic target with the given content	81	13 - γ	10-t 8-c
6	A93	93 Al_2O_3 + 7 1.6YSZ	90 γ+1 t + 9 c		86	15 - γ	8-t 6-c

S_{BET} – specific surface area, dx – the average crystallite size of Al_2O_3 defined by the X-ray analysis, dx_2 – the average crystallite size of the additional phases, γ and δ - alumina phases, m, t and c – monoclinic, tetragonal, and cubic phases of ZrO_2(YSZ), a and r – TiO_2 modifications anatas and rutile.

Table 2. The characteristics of the composite nanopowders.

The notable diffusion background in the 2Θ < 40° diffraction angles indicates the presence of sufficient fraction of amorphous alumina in LE nanopowders. Besides this the tetragonal modification of zirconia, as well as the cubic one, had the periods of the crystal lattice that were substantially smaller than that of the corresponding given yttria concentration. The same effect was seen on other Al_2O_3+YSZ composite powders earlier and it is connected with the dissolving of alumina in zirconia (Srdic´ et al., 2008). Knowing that ionic radius of aluminum is 20 % smaller that the radius of Zr and assuming the effect of elementary cell volume decrease to be linear at Zr atoms substitution by Al atoms, the content of alumina in each zicronia phase was estimated. In particular for A45 powder the alumina content was 0.9 and 23 mol.% in tetragonal and cubic zirconia modifications correspondingly. Thus the LE method leads to a formation of zirconia modifications containing a substantial amount of dissolved alumina along with yttria. The substantial amount of cubic phase at low concentration of yttria - 1.6 mol.% indicate on the stabilizing role of alumina dissolved. Small X-ray size of zirconia that are notable smaller than the particle size of the individual 2.8YSZ (table 2) could be connected with the influence of alumina dissolved.

Thus the absence of some Mg containing phases as well as the presence of Al_2O_3+ZrO_2 solid solution indicate the more uniform distribution of the components in AM1, A45 and A93 nanopowders.

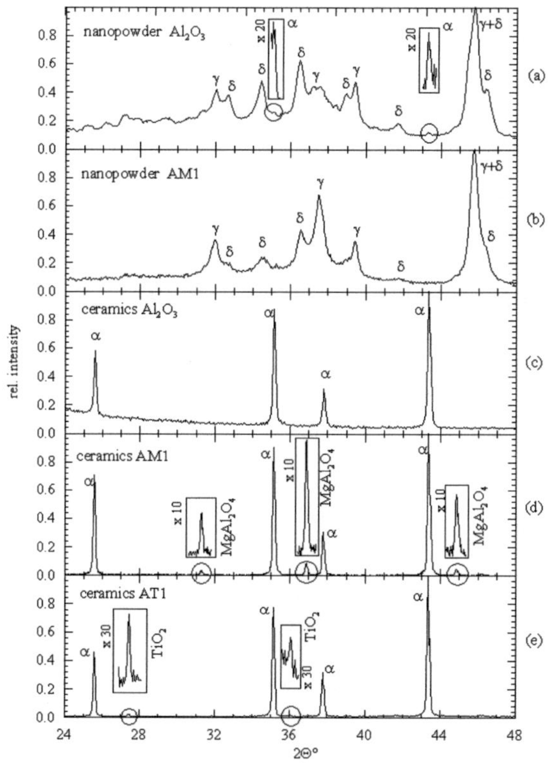

Fig. 2. XRD data (Cu Kα) for EEW nanopowders (a) - Al_2O_3, (b) - AM1 and sintered ceramics Al_2O_3 (c), AM1 (d), AT1 (e).

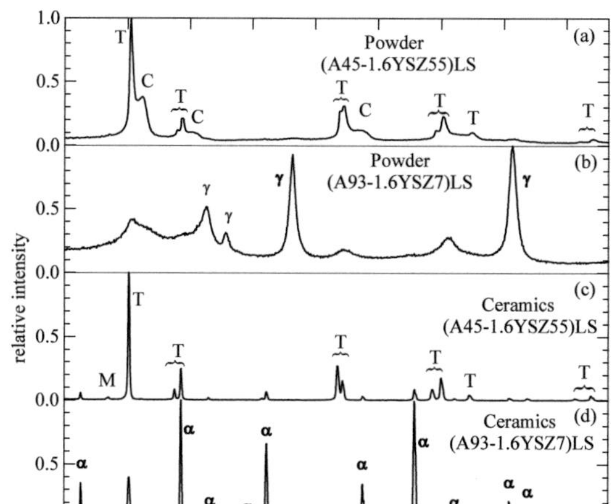

Fig. 3. XRD data (Cu Kα) for LE nanopowders (a, b) and sintered ceramics (c, d).

3. Experimental methods

3.1 MPC, the magnetic-pulsed compaction method

At nanosized particles the properties that prevent the consolidation of the particles into homogeneous compact material become sufficient: the specific bound energy as well as the amount of agents adsorbed and aggregating ability increases. That is why the nanosized powders are characterized with unsatisfactory compressibility and hence, the traditional static methods don't allow to reach high compacts density (Vassen & Stoever, 1992). This leads to a formation of large pores and rapid grain growth during sintering (Chen & Mayo, 1993). The nanopowder compaction and recrystallization difficulties upon sintering limited substantially the possibilities of the traditional compaction methods as well as sintering for fabrication of nanostructured materials and motivated for the searching of new approaches. For this class of nanomaterials hot isostatic pressing (HIP) method appeared to be the most perspective for the fabrication of high dense bulk zirconia ceramics with the grain size less than 50 nm (Hahn & Averback, 1992). Though from the point of view of shaped variety and cost decrease the traditional technology (cold pressing with the following sintering at normal pressure) is still the best one.

In order to fabricate dense compacts from nanosized powders the use of intensive pulsed compression methods appears to be quite attractive. The fast movement of the powder medium makes it possible to overcome the interparticle friction forces effectively. That is why high (> 0.6) relative densities can be achieved (Graham & Thadhani., 1993).

In present work the uniaxial magnetic-pulsed press (Institute of Electrophysics, Ekaterinburg, Russia) was used (Ivanov et al., 1999). It allows to produce pulsed pressure up to 1000 kN. The MPC method is based upon the principle of the throwing the conductor out of the pulsed magnetic field zone. The throwing force is the result of the interaction of the currents in the conductor and the magnetic field. In our research we've used well-known scheme of the conductor's acceleration (Fig. 4). The pulsed generator with the capacitive storages have been applied as the sources of the pulsed power currents and the magnetic fields. The capacity of energy, the voltage of the storage, and the characteristic pulse duration of the device figured on the scheme are 20 kJ, 5 kV, 120 - 360 μs.

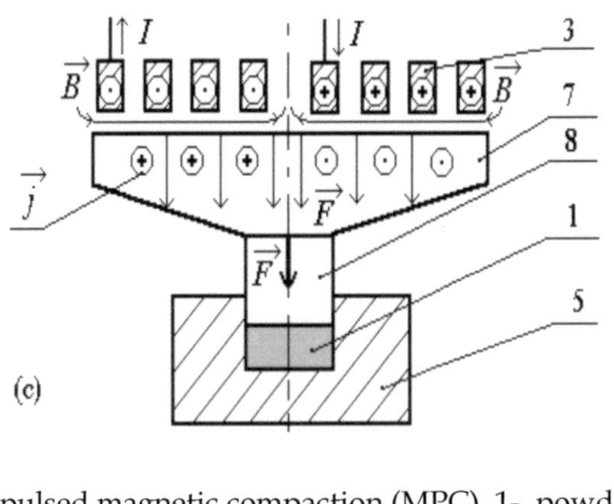

Fig. 4. Scheme of the pulsed magnetic compaction (MPC). 1- powder; 3- inductor; 5- mould; 7- concentrator; 8- piston.

The pulse of the pushing magnetic force on the side of the inductor 3 is apprehended by the hard current-conducting concentrator 7. The concentrator transmits the mechanical pulse to the piston 8 made of the hard-alloy which compresses the powder inside the mould. The high pulsed pressures in the mould (> 1 GPa) are achieved because of the fact that the area of the piston cross section is smaller than the area of the concentrator surface faced with the inductor.

By changing the pulsed pressure amplitude the compacts with relative density up to 0.7 were obtained. In present work the disk-shaped compacts 15-30 mm in diameter and 1.0-3.5 mm thick were used. It should be noted that the densities of the compacts from nanopowders achieved at least 10 - 15% higher than that of traditional "cold" static methods (Ivanov et al., 1997). The fracture image of the MP pressed alumina and YSZ nanopowders is shown on figure 5. The homogeneous location of grains with small amount of pores can be pointed out. The X-ray analysis indicate the intensive mechanical activation of the compacted powders. First of all, the additional peak widening appears. It is connected with the microdistortion of the crystalline lattice of about 0.1%. Second of all, the increase of the stable modifications is observed. For instance, in MP pressed weakly aggregated EEW alumina powder the notable amount (around 1-2 wt.%) of θ– and α– modifications appear in EEW alumina powder, doped with MgO (AM1) the amount of δ-alumina increases on 5 - 6 wt.%. At the same time the amount of less stable γ–alumina decreases (fig 2). The same effects were discovered earlier at EEW TiO_2 nanopowder compaction. Here the amount of more stable and dense rutile increased on about 7% at the decrease of the anatase amount (Ivanov et al., 1995). At that the specific area of the compacts decreases on 10 - 20% comparing to the initial powder. This fact indicated the sufficient increase of the interparticle contacts in the compacts. This leads to the increase of the starting sintering speed.

In general such changes promote the intensification of the following sintering process.

(a) (b)

Fig. 5. MPC green body fracture REM images of nanopowders (a) Al_2O_3, (b) YSZ.

3.2 Sintering and attestation methods

The shrinkage dynamics upon sintering was investigated at constant heating rate at the temperatures up to 1460°C in air by the dilatometric analysis using Netzsch DIL 402C. During sintering the quenching was used for the determination of phase content and grain size. X-ray diffraction method (Brucker D8 DISCOVER diffractometer, Cu Kα radiation) was used for phase content, CSR - coherent scattering region or the dimension of the crystal phases d_x determination in the initial powders, green bodies and ceramic samples. The d_x size was determined by the widening of the diffraction peaks using the Sherrer-Selyakov method. TEM (Jeol JEM 2100) was used for the nanopowders particles analysis. The structure of the ceramics fractures was studied via AFM (NT-MDT Solver P47) and REM (LEO982).

The density of the samples was determined by weighing in water, the specific surface area – by the adsorption of the N_2 or Ar (Micromeritics Tristar 3000).

Microhardness (Hv) and fracture toughness (K_{IC}) of the ceramics was determined by Vickers indentation method at 2 and 7 N load correspondingly (Micro Materials Ltd Nanotest 600).

The relative wear resistant was tested in water abrasive suspension of the aluminosilicate with the 5-50 microns particles as described in (Bragin et al., 2004).

4. Results and discussion

4.1 Sintering of nanopowders Al_2O_3 and influence of additives of MgO, TiO_2

Upon studying the alumina ceramics fabrication from the nanosized powders, containing polymorphs g and t, it appears to be actual to reveal the role of the multistep polymorph transition $\gamma + \delta \rightarrow \theta \rightarrow \alpha$. The influence degree on the whole sintering process is sufficient as the last stage is accompanied by the rapid grain growth of the appearing α - Al_2O_3 (Freim et. al., 1994). At that the features of the polymorph transition are essentially dependent from the quality of the initial powder (disperse and phase content, particles shape etc.) (Lange, 1984), density of its packaging [6] and the nature of the dopants that are inserted in alumina intentionally (Smoothers & Reynolds, 1954).

The influence of the dopants on the polymorph alumina transition upon nanopowders sintering, compacted up to high relative density (around 65%) is illustrated on figure 6. The addition of titania (fig. 6. a) decreases the temperature of the starting point of the polymorph transition relative to pure alumina, meaning the decrease of the temperature of the starting point of the shrinkage of the material (comparison a, b, and c on fig 6.). Moreover, the polymorph transition runs in this system at sufficiently lower densities, showing the decrease of the activation barrier comparing to pure alumina. In contrast to titania (fig. 6. a) magnesia increases the temperature border of the polymorph transition, increases its speed sharply and shrinkage speed in general (fig. 6. b).

Thus the form of the shrinkage curve of the compacts from the nanopowders of the metastable forms of alumina is determined by the processing of the polymorph transitions of alumina to stable α-Al_2O_3. At the staring stage ($t < T_A$) the material consists of the γ, δ, and θ phases mixture. The polymorph transition $(\gamma + \delta + \theta) \rightarrow \alpha$ starts at T_A temperature and can be T_B considered as the temperature of the polymorph transition completion. At $t > T_B$ the material consists of α-alumina and at T_C the α–alumina shrinkage starts. Thus the shrinkage curve showed in fig. 7 is a sum of two S-shaped curves: the shrinkage curve defined by the polymorph transition starting at T_A and shrinkage curve upon sintering (the decrease of the porosity) staring at T_C. The T_A values of the polymorph transition are shown in table 3. The increase of T_A leads to the highest shrinkage speed in the T_A-T_B region of the AM1 nanopowder.

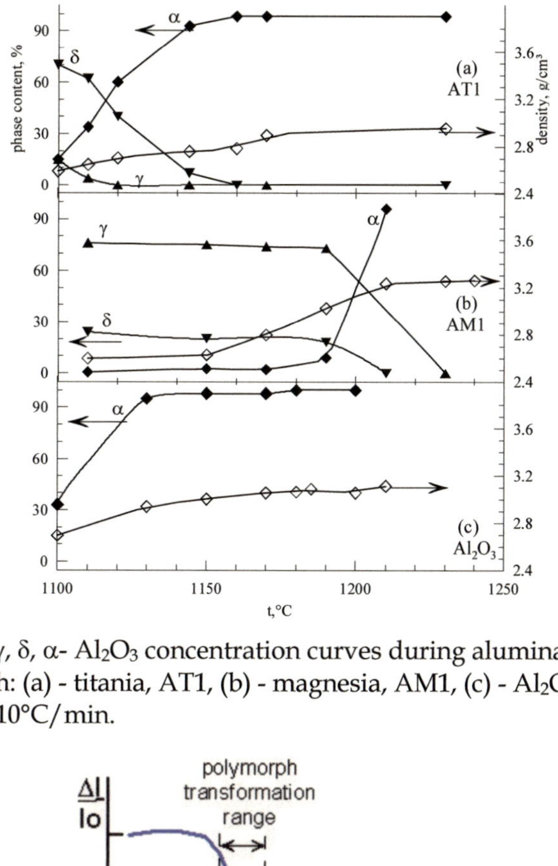

Fig. 6. Shrinking and γ, δ, α- Al_2O_3 concentration curves during alumina nanopowders sintering. Doping with: (a) - titania, AT1, (b) - magnesia, AM1, (c) - Al_2O_3 without dopants; constant heating rate 10°C/min.

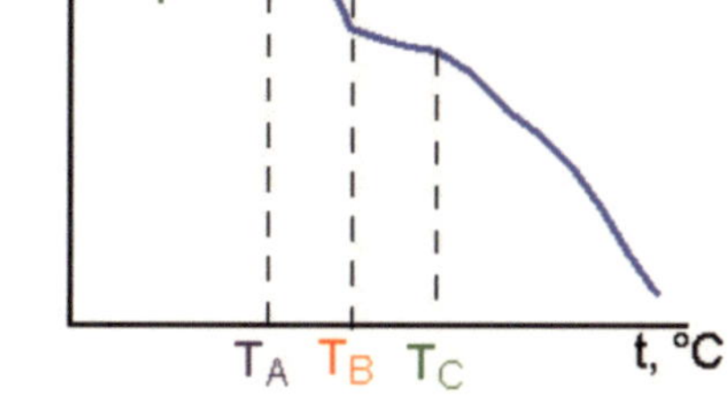

Fig. 7. Generalized sintering curve of alumina nanopowders.

Al_2O_3	AT1	AM1	A40	A85	A45	A93	t-YSZ
1050	1090	1140	1040	1170	1140	1150	850

Table 3. The temperatures T_A (±20°C) of the shrinkage beginning of nanopowders at constant heating rate 5°C/min.

The sufficient influence of the second phase nature on the grain structure evolution upon sintering of the ceramics is found (fig. 3.8). The size of the primary α–crystallites that are formed during sintering of the AM1 (MgO doped Al_2O_3) ceramics is comparable to the size of the crystallites of the initial metastable alumina modifications (fig. 3.8 b). The abnormal size change of the α-alumina crystallites in the system doped by TiO_2 is found (fig. 3.8 a). The size of the primary α–crystallites, as in the case of undoped alumina, is sufficiently larger than the size of the initial phases crystallites. Although in the temperature region of the mass formation of α–modification a large amount of the crystallites comparable with the crystallites of the initial modifications in terms of size appear. The following sintering is accompanied by a monotonic grain growth. The effect of the decelerated of the α–alumina crystallites upon magnesia doping (AM1, figure 3.8 b) is found. At equal densities the grain size of the alumina ceramics doped with titania (figure 3.8, a) is much larger comparing to AM1.

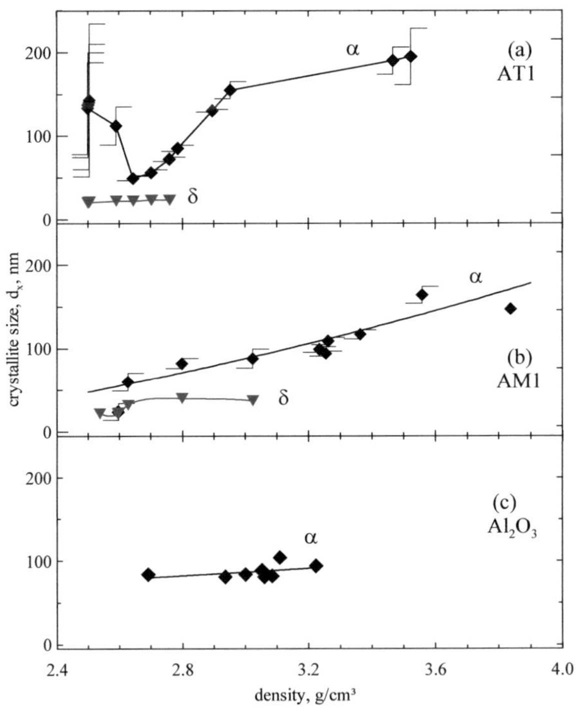

Fig. 8. Average X-ray crystallite size evolution upon sintering of the alumina ceramics, doped with: (a) - titania, AT1, (b) - magnesia, AM1, (c) - Al_2O_3 without dopants; constant heating rate 10°C/min.

At our opinion the positive role of the magnesia addition in fabrication of the dense and hard alumina-based ceramics is connected with the nature of the magnesia influence on the diffusion processes upon alumina sintering, as well as with the technique of the additive insertion. It's important to note that MgO is dissolved in alumina nanopowder at atomic level and it doesn't form separate phases. The second $MgAl_2O_4$ spinel phase along with the α–alumina appears only during sintering of the AM1 compacts. It is seen from the comparison of the X-ray patterns of the initial powder (fig. 2, b) and sintered ceramics (fig. 2, d). On the contrary in AT1 ceramics there were no signs of mixed combination with alumina and titanium found. The α–alumina and TiO_2- rutile peaks are clearly seen on x-ray patterns (fig. 2, e).

The ceramics structure data, from analysis of the fracture images obtained by AFM in "height" and "Mag-cos" modes (figures 9 and 10) prove the X-ray data. In all cases the material consists of polycrystalline blocks, which shape and size depend on the dopant. The

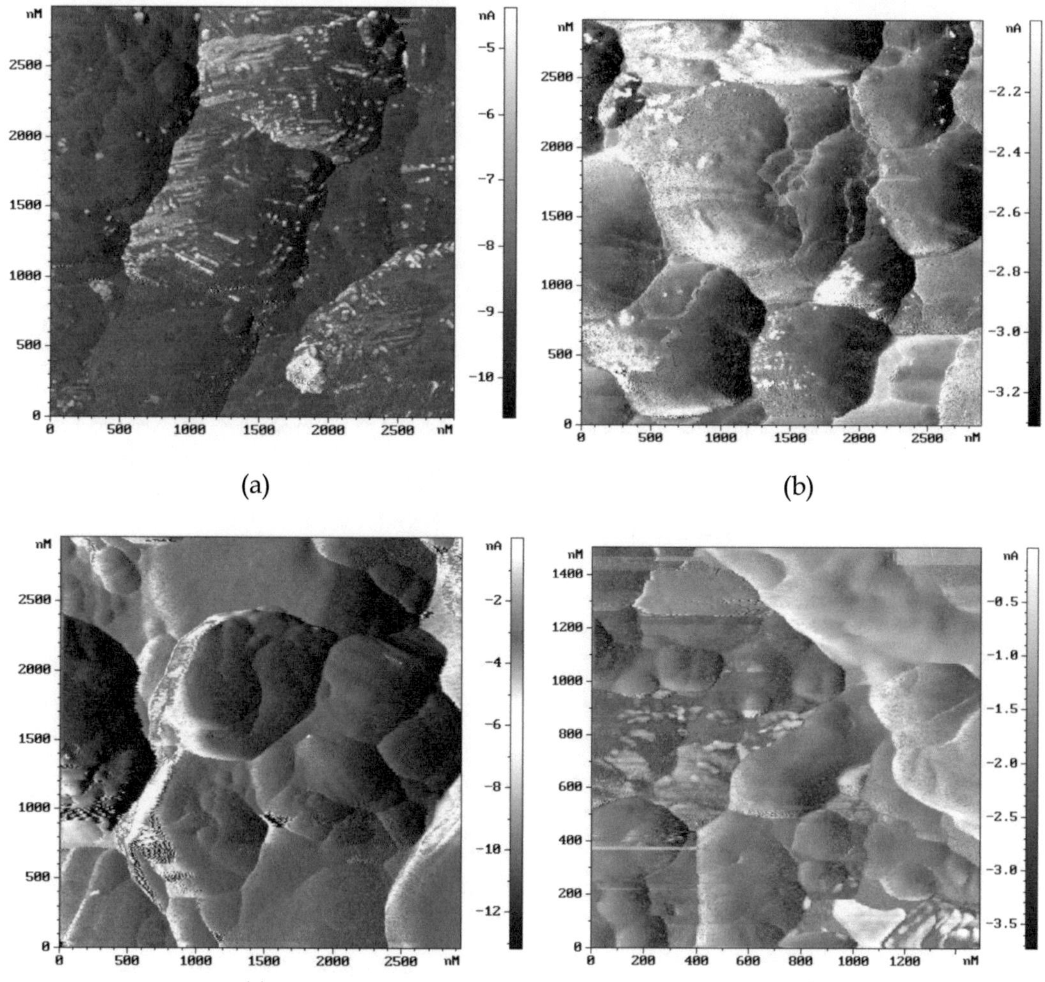

Fig. 9. Ceramic fracture AFM images of (a) - AM1 (τ_s = 6 min); (b) AM1 (τ_s = 30 мин); (c) AT1 (τ_s = 6 min); (d) AZ10 (τ_s = 60 min). ("Mag-cos" mode).

blocks's average size, d, given in table 4, and obtained by the intersection method of the AFM images clearly demonstrate the inhibitory influence of the MgO and ZrO_2 on the recrystallization processes in alumina matrix. It's worth mentioning the originality of MgO dopant which is uniformly dissolved in the initial metastable alumina and its segregation as $MgAl_2O_4$ nanocrystals on the boundaries of the matrix α–alumina crystals takes place at the polymorph transition to α-alumina only. On AFM images, taken in "Mag-cos" mode (fig. 3.9, a) the uniformly distributed contract light regions with around 20 nm in size are clearly seen.

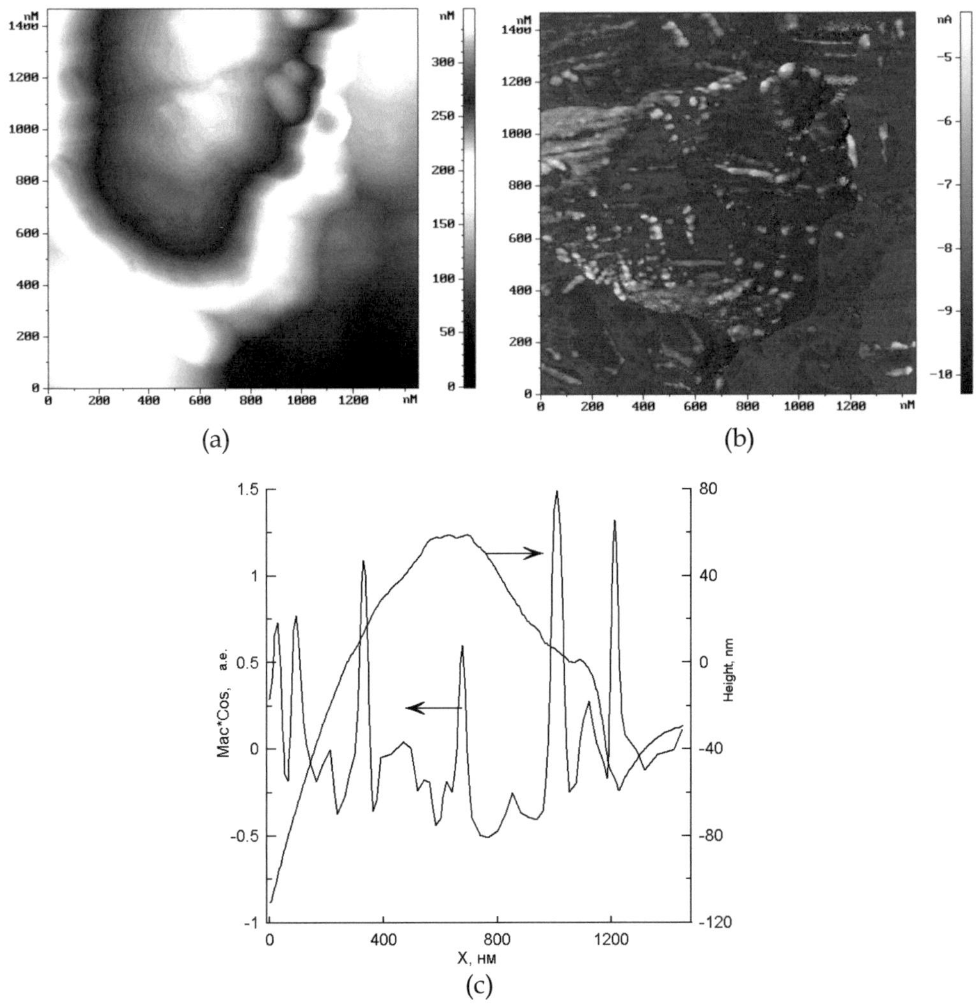

Fig. 10. AM1 ceramic fracture AFM images in (a) "height" and (b) "Mag-cos" modes with the corresponding profiles at Y=800 nm (c).

Probably these regions correspond to the second phase - $MgAl_2O_4$, defined by X-ray. It is proven by the combined structural analysis in "height" and "mag-cos" modes (figure 10). Clear peaks on the "mag-cos" curve (fig. 10, c) and corresponding smooth relief contour show the invariation of the block surface elastic properties and the presence of the doped phase. This peculiarity is present upon the decrease of the sintering dwell time of such ceramics (fig. 9, b). Thus the doping by MgO allowed obtaining fine and homogeneous structure in alumina ceramics with uniformly dispersed $MgAl_2O_4$, nanophase over the grain surface leading to grain growth ingibition.

In case of the TiO_2-doped ceramics the largest blocks are observed (fig. 9). Its surface, in contrast to the mentioned above types of ceramics, is smooth. This fact doesn't allow to state the substantial difference between the crystal and block size. The accumulations of the doped phase are non-uniformly distributed over the volume. They are localized as about 100 nm thick interlayers between blocks (see light regions on fig. 9). Some isolated grained of the same size can be also seen.

4.2 Shrinkage peculiarities upon the sintering of the alumina nanopowders based ceramics doped with ZrO_2

A wide range of Al_2O_3+ZrO_2 compositions was studied (table 2). Pure zirconia as well as partially stabilized zirconia in tetragonal modification - as solid solution with Y_2O_3 (t-YSZ) with concentration of 1.6 and 2.8 mol% was used. The notable difference of the shrinkage curves of the nanopowders mixtures from the composite LE -nanopowders can be seen (fig. 11). It should be noted that LE nanopowders differ by sufficiently nonequilibrium material state, namely large fraction of alumina as solid solution in zirconia and in amorphous state. That is why its shrinkage behavior sufficiently differs from other nanopowders. It can be seen, that the shrinkage trend of the latter (A45 and A93 pp. 5 and 6 of table. 2) though the large difference of the Al_2O_3/t-YSZ ratio. On the other hand some differences of the nanopowders with close composition (A93 and A85) are seen. The shrinkage of the A85 nanopowder, obtained by the mechanical mixing of two powders (p.3 table 2), is similar to the shrinkage of the metastable alumina with the only difference in elevated for 100°C sintering temperature, as the shrinkage behavior of the A93 LE nanopowder is close to the shrinkage of t-YSZ, with 200°C temperature up shift (table 3).

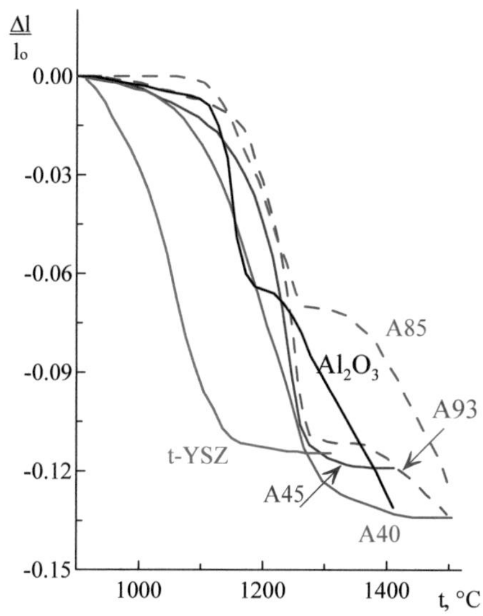

Fig. 11. Shrinking curves of nanopowders Al_2O_3 + t-YSZ (constant heating rate 5°C/min).

It is seen that the shrinkage curve of A40 and A45 compositions doesn't have the typical fracture, peculiar to the metastable alumina sintering. Bearing in mind the difference of the starting shrinkage temperatures of alumina and t-YSZ, it can be assumed that by the beginning of the $\gamma \rightarrow \alpha$ polymorph transition in alumina, the zirconia matrix is already formed. That is why the effect of the alumina crystal volume change is not notable on the shrinkage curve.

4.3 Peculiarities of the structural-phase state of the composite submicron alumina matrix ceramics with the YSZ and ZrO_2 additives

It appeared that in alumina exceeded ceramics the addition of zirconia doesn't lead to a substantial alumina grain growth inhibition effect upon sintering (fig 12). The average X-ray crystallite size of α-Al_2O_3 increases monotonously out of the confidence range of the X-ray method at 0.98 dense ceramics and denser. The nanopowders synthesis route appeared more important. It is seen on figure 12 that the crystallites size of α-Al_2O_3 in the ceramics, obtained from composite LE nanopowder (A93 type) is substantially larger than that of the ceramics, sintered from A85 nanopowder, obtained by the mixing of the individual nanopowders: EEW Al_2O_3 and LE 2.8YSZ.

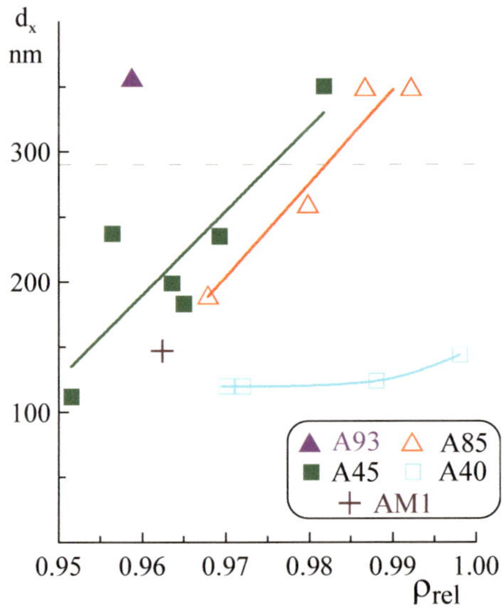

Fig. 12. Evolution of the average α-Al_2O_3 crystallite size during sintering of ceramics Al_2O_3 + t-YSZ. (constant heating rate 5°C/min) Dotted line shows the border of the confidence range of the X-ray method.

The notable effect of the grain growth limitation is obtained in the A40 ceramic samples where the volume fractions of Al_2O_3 and T-YSZ were approximately the same. While the density was reaching the theoretical limit the X-ray crystallite size didn't exceed 140 nm (fig 12 and table 4). In this case the limitary crystalline size can be limited by the homogeneity of the Al_2O_3 and T-YSZ initial powders mixing level reached. Perhaps the nanopowders mixing took place at aggregates level and the limitary crystal size in ceramics was determined by the quantity of the material in the volume of the single aggregate.

The visual interpretation of the X-ray peaks widening is given on figure 13 for the corundum doublet peak {024}. It is seen that the width of the reference peaks of coarse grained ceramics is the smallest one and two peaks are clearly separated. The corresponding peaks for the composite ceramics are merged together and are barely observable due to the presence of the shelf on the right side of the peak merged. At almost equal ceramics relative density (0.965) the peaks of the composite containing approximately equal volume fraction of tetragonal zirconia (2.8YSZ) and Al_2O_3 are the widest ones. This shows the finer structure of the ceramics with such composition.

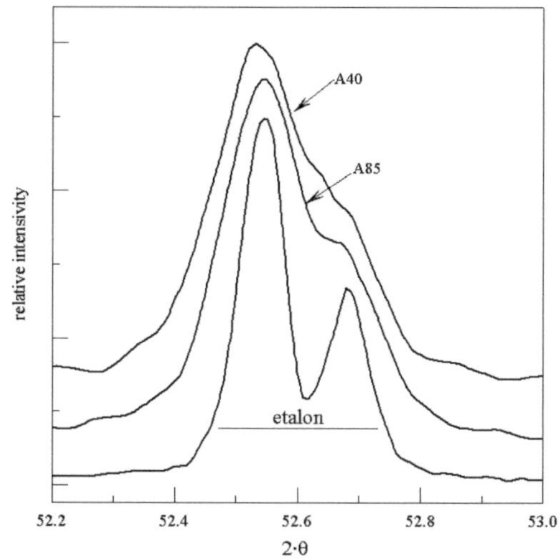

Fig. 13. X-ray corundum doublet reflex {024} (Cu K_{α}) for two composite ceramic Al_2O_3+YSZ, in comparison with rough crystalline ceramics Al_2O_3.

The second phase (Al_2O_3) didn't influence substantially on the grain growth of t-YSZ. While reaching the theoretical density the average X-ray crystallite size in all types of ceramics, including individual 2.8YSZ, becomes almost identical, around 70 - 90 nm.

Two consequent states of the polycrystalline structure at 0.92 and 0.97 densities are shown on the AFM images of the ceramics fractures (fig. 14). As the ceramics density increases (independently of the powder synthesis route) the general trend of the relief development is tracked. At relative density of 0.92 the relief consists of large blocks on the surface of which the boundaries of smaller grains are clearly seen (fig. 14, a and b). At relative density of 97% small fragments located on the surface of large blocks become less notable (fig. 14, c and d). The fraction of fractures that take place over the grain volume increases as the relief elements (formed upon crystal fracture) appear. There are practically no separated small particles in the bulk of the material. The "roundness" of the boundaries disappears. At that the structural changes that are observed in "Mag-cos" mode are more notable (fig. 15 a - d).

The rapid decrease of the intergranular boundaries contrast is observed. At starting sintering stages (relative density of 0.92, fig. 15, a) the boundaries between grains look like light lines. Such "brightening" of the boundaries is connected with the AFM apparatus peculiarities at the analysis surface with large slope. Large height difference on the grain boundaries allows to conclude that the fracture of the 0.92 dense ceramics takes place along the grain boundaries.

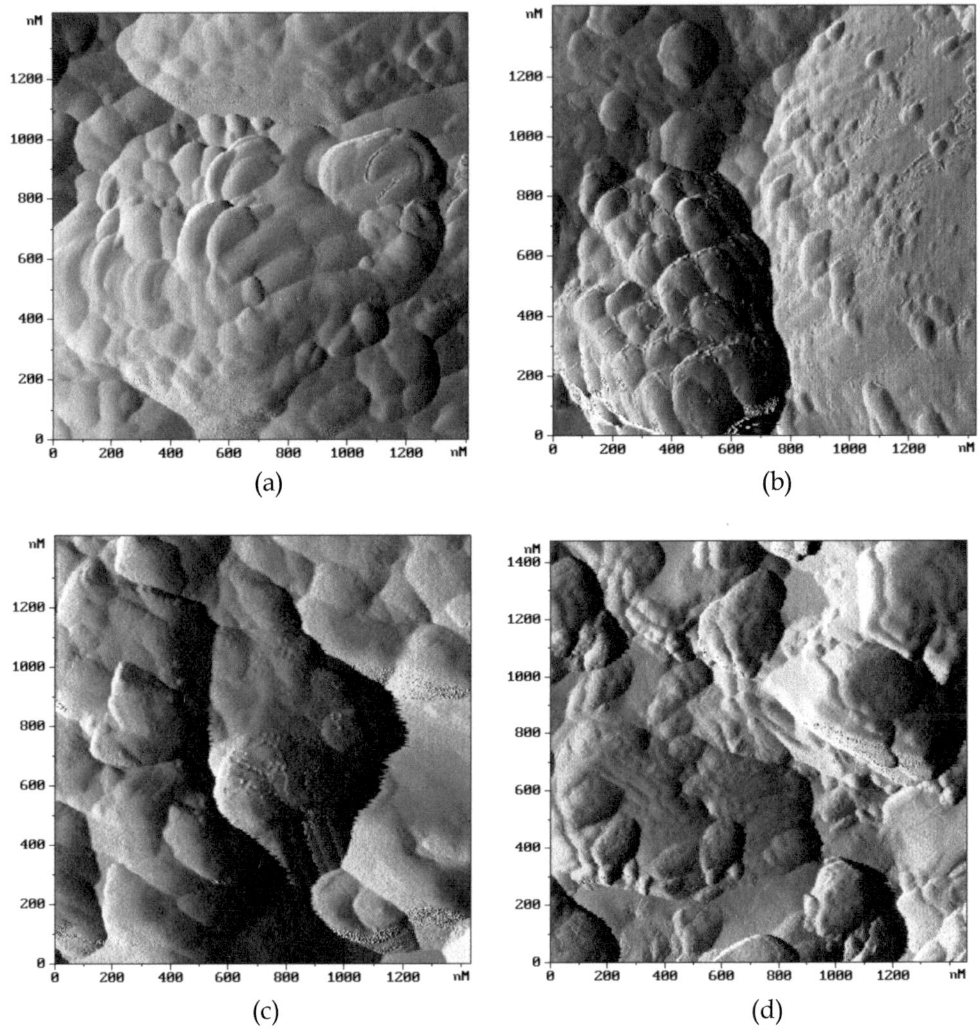

Fig. 14. Fracture AFM images in "height" mode of Al_2O_3 + t-YSZ ceramics at relative densities of 0.92: (a) - A93, (b) - AZ10 and 0.97: (c) - A93, (d) - A85.

The images of the 0.97 dense ceramics taken at "Mag-cos" mode (fig. 15, b) differ a lot: large white and small dark regions are distributed randomly. As the volume fraction of $\alpha-Al_2O_3$ phase is larger than t-YSZ fraction in A93, A85 and AZ10 ceramics, one can assume that light regions are the large corundum blocks and dark regions are the insertions of zirconia. It should be pointed out that "Mag-cos" images of A93 ceramics sintered from laser powder (fig. 15, b) are the same as for A85 ceramics sintered from the mechanical powder mixtures (fig. 15, c).

In case of AZ10 ceramics (fig. 14 d and 15 d) the non-uniformity of the block's shape and size should be pointed out. Probably such huge difference from A85 (having almost the same composition and prepared by the same route) is connected with the non-uniform distribution of ZrO_2 dopant in the initial AZ10 powder.

Fig. 15. Al_2O_3 + t-YSZ ceramic fracture AFM images in "Mag-Cos" mode (a) - A93 (relative density 0.92), (b) - A93 (c) - A85, (d) - AZ10 (relative density 0.97).

AFM images of A40 ceramics fracture in "Mag-cos" mode (fig. 16) show the equidimensional dark and light regions, distributed over the image area more uniformly compared to the fracture images of other ceramics (fig. 15, b and c). Thus in this ceramics the oxide grains are mixed quite well.

It looks interesting to compare the grain size obtained from AFM images analysis and from X-ray data (table 4). The X-ray grain size in A85 and A93 ceramics is larger than the confirmative range of the method. The X-ray analysis showed that in A40 and A45 ceramics the X-ray crystallite size is small being twice as small as obtained from AFM data. Moreover, according to AFM data the grain size in the ceramics sintered from nanopowders obtained by different methods doesn't differ and is 280±10 nm. The difference in the values is logical as AFM shows the size of the "blocks" which can consist of a large number of crystals with small X-ray crystallite size. The cantilever tip has a radius of about 5 - 10 nm. Therefore if the boundary width between crystals is smaller than this value it's not registered by this method.

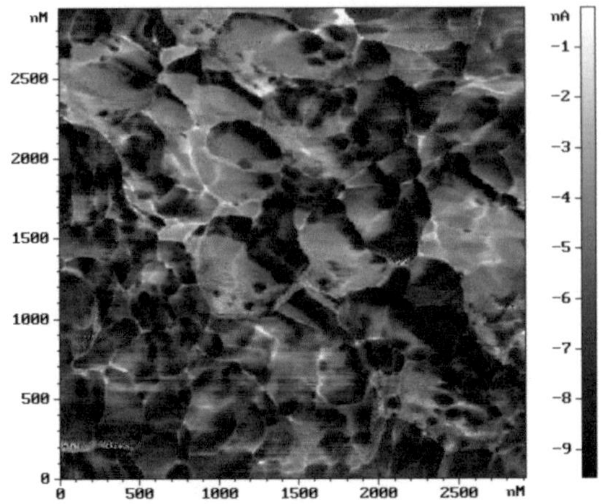

Fig. 16. A40 ceramic fracture AFM image in "Mag-Cos" mode, relative density 0.99.

4.4 Hardness and fracture toughness of submicron alumina-based ceramics

For the fabrication of the full dense ceramics for each powder type the corresponding temperature and dwell time were picked for at least 65% green bodies. The main characteristics of the ceramic samples are shown in table 4. All samples contained stable modifications. In the ceramics sintered from composite LE nanopowders (A45 and A93) up to 3% of monoclinic zirconia was found.

The microhardness of all ceramic samples was in 17-21 GPa range. The best hardness of 21 GPa had AM1 (MgO doped alumina) ceramics. This value is comparable to the hardness of monocrystall alumina - leucosapphire 20.9 GPa. High ceramics properties (table 4) prove the assumption that particle aggregates of the powders obtained by MPC and LE methods are weak and don't influence the compaction process with the following sintering.

The microhardness of the ceramics containing t-YSZ depends on its amount and increases from 17 up to 20 GPa with the decrease of t-YSZ amount from 60 down to 7 wt.%. It should be noted that for A93 ceramics sintering the highest temperature of 1490°C was used. At that its relative density being 0.961 was the lowest one in the row of the ceramics studied.

The microhardness of the ceramic samples with large fraction of YSZ (A40 and A45) being 17 GPa is close to the values of the individual 2.8YSZ. This is defined probably by the presence of YSZ matrix where the corundum grains are distributed. Fracture toughness of the ceramics containing t-YSZ was in 5 - 6 MPa $m^{1/2}$ range.

Wear resistance is the most important integral property of the material, showing the package of its mechanical properties. The comparative analysis of the wear resistance of 3 types of ceramics (ZrO_2, MgO or TiO_2 doped alumina) showed that the specific wear resistance correlates with the microsturcture characteristics and with the hardness of the ceramics. The samples with finer structure and higher hardness have better were resistance. Thus in order to achieve high wear resistance of the ceramics it is necessary to decrease grain size at high density.

powder type	dopa-nt	t_S, °C	τ_S, min	ρ, g/cm³	H_V, GPa	K_{1C}, MPa·m½	d_X α-Al_2O_3 nm	d_X ZrO_2 nm	d_{ACM}, nm	ε g/(kW·h)	Ra nm	phase content
Al_2O3	-	1410	2	0.941	21		> 200	-	-			α-Al_2O_3
AM1	MgO	1410	5	0.976	14		180	-	-			96 - α-Al_2O_3 + 4 - $MgAl_2O_4$
		1450	6	0.969	21	4	180	-	200-300	0.04	29	
		1450	30	0.973	20	4	> 200	-	300-500	0.06	43	
AT1	TiO_2	1450	6	0.961	17	3	> 200	-	200-800	0.13	130	99 - α-Al_2O_3 +
		1450	60	0.972	17	4	> 200	-	300-1000	0.14	180	1 - R-TiO_2
AZ10	ZrO_2	1450	6	0.920	13	4	150	-	200-300			90 - α-Al_2O_3
		1450	60	0.977	19	5	> 200	70	300-600	0.07	92	7,5 - t-ZrO_2 2,5 - m-ZrO_2
A45	1,6YSZ	1410	120	0,982	17	5	> 200	80	-			45α + 52t + 3m
A40	2.8YSZ	1410	15	0.997	17	5	140	70	300-400			40α + 60 t
A93	1.6YSZ	1490	0	0.961	20	5	> 200	55	300-400			93α + 7t+m
A85	2.8YSZ	1450	60	0.980	18	6	> 200	70	300-400			85α + 15t

t_S i τ_S - sintering temperature and dwell time, ρ - density, H_V - microhardness, K_{1c} - fracture toughness, d_x- average X-ray crystallite size defined by the X-ray analysis, d_{AFM} – average grain size according AFM data, ε - specific wear; R_a - roughness of attacking surface; R – rutile, t and m – tetragonal and monoclinic forms of YSZ.

Table 4. The characteristics of the ceramics based on Al_2O_3.

The comparison of the wear resistance of AM1 ceramics with commercial XC22 corundum ceramics showed more than 3 times higher resistance of AM1 ceramics. This can substantially increase the life time of the devices without fixing.

The investigation data of the external surface of the ceramic samples after the wear testing via the optical interferential microscope with 2000x magnification logically complement and correlate with the information about the internal microstructure and the resistance of the ceramics to the wear action. The main peculiarities of the sample's surface structure are demonstrated by the projected images of the relief (fig. 3.18). The standard roughness of the attacking surface of all samples is shown in table 4. The attacking surface of the most wear resistant AM1 ceramics has the most homogenous surface structure at the absence of macro defects and the lowest roughness R_a = 29 nm (fig. 17, a). For AM1 ceramics with prolonged sintering dwell time the carryover of the material during wear characterizes with the round-shaped cavities up to 5 microns in size and up to 0.3 microns deep (fig. 17, b) at low average roughness. In case of AT1 ceramics, sintered for 6 minutes, the carryover tracks have 10 microns jagged shape (fig. 17, c). Upon wearing of AT1 ceramics with prolonged sintering dwell time (60 minutes) (fig. 17, d) the increase of fracture non-uniformity and grain size of the

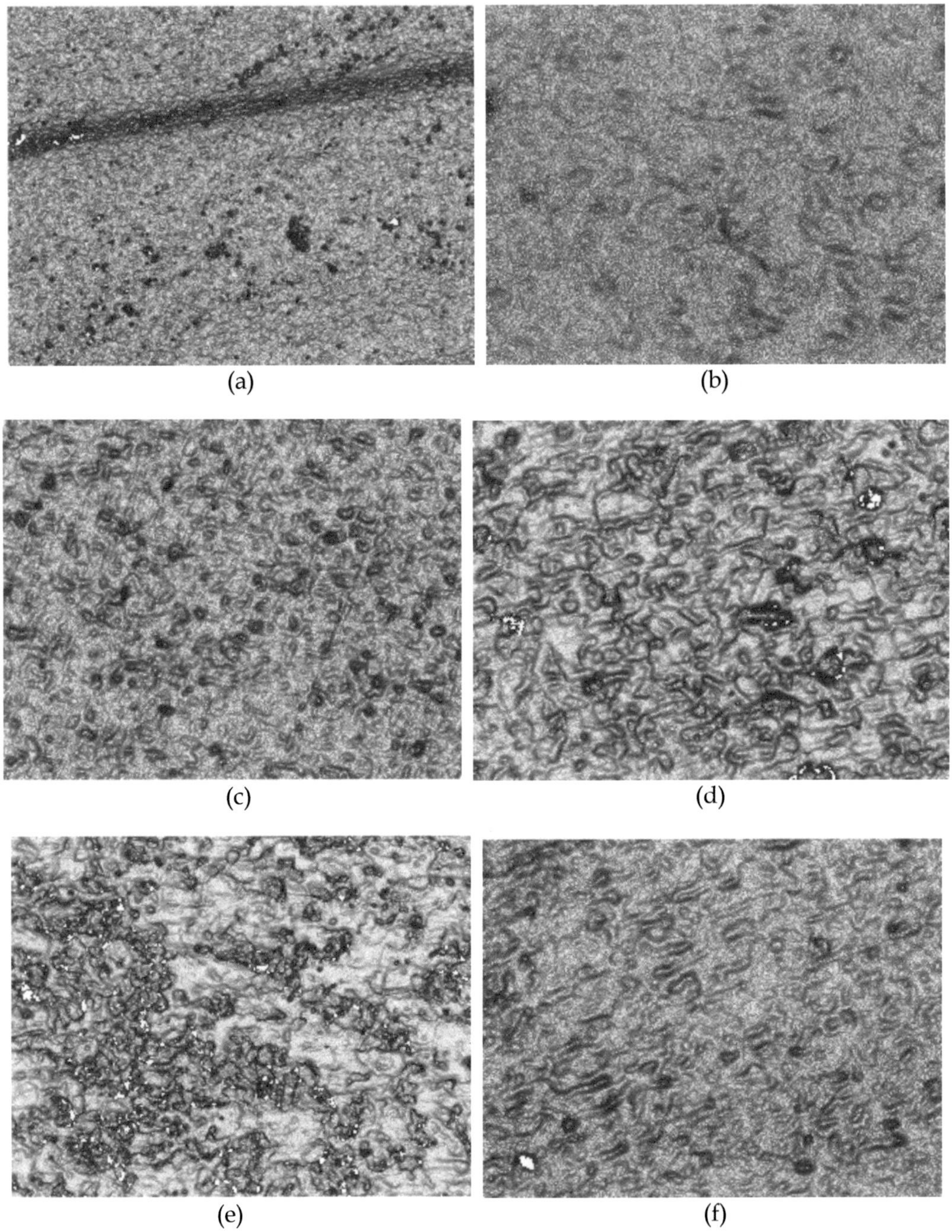

Fig. 17. The characteristic images of the working surfaces of the submicron ceramics after the wear test: the attack surface (2000x): (a) - AM1 (τs = 6 min), (b) - AM1 (τs = 30 min), (c) AT1 (τs = 6 min), (d) - AT1 (τs = 60 min), (e) - AZ10 (τs = 60 min) ; (f) sliding surface (1000x) AM1 (τs = 6 min. Image area at 2000x - 70x50 microns, at 1000x 40x110 microns.

erosion cavities up to 20 microns at the depth of 1.5 microns was observed. The damage dimensions of the AZ10 ceramics surface is much less and its low roughness $R_a = 92$ nm (fig. 17, d) correlates with high wear resistance comparing to AT1 ceramics. Though two types of ceramics doped with TiO_2 and ZrO_2 had sufficient difference in wear factor (ε, table 4) they are characterized with notable relief trend corresponding to the direction of the abrasive flow.

The wear of the working surface of the sample (in addition to the abovementioned peculiarities) is characterized by the 2-3 times elevated roughness and gutter formation, oriented along the moving direction. In particular fig. 17 illustrates this peculiarity for AM1 (dwell time - 6 minutes) ceramics with the characteristic gutter width of 15 microns and depth of 1.5 microns correspondingly.

The abovementioned data show that among the types of ceramics with α-Al_2O_3 matrix the composition containing homogeneously dispersed $MgAl_2O_4$ second phase is the best one. At that the homogeneity of MgO dopant was provided during the nanopowder synthesis via EEW of Al-Mg alloy. During sintering the aluminum-magnesium spinel second phase segregates and covers the surface of the formed crystallites α-Al_2O_3 (fig. 10) uniformly. This inhibits the mass movement processes and eliminates the grain growth of α-Al_2O_3 crystals effectively. Being anisotropic, corundum creates the inhomogenety of properties (hardness and microstructure stress) in polycrystal body. The decrease of the grain size leads to the decrease of the effect of the properties change moving from one grain to another. This leads to the increase of wear resistance upon the decrease of structure scale.

MgO doped alumina ceramics showed the best characteristics. In this ceramics the homogeneous distribution of the dopant was provided by the nanopowder synthesis method.

Small grain size was obtained due to high starting density of the green bodies. The use of MPC of weakly aggregated nanopowders made it possible. At that the larger contact area of the particles and high homogeneity level of the compact are reached. At elevated level of defects of intergranular boundaries and large amount of crystallization centers of α-Al_2O_3 this creates height starting position for the polymorph transition and for the shrinkage. Also it allows to conduct sintering at relatively low temperatures in 1400-1450°C range.

5. Conclusions

1. The usage of weakly aggregated nanopowders of unstable alumina phases, compacted by MPC method up to high density more than 0.65 (relative to theoretical one) allows obtaining dense ceramics with submicron structure at relatively low sintering temperatures in the 1400-1450°C range and up to 30 minutes dwell time.
2. The wear resistance of the ceramics correlates with the structural scale, density and dopant nature. In the row of the TiO_2 - ZrO_2 - MgO additives the wear characteristics of alumina increases sufficiently. The MgO doped ceramics, which is characterized with the smallest grain size of around 300 nm upon high density and hardness appears to have the best parameters.
3. The sintering at reduced down to 1450°C temperatures allowed to obtain the alumina-based ceramics which is 2.5-3 times more resistant to abrasive-erosive wear comparing with the best industrial ceramics with the same composition. Such ceramics combine high density, 0.97, small grain size of the alumina (<300 nm) and $MgAl_2O_4$ spinel (20 nm), high

hardness (20-21 GPa) and fracture toughness (4 MPa m$^{0.5}$). The ceramics is fabricated from metastable (γ and δ - phases) alumina nanopowder with the soluted magnesia in it.

4. It was found that the microhardness of the nanosized alumina / tetragonal zirconia (T-YSZ) composite ceramics depends from the T-YSZ amount and increases from 17 to 20 GPa with the decrease of its amount from 60 down to 7 wt.%. The fracture toughness of the samples is around 5-6 MPa m$^{0.5}$. At that the best elimination of the crystallites growth of alumina (140 nm) was observed for the composite ceramics with the equal amount of α-Al_2O_3 and T-YSZ phases and sintered at 1410°C up to the relative density of 0.997.

6. References

Bragin V.B., Ivanov V.V., Ivanova O.F., Ivin S.Yu., Kotov Yu.A., Kaygorodov A.S., Kiriakov S.I., Medvedev A.I., Murzakaev A.M., Postnikov V.S., Neshkov P.F., Khrustov V.R. & Shtoltz A.K. (2004). Wear resistance of a fine structured ceramics based on $A1_2O_3$ doped with magnesium, titanium or zirconium. *Perspectivnye materialy,* No.6. - pp. 48 - 56. (in russian)

Chen D-J. & Mayo M.J. (1993). Densification and grain growth of ultrafine 3 mol % Y_2O_3-ZrO_2 ceramics. Nanostruct. Materials, Vol. 2, – pp. 469 - 478.

Freim J., Mckittrick J., Katz J. & Sickafus K. (1994). Microwave Sintering of Nanocrystalline γ- $A1_2O_3$. Nanostructured Materials, Vol. 4, No 4, pp. 371 - 385.

Graham R.A. & Thadhani N.N. (1993). Solid State Reactivity of Shock-Processed Solids. In Shock Waves in Materials Science, Sawaoka A.B., pp. 35 - 99, Springer-Verlag.

Hahn H. & Averback R.S. (1992). High Temperature Mechanical Properties of Nanostructured Ceramics. Nanostruct. Materials, Vol.1, - pp. 95 - 100.

Ivanov V.V., Kotov Yu.A., Samatov O.H., Boehme R., Karov H.U. & Schumacher G. (1995). Synthesis and dynamic compaction of ceramic nano powders by techniques based on electric pulsed power. Nanostruct. Materials. Vol. 6, No 1 - 4, pp. 287 - 290.

Ivanov V.V., Paranin S.N., Vikhrev A.N. & Nozdrin A.A. (1997). The efficiency of the dynamic method of sealing nanosized powders, Materialovedenie, No 5, pp. 49 – 55 (in Russian)

Ivanov V., Paranin S., Khrustov V. & Medvedev A. 1999. Fabrication of articles of nanostructured ceramics based on Al2O3 and ZrO2 by pulsed magnetic compaction and sintering, Proc. of 9th World Ceramic Congress Cimtec - Ceramics: Getting into the 2000's, - Florence, Italy,. - Part C. - pp. 441 - 448.

Lange F.F. (1984). Sinterability of Agglomerated Powders, Journal of Amer. Cer. Soc, Vol. 67, No 2, – pp. 83 – 89.

Kotov Yu.A., Osipov V.V., Ivanov M.G., Samatov O.M., Platonov V.V., Azarkevich E.I., Murzakaev A.M. & Medvedev A.I. (2002). Properties of oxide nanopowders prepared by target evaporation with pulse-periodic CO_2 laser. Technical Physics, Vol. 47, No. 11. pp. 1420-1426.

Kotov Yu.A. (2003). Electric explosion of wires as a method for preparation of nanopowders. Journal of nanoparticle research. Vol.4, pp. 539 - 550

Smothers W.J. & Reynolds H.J. (1954). Sintering and grain growth of alumina. Journal of. Amer. Cer. Soc, V. 37. - No 12, pp. 588 - 595.

Srdic´ V.V., Rakic´ S & Cvejic´ Z. (2008). Aluminum doped zirconia nanopowders: Wet-chemical synthesis and structural analysis by Rietveld refinement. Materials Research Bulletin, Vol.43, - pp. 2727 – 2735

Vassen R. & Stoever D. (1992). Compaction Mechanisms of Ultrafine SiC Powders. Powder Technology, Vol.72, pp.223 - 226.

4

Advanced Sintering of Nano-Ceramic Materials

Khalil Abdelrazek Khalil[1,2,*]
[1]*Mechanical Engineering Department, College of Engineering, CEREM, King Saud University, Riyadh*
[2]*Department of Material Engineering and Design, Faculty of Energy Engineering, South Valley University, Aswan,*
[1]*Saudi Arabia*
[2]*Egypt*

1. Introduction

In the past few years, the use of ceramic materials has significantly increased in various applications due to the unique characteristics of these materials in comparison with metals and polymers. The advantageous properties of ceramic materials are hardness, rigidity, abrasive toughness and low density. Ceramics are a class of materials broadly defined as "inorganic, nonmetallic solids". They have the largest range of functions of all known materials. The last decades have seen the development of the enormous potential of functional ceramics based on unique dielectric, ferroelectric, piezoelectric, pyroelectric, ferromagnetic, magnetoresistive, ionical, electronical, superconducting, electrooptical, and gas-sensing properties. Similar scientific developments also have taken place in structural ceramics. Thermal, chemical, and mechanical stability of many oxide and nonoxide compounds laid the foundation for improved processing, which led to an improved level of microstructure design and defect control. This in turn resulted in never-before-seen improvements in mechanical performance and in the reliability of the properties of components and devices.

In addition, superior combinations of thermal, insulating, and mechanical properties have become the basis of huge applications in the packaging of microelectronics and power semiconductors. Therefore, ceramic materials have now become the cornerstone of such advanced technologies as energy transformation, storage and supply, information technology, transportation systems, medical technology, and manufacturing technology. In addition to these trends, present-day environmental regulations and awareness and the recycling of materials will affect the use of materials and require less expensive production processes. Following technological trends, the needs for future basic research in the field of ceramics can be divided into four major areas:

(1) materials and materials properties research in order to widen the area's scope and match its needs for future applications, (2) research to increase the knowledge of economical and

* Corresponding Author

ecological production processes for materials, components, and devices, (3) miniaturization and integration, and (4) modeling and numerical simulation, which would complement or even act as a substitute for present areas of experimental work, thus not only directing research to defined questions, but also reducing practical work and time periods typically combined with product development.

Nanocrystalline ceramic materials have been the subject of interests and focus of research programs around the world for the past two decades owing to the expectations that the mechanical behavior may improve significantly when grain sizes reduce to nanometer scale. However, although numerous technologies are available for making nanosized ceramic powders, obtaining true nanocrystalline ceramic (average grain size <100 nm) has been a great challenge due to the difficulties of controlling grain growth during sintering. Thus, the use of conventional methods of powder consolidation often result in grain growth in the compact or surface contamination due to the high temperatures and long sintering duration involved. It is therefore essential to minimize grain growth through careful control of the consolidation parameters, particularly sintering temperature and time. Evaluation of the mechanical properties of nanocrystalline ceramic materials is also difficult because there are little published data that are based on specimens with truly nanoscale grain sizes.

In this chapter, the challenges and results of sintering nanocrystalline ceramic powders will be examined as well as the various technologies for producing nanosized ceramic powders. The key challenge to the production of bulk nanocrystalline ceramic materials is to control the rapid grain growth during the early stage of sintering. This chapter provides an overview of the development of nanocrystalline ceramic materials. The review will first summarize different methods of advanced sintering of nanosized ceramic powders. The review of the advanced sintering and consolidation of the nanosized ceramic powders will emphasize the challenges and the progress toward achieving nanoscale grain sizes, or grain sizes that are as fine as possible, at sintered states. In the last section of this review, focusing on the high frequency induction heat sintering of nanosized ceramic powders will be given in details and summarized.

2. Conventional sintering process

Dense nanostructured ceramic materials are usually obtained by pressing and conventional sintering of nanopowders using pressure assisted methods, such as hot pressing, hot isostatic pressing, sinter forging, etc. (M.J. Mayo, 1997; J.R. Groza, 1999; Dj. Veljovic et, al., 2007). The high sintering temperatures and long sintering times required for the consolidation of ceramic powders often result in extreme grain coarsening and decomposition of the ceramic, which is characteristic for conventional sintering methods and results in the deterioration of the mechanical properties of ceramics (Y.W. Gua et al., 2004 & C.Y. Tang et al., 2009). Hot pressing of some ceramic materials was found to allow the occurrence of densification at temperatures much lower than during conventional sintering (R. Halouani et al., 1994 & Dj. Veljovic et al., 2009). The advantages of the hot pressing technique are the enhancement of the densification kinetics and the limiting of grain growth, while the disadvantages are the limited geometry of the end product and the expensive equipment required. Furthermore, in conventional HP techniques, the powder container is typically heated by radiation from the enclosing furnace through external heating elements and convection of inert gases if applicable. Therefore, the sample is heated as a consequence of the heat transfer occurring by conduction from the external surface of the container to the powders. The resulting heating rate is then typically slow and the process can last hours. In addition, a lot of heat is wasted as the whole volume of space is heated and the compact indirectly receives heat from the hot environment.

3. Advanced sintering process

Consolidation of nanocrystalline powders is a very difficult problem. A peculiarity of sintering nanocrystalline powders is competition between the processes of densification and microstructure coarsening, which occur in parallel. In order to consolidate a material with a density close to theoretical and a grain size as small as possible, various techniques have been developed that accelerate the former and decelerate the latter factors. Of these, methods involving fast heating, high pressure, and addition of various agents that accelerate shrinkage and inhibit grain growth are the best known. To overcome the problem of grain growth, unconventional sintering and densification techniques have been proposed. Advanced sintering was found to show great potential in ceramics processing (S. Vijayan & H. Varma, 2002). These include the use of grain growth inhibitors in solid solution or forming discrete second phases, high-pressure densification, spark-plasma sintering and related techniques, shock densification, high-frequency induction heating and magnetic pulse compaction (Allen et al., 1996; Kim and Khalil, 2006; Godlinski et al., 2002; Krell et al., 2003; Jiang et al., 2007).

3.1 Microwave sintering

Microwave sintering of ceramics in general has been known for the past three decades. It has several advantages with regard to processing like being rapid and possibility of selective heating, imparting improved properties to the processed materials by way of inhibiting grain growth in addition to reducing the processing time and energy required (Clark D & Sutton WH, 1996; Katz JD, 2005; Agarwal DK, 2005). Use of microwave technology in material science and processing is not rather new. The areas where this technology has been applied include: process control, drying of ceramic sanitary wares, calcination, and decomposition of gaseous species by microwave plasma, powder synthesis, and sintering (D. Agrawal & Sohn, 2006). Microwave processing of materials was mostly limited until 2000 to ceramics, semiconductors, inorganic and polymeric materials. There was a misconception between researchers that all metals reflect microwave or cause plasma formation, and hence cannot be heated, except exhibiting surface heating due to limited penetration of the microwave radiation. The researchers did not notice that this relation is valid only for sintered or bulk metals at room temperature, and not for powdered metals and/or at higher temperatures (D. Agrawal & Sohn, 2006). No wit has been found that the microwave sintering can also be applied as efficiently and effectively to powdered metals as to many ceramics.

This technique provides a series of benefits, such as great microstructure control, no limit of the geometry of the product, improved mechanical properties of the materials and reduced manufacturing costs due to energy savings, lower temperatures of sintering and shorter processing times. Microwave energy is a form of electromagnetic energy with the frequency range of 300MHz to 300 GHz. Microwave heating is a process in which the materials couple with microwaves, absorb the electromagnetic energy volumetrically, and transform into heat. This is different from conventional methods where heat is transferred between objects by the mechanisms of conduction, radiation and convection. In conventional heating, as mentioned before, the material's surface is first heated followed by the heat moving inward. This means that there is a temperature gradient from the surface to the inside. However, microwave heating generates heat within the material first and then heats the entire volume (P. Yadoji et al., 2003). This heating mechanism is advantageous due to the following facts: enhanced diffusion processes, reduced energy consumption, very rapid heating rates and considerably reduced processing times, decreased sintering temperatures, improved physical and mechanical properties, simplicity, unique properties, and lower environmental hazards. These are features that have not been observed in conventional processes (P. Yadoji et al., 2003; D. Agrawal, 1999; D.E. Clark & D.C. Folz, J.K., 2008; C. Leonelli, P. et al., 2008; R.R. Menezes, 2007). The microwave furnace is shown in Figure 1.

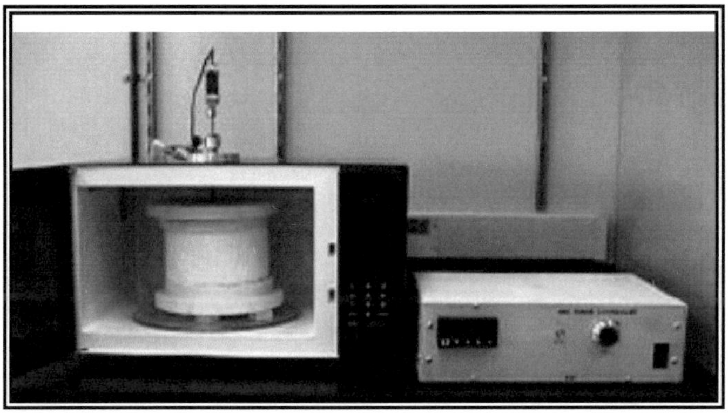

Fig. 1. Microwave furnace with thermal pod, thermocouple, and controller.

In addition to a microwave furnace, the critical components for microwave sintering include an insulation box and susceptors. The insulation box consists of a small chamber fabricated from low-density rigid insulation board. Low density and very low dielectric loss are required for the box to make it microwave transparent. Microwaves pass through the material with little interaction, allowing the contents to heat. The box, in essence, acts as an oven within the microwave chamber or applicator, as it allows microwaves to pass though but contains the heat generated by the contents. Many ceramic materials do not absorb microwaves (2.45 GHz) well at room temperature (Committee on Microwave, 1994). Susceptors are useful for initial heating of these ceramics. Susceptors are made of a material that absorbs microwaves at room temperature and act as heating elements, which "boost" the temperature until the dielectric loss in the ceramic is high enough that the ceramic couples directly with the field. For example, using silicon carbide susceptors, zirconia will heat primarily by radiation from the SiC, until it reaches approximately 600°C, whereby the zirconia couples preferentially and heats volumetrically.

Microwave sintering of cemented carbides like WC–Co also has been investigated since 1991 after the pioneering work of Cheng (Cheng J, 1991; Cheng JP et al., 1997; Breval E et al., 2005) and thereafter by Porada (Gerdes T & Porada MW, 1994; Rodinger K et al., 1998; Kolaska H et al., 2000). Breval et al. (Breval E et al., 2005) investigated on the microwave sintering of 0.1–1 μm sized WC particles with cobalt as the binder and compared the results with conventional sintering of the same powders. They reported that the microwave sintered sample hardly exhibits any growth and the cobalt phase does not reveal any dissolution of tungsten whereas in the conventionally sintered sample, nearly 20% of W had dissolved in the cobalt binder phase. They found that the microwave sintered sample always showed improved mechanical properties when compared to the conventionally sintered one. Porada and her group showed that the microwave reaction sintering of W, C and Co powders yielded sintered WC–6Co compacts with fine and uniform microstructure (with an average grain size of 0.6 lm) which exhibited a 10% increase in hardness values in comparison to tools made by a conventional route (Gerdes T & Porada MW, 1994; Rodinger K et al., 1998; Kolaska H et al., 2000). Recently, microwave post-treatment of WC–Co cutting tools was also investigated by Ramkumar et al. Ramkumar J et al., 2002 & Aravindan S et al., 2005). These authors reported that a favorable response was obtained on microwave treatment of cemented WC drill inserts and the hardness had increased from 1372–1525 to 1700–1900 kg/mm^2 through a microwave treatment.

3.2 Spark plasma sintering

A spark sintering method was investigated and patented in the 1960s and used to compact metal powders, but due to high equipment cost and low sintering efficiency it was not put to wider use (Inoue, K. (US Patent NO. 3 241 956). The concept was further developed during the mid 1980s to the early 1990s, and a new generation of sintering apparatus appeared named Plasma Activated Sintering (abbreviated PAS) and Spark Plasma Sintering (abbreviated SPS). Common to these systems is the use of pulsed direct current to heat the specimens. These sintering techniques currently attract growing attention among productions engineers as well as materials researchers. Whether plasma is generated has not been confirmed yet, especially when non-conduction ceramic powders are compacted. It has, however, been experimentally verified that densification is enhanced by the use of a pulsed DC current or field (Mishra et al., 1998). This family of techniques is in academia also named as pulsed electric current sintering (PECS) (Yoshimura et al., 1998; Murayama, 1997; Zhou et al., 2000) or electric pulse assisted consolidation (EPAC) (Mishra et al., 2000). SPS allows compaction of ceramic and metal powders at low temperature and in short time (within minutes).

The basic configuration of a SPS unit is shown in Figure 2. It consists of a uniaxial pressure device, where the water-cooled punches also serve as electrodes, a water-cooled reaction chamber that can be evacuated, a pulsed DC generator, pressure-, position- and temperature-regulating systems. Spark Plasma Sintering resembles the hot pressing process in several respects, i.e. the precursor powder (green body) is loaded in a die, and a uniaxial pressure is applied during sintering process. However, instead of using an external heating source, a pulsed direct current is allowed to pass through the electrically conducting pressure die and, in appropriate cases, also through the sample. This implies that the die also acts as a heating source and that the sample is heated from both outside and inside. The use of a pulsed direct current also implies that the samples are exposed to a pulsed electric field during the sintering process. In the SPS method, the powder sample in a graphite die is gradually pressed in a vacuum, and heated by a pulse current (Joule heating). The sintering is considered to proceed very quickly due to the spark plasma induced by the large pulse current. This can be attributed to the in situ particle surface activation and purification by the spark plasma generated during the process. Therefore, heat and mass transfer between the particles induced by electric current can be rapidly accomplished (R. Chaim et al., 2007; H. Borodianska et al., 2008; O. Vasylkiv et al., 2008; H. Borodianska et al., 2009). The very short duration and relatively low homologous temperature involved in the technique make it very attractive for densification and preservation of the nanocrystalline character in ceramics.

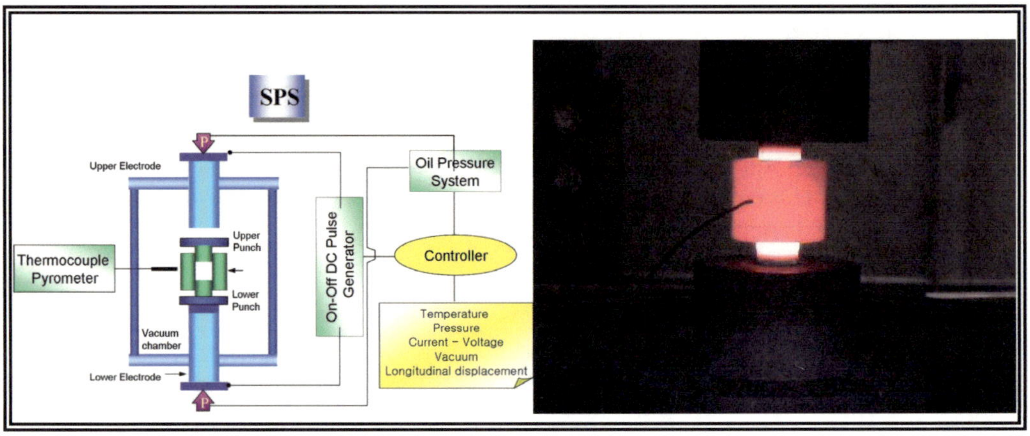

Fig. 2. Basic configuration of a typical SPS system and photo of the heated die.

Numerous experimental investigations point to the possibility of using SPS to consolidate parts to nearly full density while retaining the fine grain size and phase homogeneity composites (P. Angerer, et al., 2004; K. Kakegawa et al., 2003; K.A. Khor et al., 2004). This property of SPS is of undoubted significance in the case of manufacture of bulk nanostructured parts, for which the control of grain growth is one of the major problems (H. Borodianska et al., 2008; O. Vasylkiv et al., 2008; H. Borodianska et al., 2009). The formation of plasma at particle contacts makes it possible to obtain a fine-grained structure commensurable with the initial powder particle size. However, at the same time the grain-boundary framework in SPS-derived nanoceramics is often underdeveloped (H. Borodianska et al., 2009). The enhanced kinetics of different thermally activated processes where SPS was used, such as densification (Z.A. Munir et al., 2003; M. Nygren & Z. Shen, 2003; O. Vasylikiv et al., 2009; K. Morita et al., 2008), reactive sintering (W.W. Wu et al., 2007; J. Zhang et al., 2007; R. Licheri et al., 2008; J.G. Santanach et al., 2009), joining (Y.J. Wu, et al., 2003; W. Liu & M. Naka, 2003; C. Elissalde et al., 2007), liquid and solid state crystal growth (M. Omori, 2000; J.K. Park, 2006), raised questions whether the atomistic mechanisms active by conventional heating change due to the nature of the SPS process. In this respect, enhanced densification kinetics observed in YAG ($Y_3Al_5O_{12}$), was analyzed in terms of particle surface softening due to the plasma (R. Chaim et al., 2006). Such a behavior originated most probably from the extremely high creep resistance of YAG prohibiting plastic deformation, and its high electrical resistance enhancing charging and discharge at the particle surfaces. Therefore, densification of conducting or semiconducting ceramics by SPS may shed further light on the interrelations between the ceramic mechanical and electrical properties and its densification behavior by SPS. So, SPS is most promising and has already proved successful in consolidation of various nanoceramics and composites (P. Angerer, et al., 2004; K. Kakegawa et al., 2003; K.A. Khor et al., 2004). SPS is widely used nowadays as it offers the possibility of performing a rapid consolidation of difficult-to-sinter ceramic and ceramic composites at reduced temperatures.

3.3 High frequency induction heat sintering

The novel technique of high-frequency induction heat sintering (HFIHS) has been shown to be an effective sintering method that can successfully consolidate ceramics and metallic powders to near theoretical density. The (HFIHS) process is a sintering method for the rapid sintering of a nanostructured hard metal in a high-temperature exposure along with pressure application. It is similar to hot pressing, which is carried out in a graphite die, but heating is accomplished by a source of high frequency electricity to drive a large alternating current through a coil. This coil is known as the work coil. The passage of current through this coil generates a very intense and rapidly changing magnetic field in the space within the work coil. The workpiece to be heated is placed within this intense alternating magnetic field. The alternating magnetic field induces a current flow in the conductive workpiece.

The basic configuration of a HFIHS unit is shown in Figure 3. It consists of a uniaxial pressure device, a graphite die (outside diameter, 45 mm; inside diameter, 20 mm; height, 40 mm). A water-cooled reaction chamber that can be evacuated, an induced current (frequency of about 50 kHz) pressure-, position- and temperature-regulating systems are also presented. HFIHS resembles the hot pressing process in several respects, *i.e.*, the precursor powder is loaded in a die, and uniaxial pressure is applied during the sintering process. However, instead of using an external heating source, an intense magnetic field is applied through the electrically conducting pressure die and, in some cases, also through the sample. Thus, the die also acts as a heating source, and the sample is heated from both

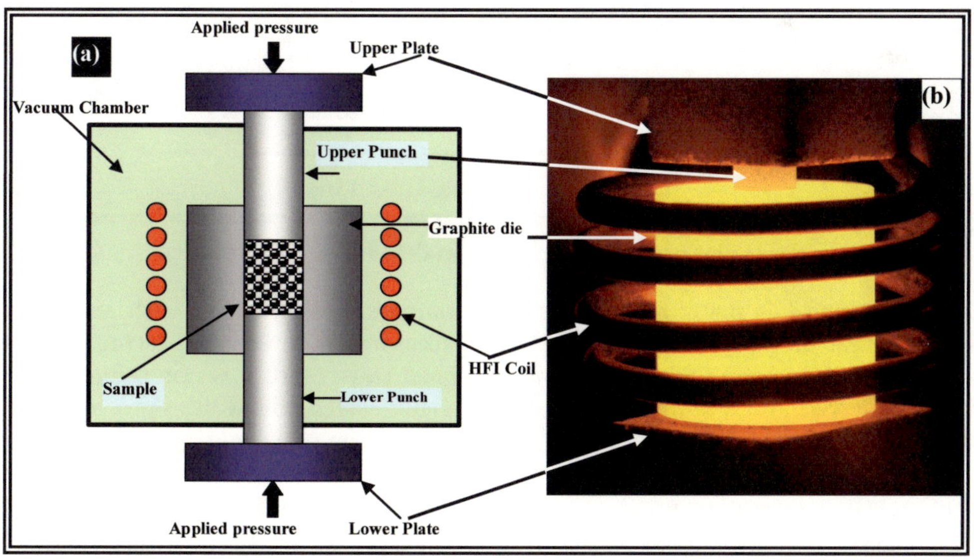

Fig. 3. (a) Schematic diagram of high-frequency induction heated sintering apparatus, (b) Photo of the heated die.

the outside and inside. Temperatures can be measured using a pyrometer focused on the surface of the graphite die. The system is first evacuated to a vacuum level of 1 × 10^{-3} Torr, and uniaxial pressure is applied. An induced current is then activated and maintained until densification, indicating the occurrence of sintering and the concomitant shrinkage of the sample, is observed. Sample shrinkage is measured by a linear gauge that measures the vertical displacement. The typical parameters for the process are presented in Table 1, and the four major stages of the HFIHS and densification processes are shown in Fig. 4.

Parameter	Applied value	Parameter	Applied value
Vacuum level	1 × 10^{-3} Torr	Applied pressure	10 - 300 MPa
Heating rate	100 - 1200 °C/min	Cooling rate	500 °C/min
Total power capacity	15 kW	Output of total power	0 - 100%
Resistance heating frequency	50 kHz	Duration time	~10 min

Table 1. Processing conditions of high-frequency induction heated sintering process

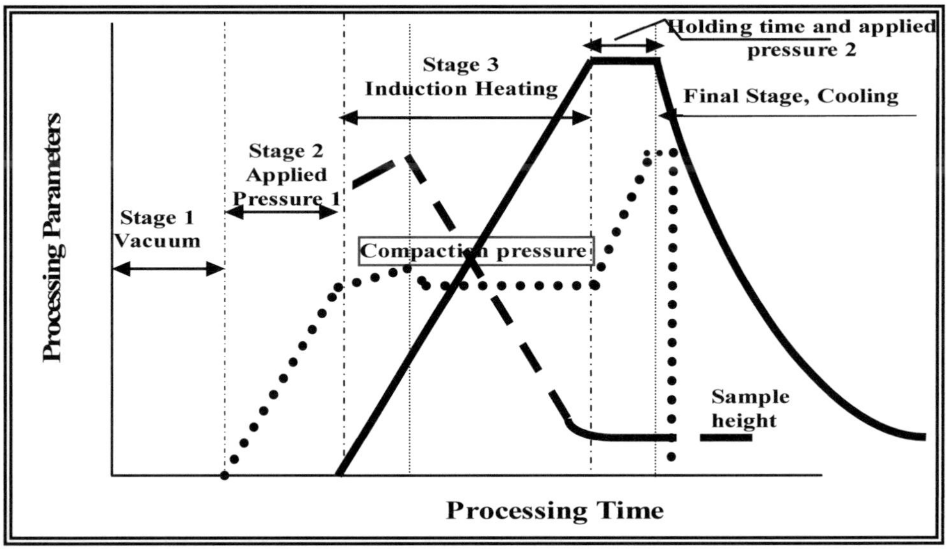

Fig. 4. The four major stages of sintering process in the HFIHS apparatus.

Induction heating has many advantages over competitive techniques, such as radiant or convection heat and laser technologies. This process is a non-contact technique, which provides localized heating through custom-designed coils. Since it is non-contact, and the heat is transferred to the product via electromagnetic waves, the heating process does not contaminate the material while being heated. The process is energy efficient process converts up to 80% of the expended energy into useful heat to save costs. It allows the right amount of heat to be applied exactly where it is needed for an exact period of time, ensuring controlled, accurate performance, increased production and reduced distortion.

Considerable research has been performed on using the novel technique of HFIHS (Khalil et al., 2006, 2007, 2010, 2011, 2012). The HFIHS process involves the rapid sintering of a nanostructured hard metal in a very short time with high-temperature exposure and the application of pressure. This process is advantageous because it allows for rapid densification to near the theoretical density of the associated materials and inhibits grain growth in nanostructured materials. Another important reported aspect of the HFIHS process is the role of rapid heat transfer to the product via electromagnetic waves. In previous papers HFIHS (Khalil et al., 2006, 2007a, 2007b, 2007c), the authors have reported the investigation of the consolidation and mechanical properties of different nanostructured ceramic materials by HFIHS. The grain size of the sintered material was greater than 100 nanometers. Furthermore, the specific microscopic effects of HFIHS were not determined.

Khalil and S. W. Kim (Khalil & S. W. Kim, 2007) have studied sintering of Al_2O_3-8YSZ by HFIHS. They found that, HFIHS was effective in the preparation of fine-grained, nearly fully dense of Al_2O_3-8YSZ ceramics from the powder with a smaller particle size by optimizing the overall processing parameters. The samples were densified by heating to a sintering temperature in the range of 1300 to 1500°C, and then fast cooled to 500°C within short time. The density of the samples kept increasing with the rising of the sintering temperature; on the other hand sintering pressure has a relatively small influence on the density. For these composites, a relative density of more than 99% theoretical density was achieved after sintering at 1400°C. Low density was found when a higher heating rate, 700°C/min, was applied during the HFIHS process. Microstructure inhomogeneity, where the edge was denser than the inside of the sample, appeared with a high heating rate. A difference in temperature between the surface and the center of the sample exists, and it depends on the heating rate. When the heating rate was 200°C/min, the inside could be sintered almost as dense as the edge of the sample. Al_2O_3-(ZrO_2+ 8%mol Y_2O_3) nanopowders with 20 vol% - (ZrO_2+ 8%mol Y_2O_3) were consolidated very rapidly to full density using high frequency induction heating sintering (HFIHS).

The variations of shrinkage displacement and temperature with time for various maximum sintering temperatures under a pressure of 60 MPa are shown in Figure 5. In all cases, the application of the current and subsequent increase in temperature resulted in initial thermal expansion followed by shrinkage due to consolidation. The onset of shrinkage occurred at temperatures in the range of 600 to 800°C for all samples due to rearrangement of the powders as well as plastic deformations. As the temperature increased, the shrinkage displacement increased gradually. The rate of shrinkage decreased as a temperature of the maximum sintering temperature was reached. It is clear that, the rate of shrinkage displacement became zero when the maximum sintering temperature was reached. At a temperature of 1400°C, the specimen attained minimum height (maximum shrinkage) before reaching the maximum sintering temperature. Although, the maximum sintering temperature was reached at 1500°C, there was no more shrinkage beyond 1400°C. It is clear from these Figures that, the maximum sintering temperature with respect to maximum shrinkage was 1400°C.

Figures 6 and 7 show microstructures of the samples sintered at a heating rate of 200 and 700°C/min, respectively. In general, sintered samples with higher relative densities exhibit greater mechanical properties. However, there is also an influence of microstructural uniformity on mechanical properties. Figure 6 (a) to (d) shows various SEM micrographs of fracture surfaces of the samples. The microstructure seems to be like a green compact for the 1300°C sample because the sintering temperature was too low for sintering. In the case of

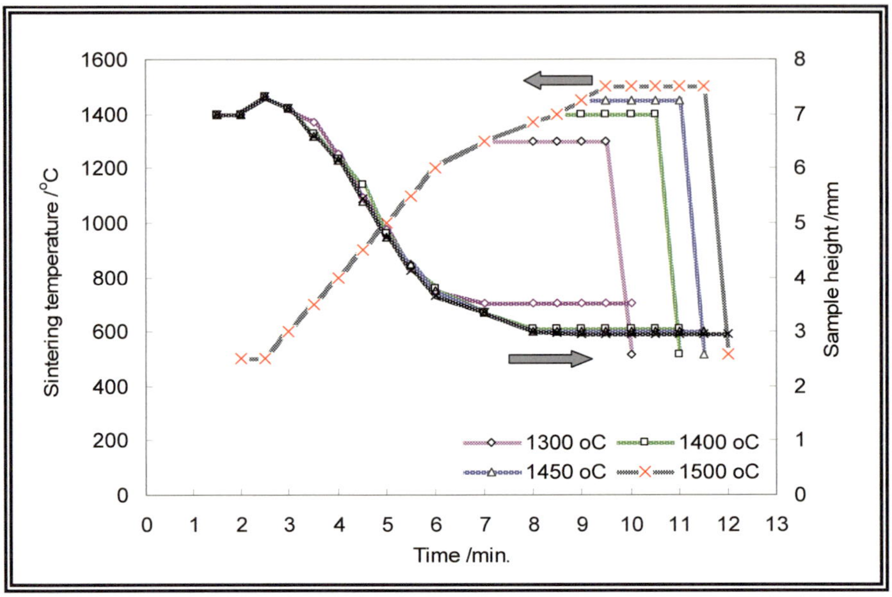

Fig. 5. Variations of temperatures and shrinkage displacements vs. heating time at various maximum sintering temperatures.

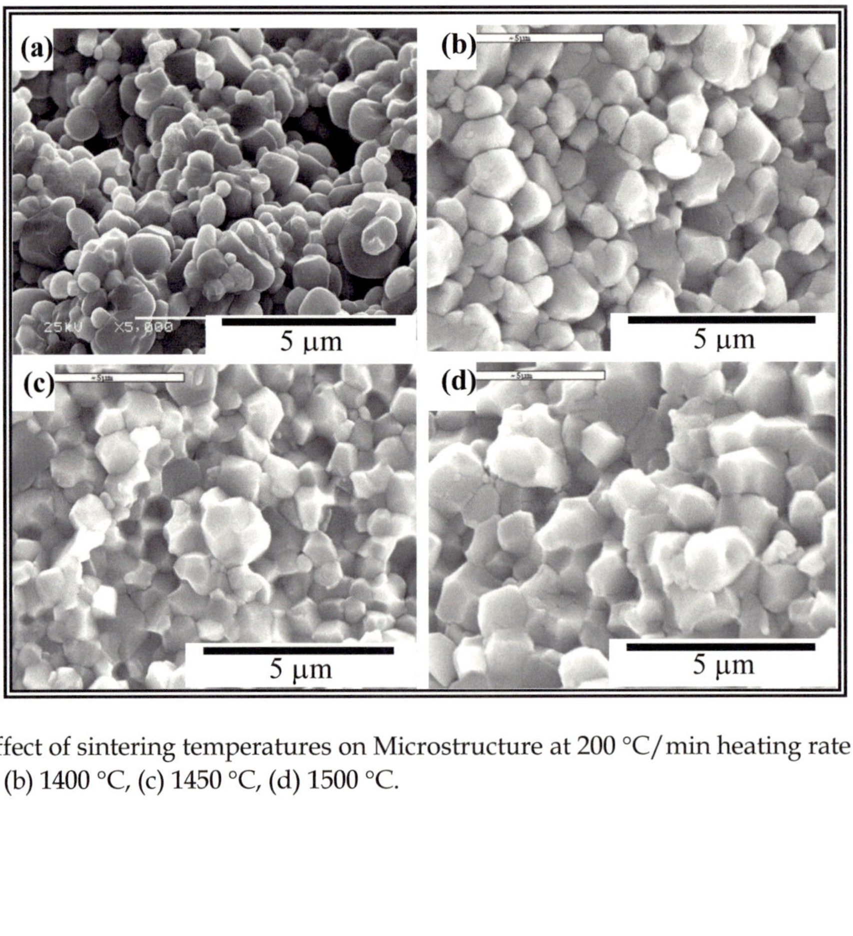

Fig. 6. Effect of sintering temperatures on Microstructure at 200 °C/min heating rate (a) 1300 °C, (b) 1400 °C, (c) 1450 °C, (d) 1500 °C.

Figure 6 (a), there exist a number of closed pores, entrapped in sample grains. Samples sintered at temperatures from 1400°**C** to 1450°C, showed highly homogeneous microstructures without agglomerates. These provided better densification, less porosity in the sample and no abnormally grown alumina grains. The intragranular fracture mode was dominant during fracture, indicating the presence of stiff grain-boundaries. This is clear in Figure 6 (c). At temperature 1500°**C**, Figure 6 (d), the grain size rapidly increased due to a high sintering temperature. Abnormal grains appeared only when the powder was sintered for a long sintering time (due to a lower heating rate, 200°C/min) and a high sintering temperature (1500°C).

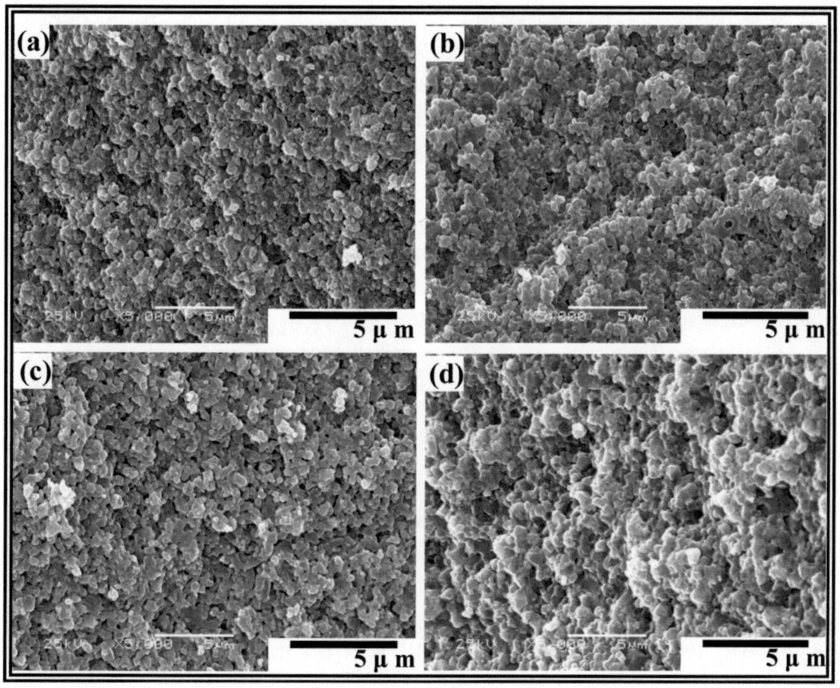

Fig. 7. Effect of sintering temperatures on Microstructure at 700 °C/min heating rate (a) 1300 °C, (b) 1400 °C, (c) 1450 °C, (d) 1500 °C.

Figure 7 (a) to (d) shows various SEM micrographs of fracture surfaces of the samples sintering at a high heating rate (700°C/min). The grain size is relatively small, less then 200 nm, compared to that in Fig. 6, but more porosities appear. We can conclude that, the heating rate of 200°C/min resulted in a homogeneous microstructure and an average grain size of about 500 nm. Essentially, the HFIHS process makes rapid sintering possible by application of a proper heating rate, and it also produces a sample with a low relative density and abnormal grains (as in samples made by a conventional method) when a lower heating rate is employed.

Experiments were conducted on synthesis and processing of nanostructured Alumina-20vol% 3YSZ composites sintered by HFIHS by S. W. Kim and Khalil (K. S. Won & Khalil, 2006). They conclude that, The Alumina micro-powder and 3YSZ nano-powders mixtures with distinct nanocrystalline characteristics were synthesized and optimized by using wet-milling technique. This technique is remarkable due to the easiness of application. The samples were densified by heating to a sintering temperature in the range of 1100 to 1400 °C, and then rapidly cooled to 500 °C. The density of the samples kept increasing with the rising of the sintering temperature. A relative density more than 99% of theoretical density of the composites was achieved after sintering at 1370 °C. Al_2O_3-3YSZ composites with small grain size, homogeneous microstructure, higher density, hardness and toughness were successfully produced at relatively low temperatures.

Mg/HAp nanocomposites were successfully synthesized using a high-frequency induction heat sintering method by Khalil and A. Almajid (Khalil &A. Almajid, 2012). They conclude that, HFIHS was effective in preparing fine crystalline, nearly fully dense Mg/HAp nanocomposites from powders with smaller crystal sizes by optimizing the overall processing parameters. The relative densities and microhardness values of the specimens increased with increasing sintering temperature, reaching values as high as 99.7 % and 60 HV, respectively, at 550 °C. Meanwhile, at a sintering temperature of 580 °C, the microhardness and relative density slightly decreased. The crystal size of the sample sintered at 500 °C was approximately 37 nm. However, higher sintering temperature resulted in an increase in crystal size. The compressive strength of the sample sintered at 500 °C was low, approximately 192.7 MPa. However, as the sintering temperature of the nanocomposites increased, the compressive strength increased. For example, the compressive strength of the nanocomposite sintered at 550 °C reached as high as 194.5 MPa. For the sample sintered at 580 °C, however, the compressive strength decreased to approximately 62.5 MPa due to an increase in grain size, as observed by FE-SEM. By increasing the applied sintering pressure from 30 to 50 MPa, the relative density and microhardness of the samples greatly increased. When the applied pressure increased from 50 to 80 MPa, the effect was small. The results of crystal size measurements reveal an increase in crystal size with increasing sintering pressure of nano-size HAp reinforcements in the magnesium matrix. High applied pressure during sintering leads to an increase in the apparent activation energy and, subsequently, to an increase in crystal size. The ductility of the composite samples was found to be dependent on the increase in crystal size during sintering. The effect of sintering time on the mechanical and microstructural properties of composite samples has been reported. In this study, the relative density and microhardness increased with increasing sintering time up to 99.7% and 60 HV, respectively, after a 3-min holding time and then decreased with increasing sintering time. Sintering for a relatively short time produces small crystal sizes, but at the same time, the relative density is low. The compressive strength was significantly improved with increasing sintering time up to 3 min and then decreased when sintering for 4 min. The highest compressive strength of 192.7 MPa for sintered Mg/HAp samples was obtained at a sintering time of 3 min, whereas maximum compressive strengths of 154 and 137 MPa for the sintered samples were obtained at sintering times of 1 and 4 min, respectively. The density and microhardness increased with increasing heating rate, reaching a maximum level at a heating rate of 400°C/min. Meanwhile, at heating rates above 400°C/min, the relative density of the specimens decreased with an increase in the heating rate. The crystal size is strongly influenced by variation of the heating rate. High heating rates produce very fine crystal sizes, nearly 34 nm at 1200 °C/min. A heating rate of 400 °C/min produces small crystals (around 38.7 nm) with desirable mechanical properties. Samples sintered at a low heating rate of 200 °C/min showed greater ductility and a low compressive strength of approximately 185 MPa. The compressive strength was slightly increased by increasing the heating rate up to 400 °C/min, reaching as high as 192.7 MPa, followed by a decrease with increasing heating rates.

M. Dewidar, (M. Dewidar, 2010) used high frequency induction heating sintering method for sintering high density compacts, containing as small grains as possible of powders. The alloy of Ti-6Al-4V was modified by addition of 2.5, 5, and 10 wt.% tungsten through powder metallurgy. The use of the high frequency induction heating sintering technique allows sintering to nearly full density at comparatively low temperatures and short holding times, and therefore suppressing grain growth. Different process parameters such as sintering temperature, and applied pressure have been investigated. The obtained compacts are characterized with respect to their densities, grain morphologies and pore distributions as well as hardness. Ti-6Al-4V/W powder precursors have been successfully compacted and consolidated to densities exceeding 98.8%. The maximum compressive strengths were obtained at sintering temperature 1000 °C for the samples containing 5% W, and at 1100 _C for the samples with 10% W. Maximum hardness was obtained 45 HRC at 1100 °C for 10% W.

4. Summary

Densification of nanocrystalline ceramic materials have been the subject of interests and focus of research programs around the world for the past two decades owing to the expectations that the mechanical behavior may improve significantly when grain sizes reduce to nanometer scale. Studies of nanopowder densification have lead to a better understanding of numerous sintering issues such as powder agglomeration, surface condition or contamination, pore role in sintering and grain growth. The resultant control of synthesis (e. g., non-agglomerated nanopowders) and processing has enabled fabrication of fully dense nanocrystalline parts particularly ceramic. However, although numerous technologies are available for making nanosized ceramic powders, obtaining true nanocrystalline ceramic (average grain size <100 nm) has been a great challenge due to the difficulties of controlling grain growth during sintering. Thus, the use of conventional methods of powder consolidation often result in grain growth in the compact or surface contamination due to the high temperatures and long sintering duration involved. It is therefore essential to minimize grain growth through careful control of the consolidation parameters, particularly sintering temperature and time. The challenges and results of sintering nanocrystalline ceramic powders have been summarized as well as the various technologies for producing nanosized ceramic powders. This chapter provides an overview of the development of nanocrystalline ceramic materials. Major improvements in the nanopowder synthesis methods and understanding of the densification process have resulted in fully dense parts by the largest multitude of sintering techniques, including conventional sintering.

5. Acknowledgement

This work was financially supported by the Center of Excellence for Research in Engineering Materials (CEREM), College of Engineering, King Saud University, project No. 430-CEREM-04.

6. References

Allen, A.J., Kruegger, S., Skandan, G., Long, G.G., Hahn, H., Kerck, H.M., Parker, J.C., Ali, M.N., 1996. Microstructural evolution during the sintering of nanostructured ceramic oxides. J. Am. Ceram. Soc. 79, 1201–1212.

Agarwal DK, Cheng J, Fang Y, Roy R. Microwave processing of ceramics, composites and metallic materials. In: Clark DE, Folz DC, Folgar CE, Mahmoud MM, editors. Microwave solutions for ceramic engineers. Ohio: The American Ceramic Society; 2005. p. 205–28.

Aravindan S, Ramkumar J, Malhotra SK, Krishnamurthy R. Enhancement of cutting performance of cemented carbide cutting tools by microwave treatment. In: Clark DE, Folz DC, Folgar CE, Mahmoud MM, editors. Microwave solutions for ceramic engineers. Ohio: The American Ceramic Society; 2005. p. 255–62.

Breval E, Cheng JP, Agarwal DK, Gigl P, Dennis M, Roy R, et al. Comparison between microwave and conventional sintering of WC/Co composites. Mater Sci Eng A 2005;391:285–95.

Clark D, Sutton WH. Microwave processing of hard metals. Annu Rev Mater 1996;26:299–331.

Cheng J. Study on microwave sintering technique of ceramic materials. Ph. D. Thesis. Wuhan University of Technology, China; 1991.

Cheng JP, Agarwal DK, Komarneni S, Mathis M, Roy R. Microwave processing of WC–Co composites. Mater Res Innov 1997;1:44–52.

C.Y. Tang, P.S. Uskokovic, C.P. Tsui, Dj. Veljovic, R. Petrovic, Dj. Janackovic, Influence of microstructure and phase composition on the nanoindentation characterization of bioceramic materials based on hydroxyapatite, Ceram. Int. 35 (2009) 2171–2178.

C. Leonelli, P. Veronesi, L. Denti, A. Gatto, L. Iuliano, Journal of Materials Processing Technology 205 (2008) 489–496.

Committee on Microwave Processing of Materials: An Emerging Industrial Technology, National Materials Advisory Board, Commission on Engineering and Technical Systems, and National Research Council, "Microwave Processing of Materials," Pub. NMAB-473, National Academy Press, Washington D.C., 1994.

C. Elissalde, M. Maglione, C. Estournes, Tailoring Dielectric Properties of Multilayer Composites Using Spark Plasma Sintering.J. Am. Ceram. Soc. 90 (2007) 973–976.

D. Agrawal, Microwave Sintering of Ceramics, Composite, Metals, And Transparent Materials, Journal of Materials Education 19, Nrs. 4-6, 49–58, (1997)

D. Agrawal, Sohn MICROWAVE SINTERING, BRAZING AND MELTING OF METALLIC MATERIALS, International Symposium Advanced Processing of Metals and Materials, vol. 4, 2006, pp. 183–192.

D.E. Clark, D.C. Folz, J.K. West, Processing materials with microwave energy, Materials Science and Engineering A 287 (2000) 153–158.

Dj. Veljovic´, B. Jokic´, R. Petrovic´, E. Palcevskis, A. Dindune, I.N. Mihailescu, Dj. Janac´kovic´, Processing of dense nanostructured HAP ceramics by sintering and hot pressing, Ceram. Int. 35 (2009) 1407- 1413.

Dj. Veljovic, B. Jokic, I. Jankovic-Castvan, I. Smiciklas, R. Petrovic, Dj. Janackovic, Sintering behaviour of nanosized HAP powder, Key Eng. Mater. 330–332 (2007) 259–262.

Godlinski, D., Kuntz, M., Grathwohl, G., 2002. Transparente alumina with submicrometer grains by float packing and sintering. J. Am. Ceram. Soc. 85, 2449–2456.

Gerdes T & Porada MW, Microwave sintering of metal–ceramic and ceramic– ceramic composites. Mater Res Soc Symp Proc 1994;347:531–7.

H. Borodianska, L. Krushinskaya, G. Makarenko, Y. Sakka, I. Uvarova and O. Vasylkiv Si3N4–TiN Nanocomposite by Nitration of TiSi2 and Consolidation by Hot Pressing and Spark Plasma Sintering J. Nanosci. Nanotechnol. 9[11] (2009) 6381-6389 DOI:10.1166/jnn.2009.1344

H. Borodianska, O. Vasylkiv and Y. Sakka Nanoreactor Engineering and Spark Plasma Sintering of Gd20Ce80O1.90 Nanopowders J. Nanosci. Nanotechnol. 8[6] (2008) 3077-3084 DOI:10.1166/jnn.2008.087

Inoue (K.), US Patent, N° 3 241 956 (1966).

J.R. Groza, Nanosintering, Nanostruct. Mater. 12 (1999) 987–992.

J. Gurt Santanach, C. Estourn s, A. Weibel, A. Peigney, G. Chevallier and C. Laurent, Spark plasma sintering as a reactive sintering tool for the preparation of surface-tailored Fe-FeAl2O4-Al2O3 nanocomposites Scripta Materialia 60(4), (2009), 195-198.

J.K. Park, U.J. Chung, D.Y. Kim, Application of spark plasma sintering for growing dense Pb(Mg1/3Nb2/3)O3–35 mol% PbTiO3 single crystal by solid-state crystal growth, J. Electroceram. 17 (2006) 509–513.

J. Zhang, L. Wang, L. Shi, W. Jiang, L. Chen, Scripta Mater. 56 (2007) 241–244.

Jiang, D., Hulbert, D.M., Kuntz, J.D., Anselmi-Tamburini, U., Mukherjee, A.K., 2007. Spark plasma sintering: a high strain rate low temperature forming tool for ceramics. Mater. Sci. Eng. A 463, 89–93.

Kolaska H, Porada MW, Rodinger K, Gerdes T. Composite and process for the production thereof. 2000: US patent no. 6124040.

K. Kakegawa, N. Uekawa, Y.J. Wu, Y. Sasaki, Change in the compositional distribution in perovskite solid solutions during the sintering by SPS, Mater. Sci. Eng. B 99 (2003) 11-14.

K.A. Khor, X.J. Chen, S.H. Chan, L.G. Yu, Microstructure-property modifications in plasma sprayed 20 wt.% yttria stabilized zirconia electrolyte by spark plasma sintering (SPS) technique, Mater. Sci. Eng. A 366 (2004) 120-126.

Khalil Abdelrazek Khalil and Sug Won Kim, Effect of Processing Parameters on the Mechanical and Microstructural Behavior of Ultra-Fine Al2O3- (ZrO2+8%Mol Y2O3) Bioceramic, Densified By High Frequency Induction Heat Sintering, Int. Journal of Applied Ceramic Technology, 3 [4], 322–330 (2006)

Khalil Abdelrazek Khalil and Sug Won Kim, High-Frequency Induction Heating Sintering Of Hydroxyapatite-(Zro2+3%Mol Y2o3) Bioceramics, Materials Science Forum Vols. 534-536 (2007) pp. 601-604.

Khalil Abdelrazek Khalil and Sug Won Kim, Synthesis and Densification of Ultra-Fine Al2o3-(Zro2+3%Mol Y2o3) Bioceramics by High-Frequency Induction Heat Sintering, Materials Science Forum Vols. 534-536 (2007) pp. 1033-1036.

Khalil Abdelrazek Khalil, Sug Won Kim, and Hak Yong Kim, Observation of Toughness Improvements of the Hydroxyapatite Bioceramics Densified by High-Frequency

Induction Heat Sintering, Int. Journal of Applied Ceramic Technology, 4 [1] 30–37 (2007)

Khalil Abdelrazek Khalil, Sug Won Kim, N. Dharmaraj, Kwan Woo Kim and Hak Yong Kim, Novel Mechanism to Improve Toughness of the Hydroxyapatite Bioceramics Using High-Frequency Induction Heat Sintering, Journal of Materials Processing Technology 187-188 (2007) 417–420

Khalil Abdelrazek Khalil and Sug Won Kim, Mechanical Wet-Milling and Subsequent consolidation of Ultra-Fine Al_2O_3-(ZrO_2+3%Mol Y_2O_3) Bioceramics by using High-Frequency Induction Heat Sintering, Trans. Nonferrous Met. Soc. China, 17(2007) 21-26.

Khalil Abdelrazek Khalil, Sug Won Kim and Hak Yong Kim, Consolidation and mechanical properties of nanostructured hydroxyapatite–(ZrO2 + 3 mol% Y2O3) bioceramics by high-frequency induction heat sintering, Materials Science and Engineering A 456 (2007) 368–372.

Khalil Abdelrazek Khalil & Abdulhakim A. Almajid, Effect of high-frequency induction heat sintering conditions on the microstructure and mechanical properties of nanostructured magnesium/hydroxyapatite nanocomposites, Materials and Design, in press, (2012)

Kim, S.W., Khalil, K.A.R., 2006. High-frequency induction heat sintering of mechanically alloyed alumina–yttria-stabilized zirconia nano-bioceramics. J. Am. Ceram. Soc. 89, 1280–1285.

Krell, A., Blank, P., Ma, H., Hutzler, T., van Bruggen, M.P.B., Apetz, R., 2003. Transparent sintered corundum with high hardness and strength. J. Am. Ceram. Soc. 86, 12–18.

Katz JD. Microwave sintering of ceramics. Annu Rev Mater Sci 1992;22:153–70.

M.J. Mayo, Nanocrystalline ceramics for structural applications: processing and properties, in: G.M. Chow, N.I. Noskova, Nanostructured (Eds.), Materials Science Technology, NATO ASI Series, Kluwer Academic Publishers, Russia, 1997, pp. 361–385.

Montasser Dewidar, Microstructure and mechanical properties of biocompatible high density Ti–6Al–4V/W produced by high frequency induction heating sintering, Materials and Design 31 (2010) 3964–3970

K. Morita, B.N. Kim, K. Hiraga, H. Yoshida, Fabrication of transparent MgAl2O4 spinel polycrystal by spark plasma sintering processing Scripta Mater. 58 (2008) 1114–1117.

Lin, F.J.T., de Jonghe, L.C., Rahaman, M.N., 1997. Microstructure refinement of sintered alumina by a two-step sintering technique. J. Am. Ceram. Soc. 80, 2269–2277.

Mishra, R. S., Risbud, S. H. & Mukherjee, A. K. Influence of initial crystal structure and electrical pulsing on densification of nanocrystalline alumina powder. Journal of Materials Research 13, 86-89 (1998).

Murayama, N. What can we do by pulsed electric current sintering? Seramikkusu 32, 445-449 (1997).

Mishra, R. S. & Mukherjee, A. K. Electric pulse assisted rapid consolidation of ultrafine grained alumina matrix composites. Materials Science & Engineering, A: Structural Materials: Properties, Microstructure and Processing A287, 178-182 (2000).

M. Omori, Sintering, consolidation, reaction and crystal growth by the spark plasma system (SPS), Mater. Sci. Eng. A 287 (2) (2000) 183–188.

O. VASYLKIV : "Nanoreactor engineering and SPS densification of multimetal oxide ceramic nanopowders" J. European Ceram. Soc. 28[5] (2008) 919-927

O. Vasylkiv, H. Borodianska, P. Badica, Y. Zhen and A. Tok : "Nanoblast Synthesis and Consolidation of $(La_{0.8}Sr_{0.2})(Ga_{0.9}Mg_{0.1})O_{3-\delta}$ Under Spark Plasma Sintering Conditions" J. Nanosci. Nanotechnol. 9[1] (2009) 141-149

P. Yadoji, R. Peelamedu, D. Agrawal, R. Roy, Microwave sintering of Ni-Zn ferrites: comparison with conventional sintering Materials Science and Engineering B 98 (2003) 269–278.

P. Angerer, L.G. Yu, K.A. Khor, G. Krumpel, Spark-plasma-sintering (SPS) of nanostructured and submicron titanium oxide powders Mater. Sci. Eng. A 381 (2004) 16.

R. Licheri, R. Orrù, C. Musa and G. Cao, "Combination of SHS and SPS Techniques for the Fabrication of Fully Dense ZrB_2-ZrC-SiC UHTC Composites", Materials Letters, 62, 432-435 (2008).

R. Chaim, R. Marder-Jaeckel, and J. Z. Shen, "Transparent YAG ceramics by surface softening of nanoparticles in spark plasma sintering", *Mater. Sci. Eng. A* 429, 74-78 (2006).

R. Chaim, M. Kalina, and J. Z. Shen, "Transparent Yttrium Aluminum Garnet (YAG) Ceramics by Spark Plasma Sintering", J. Eur. Ceram. Soc. 27, 3331-3337 (2007).

R.R. Menezes, P.M. Souto, R.H.G.A. Kiminami, Microwave hybrid fast sintering of porcelain bodies, J. Mater. Proc. Technol. 190 (2007) 223–229.

Rodinger K, Dreyer K, Gerdes T, Porada MW. Microwave sintering of hardmetals. Int J Refract Metals Hard Mater 1998; 16:409–16.

Ramkumar J, Aravindan S, Malhotra SK, Krishnamurthy R. Enhancing the metallurgical properties of WC insert (K-20) cutting tool through microwave treatment. Mater Lett 2002;53:200–4.

R. Halouani, D. Bernache-Assollant, E. Champion, A. Ababou, Microstructure and related mechanical properties of hot pressed hydroxyapatite ceramics, J. Mater. Sci: Mater. Med. 5 (1994) 563–568.

S. Vijayan, H. Varma, Microwave sintering of nanosized hydroxyapatite powder compacts, Mater. Lett. 56 (2002) 827–831.

Sug Won Kim and Khalil Abdelrazek Khalil, High frequency induction heating sintering of mechanically alloyed alumina-yttria stabilized zirconia nano-bioceramics, Journal of American Ceramic Society, Vol. 89, No. 4, p. 1280–1285 (2006).

W. Liu, M. Naka, In-situ joining of dissimilar nanocrystalline materials by spark plasma sintering. Scripta Materialia 48 (2003) 1225-1230.

Y.J. Wu, N. Uekawa, K. Kakegawa, Sandwiched $BaNd_2Ti_4O_{12}/Bi_4Ti_3O_{12}/BaNd_2Ti_4O_{12}$ ceramics prepared by spark plasma sintering Mater. Lett. 57 (2003) 4088–4092.

Y.W. Gua, K.A. Khora, P. Cheang, Bone-like apatite layer formation on hydroxyapatite prepared by spark plasma sintering (SPS), Biomaterials 25 (2004) 4127–4134.

Yoshimura, M., Ohji, T., Sando, M. & Nihara, K. Rapid rate sintering of nano-grained ZrO_2-based composites using pulse electric current sintering method. Journal of Materials Science Letters 17, 1389-1391 (1998).

Z.A. Munir, U. Anselmi-Tamburini, M. Ohyanagi, J. Mater. Sci. 41 (2006) 763–777.

W.W. Wu, G.J. Zhang, Y.M. Kan, P.L. Wang, K. Vanmeensel, J. Vleugels, O. Van der Biest, Scripta Mater. 57 (2007) 317–320.

Zhou, Y., Hirao, K., Toriyama, M. & Tanaka, H. Very rapid densification of nanometer silicon carbide powder by pulse electric current sintering. Journal of the American Ceramic Society 83, 654-656 (2000)

5

Development of Zirconia Nanocomposite Ceramic Tool and Die Material Based on Tribological Design

Chonghai Xu[1,2], Mingdong Yi[2], Jingjie Zhang[1], Bin Fang[1] and Gaofeng Wei[1]
[1]*Shandong Polytechnic University*
[2]*Shandong University*
PR China

1. Introduction

With the development of modern manufacturing technology, die is more and more need in high temperature, high pressure, special working conditions or complex working condition [Liu & Zhou, 2003]. The requirement in mechanical properties of die material becomes higher and higher. It is necessary to improve the new die material [Kar et al, 2004]. Structure ceramics, because of its high hardness, high temperature mechanical property, wear resistance and corrosion resistance, have been widely used [Basu et al, 2004]. However, lower fracture toughness has limited its wide applications. Moreover, the tribological characteristics also need further study [Hirvonen et al, 2006].

Tetragonal zirconia polycrystal (TZP), with lower sintering temperature and high sintering density, have got wide application in modern die industry because of the excellent mechanical properties and transformation toughening effect[Guicciardi et al, 2006]. However, the low hardness restricts their tribological applications [Zhang et al, 2009; Liu & Xue, 1996; Yang & Wei, 2000]. Titanium diboride (TiB_2) has an excellent hardness and wear resistance but with poor fracture toughness and flexural strength [Baharvandi et al, 2006]. The proper addition of TiB_2 can improve the hardness of ZrO_2 nano-composite ceramic tool and die material with the other mechanical properties still not being decreased [Basu et al, 2005].

The excellent mechanical properties of TZP ceramics are decided mainly by the transformation toughening [Hirvonen et al, 2006]. Stabilizer materials should be added into the zirconia ceramic to obtain the tetragonal zirconia at room temperature for the achievement of the transformation toughening effect [Gupta et al, 1977]. Yttria is one of the most popularly used stabilizers. The incorporation of Y_2O_3 can lower the sintering temperature and enhance the sintering density, so that both mechanical properties and wear resistance can be improved. However, the excessive addition of Y_2O_3 will cause the difficulty of the tetragonal zirconia to be transformed into the monocline one, reducing the transformation toughening effect.

Sintering is also the key process in the preparation of ceramic materials to achieve material with high performance of the composite material with defined raw materials [Hannink et al, 2000]. Nanometer grain which has high surface energy and activity can grow fast and move quickly in the sintering process. Using hot pressing technology can lower the sintering temperature and shorten the holding time. The sintering temperature and the holding time are the key sintering parameters to achieve high performance of nano-composite material [Guicciardi et al, 2006]. Moreover, the sintering parameters play an important role on the improvement of mechanical properties through the increase in the transformation toughening effect [Tu & Li, 1997].

From the friction and wear problems in the application of the existing ceramic tool and die materials, it is necessary to carry out tribological design during the material research. In the present study, a new nanocomposite ceramic tool and die material was prepared by vacuum hot pressing technique with the application of the tribological design, and the processing techniques, microstructure, mechanical properties and the friction and wear behavior was studied.

2. Experiments

ZrO_2 stabilized by 5mol% Y_2O_3(5Y-ZrO_2), TiB_2 and Al_2O_3 are the main raw materials with the average particle size of 40nm, 1.5μm and 1.5μm, respectively. Both 5Y-ZrO_2 and TiB_2 are all commercial powders. The Al_2O_3 powders were obtained by roasting the analytically pure $Al(OH)_3$ powders. Before the experiment, TiB_2 and Al_2O_3 powders were ball milled for 100 hours.

As shown in Fig. 1(a), the particle size of commercial TiB_2 powder is about 10μm, while that after ball milling is only about 1.5μm (Fig. 1(b)).

Fig. 2 and Fig. 3 are the results of X-ray energy dispersive analysis and particle size analysis, respectively. As shown in Fig. 2, the Al_2O_3 powders have nothing impurities. The average particle size of Al_2O_3 is about 1.5μm (Fig. 3).

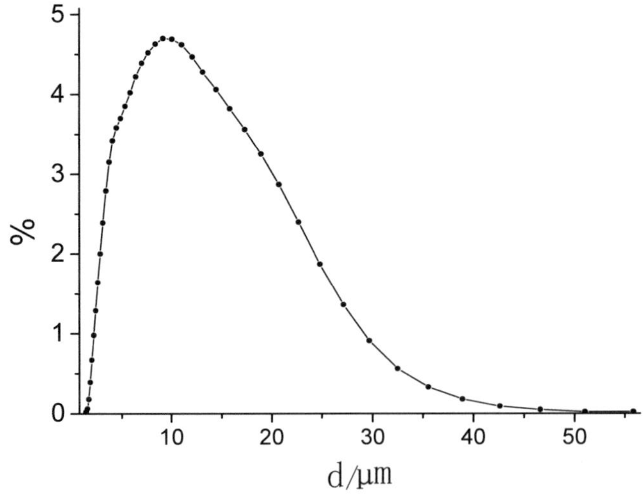

(a) commercial powders

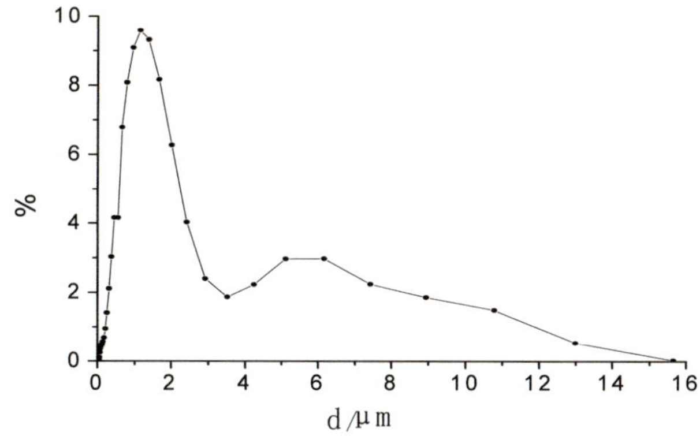

(b) after ball milling

Fig. 1. Particle size distributions of TiB_2 powder

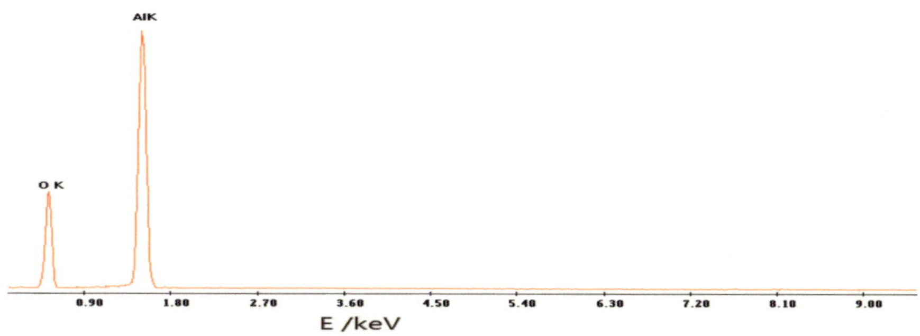

Fig. 2. EDAX of roasted Al_2O_3 powder

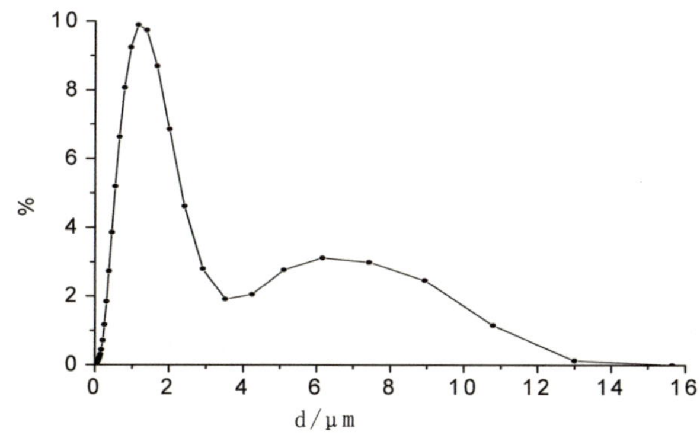

Fig. 3. Particle size distributions of Al_2O_3 powder

In the ZrO_2-TiB_2-Al_2O_3 nano composite ceramic tool and die material system, Al_2O_3 as the reinforcement phase in ZrO_2 ceramic, chemical reaction does not occur. But under the high temperature, TiO_2 may be formed by the direct reaction between TiB_2 and ZrO_2 or Al_2O_3 which is bad to the mechanical properties. The possible reactions are as follows:

$$TiB_2+ZrO_2=ZrB_2+TiO_2 \quad (1)$$

$$2TiB_2+Al_2O_3=2AlB_2+TiO+TiO_2 \qquad (2)$$

Using the data in the handbook of thermodynamic data of inorganic compounds, the standard reaction Gibbs free energy of reaction (1) and (2) at 1900K are 142.91 KJ / mol and 697.97 KJ / mol, respectively. Based on the minimum free enthalpy principle, the two reactions do not occur at 1900K. The result shows that the ZrO_2-TiB_2-Al_2O_3 nano composite ceramic tool and die material has good chemical compatibility. Fig. 4 shows the X ray diffraction before and after sintering composites.

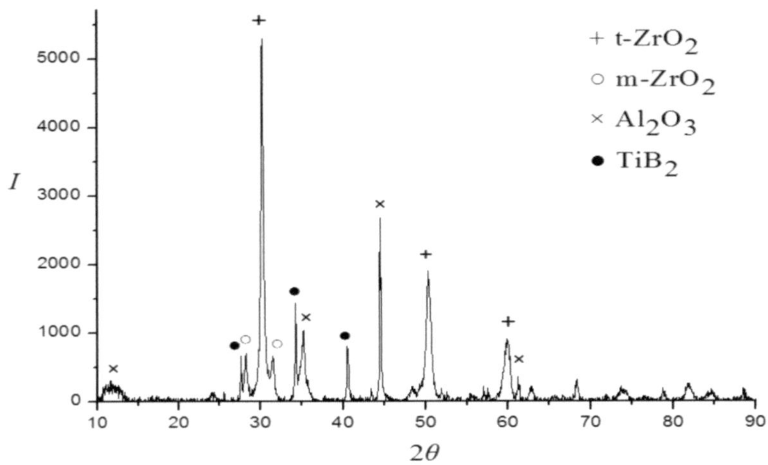

(a) before sintering

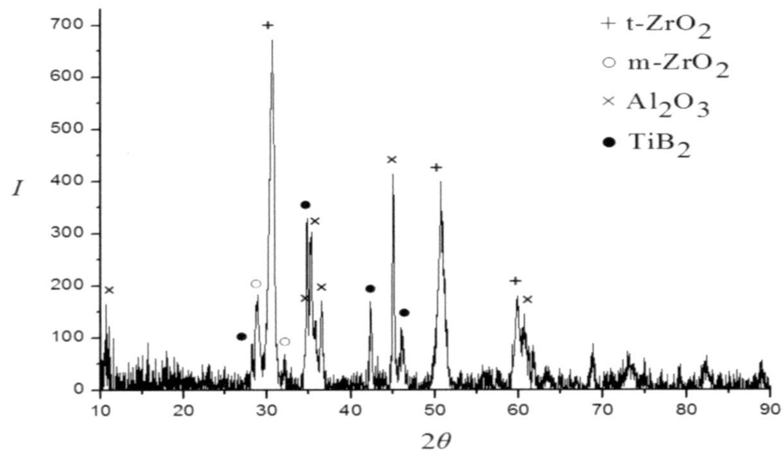

(b) after sintering

Fig. 4. XRD of the ceramic composites

As shown in Fig. 4, the phase of material had not obviously change before and after sintering. It proved that the composites have good chemical compatibility.

ZrO_2, TiB_2 and Al_2O_3 were mixed together for 48h by milling using cemented carbide balls. After milling, the slurry was dried in vacuum and screened. The mixture was hot pressed in a graphite mold at 1430°C in vacuum with the time duration of 60min and pressure of 35MPa. Composite material was made into 4mm×3mm×36mm standard sample after cutting and polishing. The flexural strength was measured by three-point bending method with a span of 20mm and loading rate of 0.5mm/min. The hardness was measured by Hv-120 Vickers hardness tester under the load of 196N for 15s. Fracture toughness was measured by the indentation method. The microstructure and phase of the composite were analyzed with environmental scanning electron microscope (ESEM, model FEI-quanta 200) and X-ray diffraction (XRD, model BRUKER D8).

The wear test was carried on the MMW-1A configuration control multi-purpose friction abrasion tester, using pin on disc form. For the actual operating conditions, 45# chilled steel rings were selected as the friction pair material. The outer diameter and the inside diameter is Ø54mm and Ø38mm, respectively with 10mm high. The hardness of the workpiese material is 44~46HRC with the surface roughness Ra=0.4μm.

According to the request of wear tester, material was made into 10mm×10mm×15mm standard sample after cutting and polishing, and the opposite surface (10mm×10mm surface) which do the friction attrition experiment was polished to the surface roughness Ra=0.1μm. After polishing, the sample is dipped into acetone and cleaned with ultrasonic washer for 5min. Finally, it is dried in vacuum for 24h.

In the experiment, the sliding dry friction was carried out without any lubricant. The normal load was 160N and the rotational speed was 200r/min. Under this condition, dry friction wear tests of the ZrO_2 nano-composites have been carried out. The friction coefficient can be obtained by directly reading in the experiment process. The data were read at 5min after the friction starts and a measured value was taken at intervals of 10min. A total of five measurements were taken to calculate the average as the final result. The wear rate was calculated by expression. The friction and wear appearance of the briquette polishing scratches surface was carried on FEI-quanta 200 environmental scanning electron microscope (ESEM).

3. Preparation of ZrO_2-TiB_2-Al_2O_3 nano composite ceramic too and die material

3.1 Components and mechanical property of ceramic material

In order to improve the comprehensive mechanical properties of ZrO_2 ceramic material, the influence of different particle size and contents of TiB_2 and Al_2O_3 powders on the microstructure and mechanical properties of ZrO_2 nano composite ceramic tool and die material is investigated. ZrO_2 nano composite ceramic tool and die material is prepared with Vacuum hot pressing technique at 1 450 °C for 60 min at 30 MPa. The results were shown in the Table 1.

As shown in Table 1, the fracture toughness of the composite is good, but the flexural strength and hardness is low. The flexural strength, the fracture toughness and the hardness of the ceramic material reaches 619MPa, 12.2MPam$^{1/2}$ and 10.71GPa, and the composites with 10 vol. % Al_2O_3 and 10 vol. % TiB_2 has the optimum comprehensive mechanical property.

Materials	Fracture toughness /MPa·m$^{1/2}$	Hardness /GPa	Flexural strength/MPa
ZB(5)A(5)	9.76	10.03	619
ZB(5)A(10)	10.59	10.20	501
ZB(5)A(15)	9.95	10.36	509
ZB(10)A(5)	10.51	10.37	617
ZB(10)A(10)	11.37	10.71	612
ZB(10)A(15)	12.20	10.19	565
ZB(15)A(5)	7.86	9.82	513
ZB(15)A(10)	7.91	10.22	524
ZB(15)A(15)	8.11	10.14	520

Table 1. Mechanical properties of the ceramic tool and die materials

Table 2 shows the mechanical properties of ZrO_2 nano composite ceramic tool and die material with different sized Al_2O_3 powders. The flexural strength and hardness of the composites with micrometer sized Al_2O_3 powders is higher than that with nanometer sized Al_2O_3 powders, but the fracture toughness is lower than the latter. In ZrO_2 nano composite ceramic tool and die material, ZrO_2 transformation toughening effect is the main toughening mechanism, the effect of Al_2O_3 on mechanical property is due to the particle reinforcement [Elshazly et al, 2008]. Nanometer sized Al_2O_3 has pinning effect on the grain boundary, restrictions the grain boundary sliding, this is propitious to make the matrix grain finer and is good for the fracture toughness. The grain size of the composite with micrometer sized Al_2O_3 is bigger than that with nanometer sized Al_2O_3, but the effect of particle reinforcement is higher than the latter, this is the main reason of the higher flexural strength.

Material	Flexural strength /MPa	Fracture toughness /MPa·m$^{1/2}$	Hardness /GPa
With micro- Al_2O_3	743	7.75	11.6
With nano- Al_2O_3	612	11.37	10.71

Table 2. Mechanical properties of the ceramic with different sized Al_2O_3 powders

Fig.5 shows the SEM morphology of ZrO_2 nano composite ceramic tool and die material with different sized Al_2O_3 powders. The microstructure of the composite with nanometer sized Al_2O_3 powders is finer than that with micrometer sized Al_2O_3 powders.

The results showed that the highest flexural strength of ZrO_2-TiB_2-Al_2O_3 nano-composite ceramic tool and die material reaches 743 MPa with 10 vol. % Al_2O_3 micrometer sized powders. The fracture toughness increased obviously along with the increase of Al_2O_3 nanometer sized powders, and reaches 11.37 MPa·m$^{1/2}$. Vickers hardness did not change obviously with different Al_2O_3 powders, while it greatly increased with the increases of Al_2O_3 content.

(a) with micrometer sized Al_2O_3 powder (b) with nanometer sized Al_2O_3 powders

Fig. 5. SEM of the composites with different sized Al_2O_3 powders

3.2 Hot pressing technology of ceramic material

3.2.1 Effects of holding time

Fig. 6 shows the effect of holding time on the flexural strength of ZrO_2 nano-composite ceramic tool and die material when sintered at 1 450 °C. It can be seen from Fig. 6 that the flexural increases first and then decreases with the increase of holding time, reaching the maximum of 878 MPa when the holding time is 60 min. Fig. 7 shows the effect of holding time on the fracture toughness and hardness of the composite material. The hardness increases with the increase of the holding time, reaching the maximum of 13.48 GPa when the holding time is 80min. Fracture toughness increases and then decreases with holding time increases, reaching the maximum of 9.91 MPa·m$^{1/2}$ when the holding time is 40min.

Fig. 8 shows the SEM morphologies of ZrO_2 nano-composite ceramic tool and die material with the holding time of 20min, 60min and 80min. It can be seen from Fig. 8, the microstructure of the composite material has changed greatly with the holding time increases, and to be the best when the holding time is 60 min. The reason is that proper holding time can improve the microstructure, ceramic material becomes further densification when the holding time is 60 min, and many nano-meter sized grains has found in the grain boundary which contribute to the trans/inter-granular mixed fracture mode occurs.

As a result, the flexural strength and fracture toughness of the composite material when the holding time is 60min are better than the other holding time.

Fig. 9 shows the effects of hot pressing temperature on the flexural strength of ZrO_2 nano-composite ceramic tool and die material with the holding time of 60min. The flexural strength first increases and then decreases with the increase of hot pressing temperature from 1420 °C to 1470 °C, and reaching the maximum of 1055 MPa at 1430 °C. Fig.10 shows the effects of hot pressing temperature on the fracture toughness and hardness. As shown in Fig. 10, the change of the hardness and the fracture toughness are nearly the same in trend as that of the flexural strength, the hardness reaching the maximum of 13.78 GPa when the hot pressing temperature of 1460 °C and the fracture toughness reaching the maximum of 10.57 MPa·m$^{1/2}$ when the hot pressing temperature is 1430 °C. It can be seen that the effect of sintering temperature on the mechanical properties of the composite material is obvious, proper hot pressing temperature can improve the mechanical properties, the flexural strength and the fracture toughness respectively reaching the maximum of 1055 MPa and 10.57 MPa·m$^{1/2}$ at 1430 °C and the hardness of this sintering temperature is lower than the maximum.

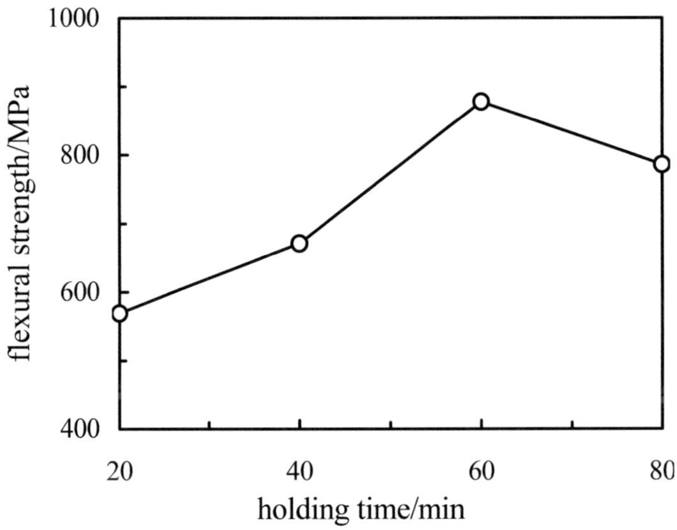

Fig. 6. Effects of holding time on the flexural strength

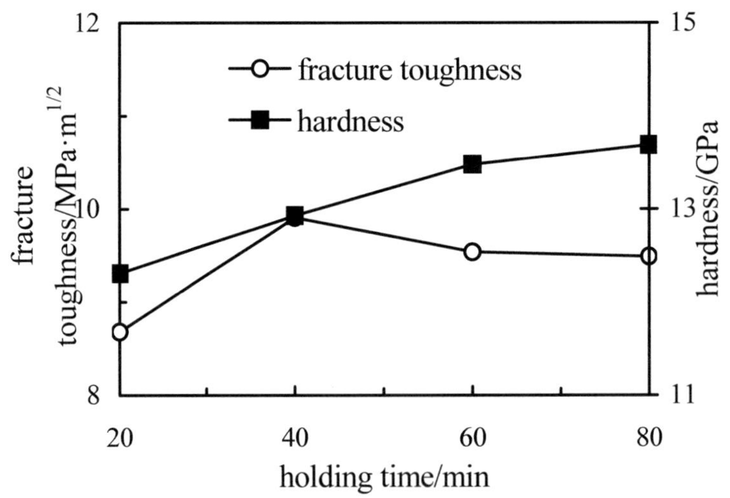

Fig. 7. Effects of holding time on the fracture toughness and hardness

(a) remain 20min at 1450 °C

(b) remain 60min at 1450 °C

(c) remain 80min at 1450 °C

Fig. 8. SEM morphologies of the ceramic tool and die material under different holding times

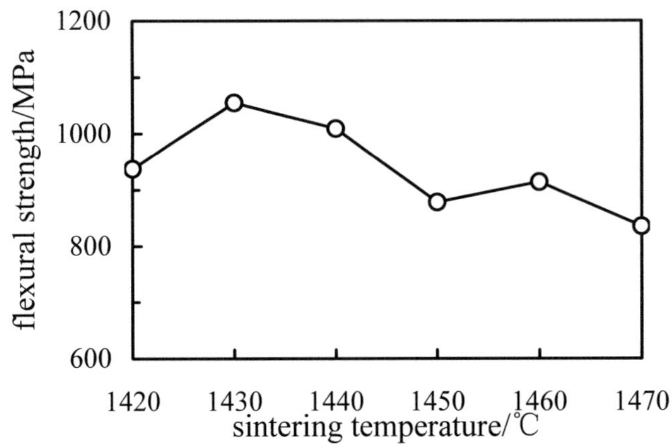

Fig. 9. Effects of hot pressing temperature on the flexural strength

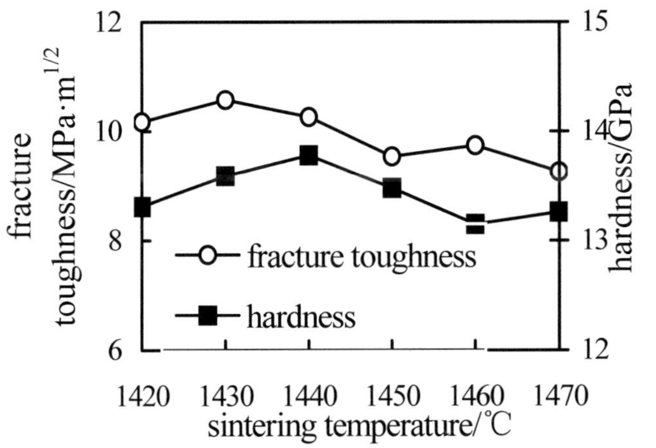

Fig. 10. Effects of hot pressing temperature on the fracture toughness and hardness

Fig. 11 shows the SEM morphologies of the ZrO_2 nano-composite ceramic tool and die material sintered at the temperature of 1420 °C, 1430 °C and 1450 °C, with the holding time of 60 min. It can be seen from Fig. 11 that the microstructure of the tested ZrO_2 nano-composite ceramic tool and die material is a typical kind of the intragranular/intergranular microstructure and the fracture mode is the mixture of both transgranular/intergranular fracture, and the mixture of both transgranular/ intergranular fracture occurred at 1430 °C is better than that which was sintered at 1420 °C, this is the main reason for the high mechanical properties. Compared to the Fig.11(c), the microstructure of the composite materials at 1420 °C and 1430 °C are obviously finer than that which was sintered at 1450 °C. The abnormal grains which can be found in Fig.11 (c) could affect the mechanical properties of the composite material.

Therefore the appropriate sintering temperature of ZrO_2 nanocomposite ceramic tool and die material is 1430 °C.

(a) 1420 °C

(b) 1430 °C

(c) 1450 °C

Fig. 11. SEM morphologies of the ceramic tool and die material under different hot pressing temperatures

3.3 Sintering process of ceramic material

Two kinds of ceramic materials were prepared by hot press technology at 1430 °C in vacuum with the time duration of 60 min and hot pressing pressure of 35 MPa. Besides, a time duration of 120 min at 1100 °C was added in composite 2. The mechanical properties of the composites are shown in Table 3.

Material	Flexural strength /MPa	Fracture toughness /MPa·m$^{1/2}$	Hardness /GPa
Composite 1	765	8.18	11.6
Composite 2	878	9.54	13.48

Table 3. Compositions and mechanical properties of the ceramic tool and die materials

It indicates that 1100 °C is approach the transformation temperature when the crystal structure of zirconium dioxide is transformed from monoclinic to tetragonal. This process also follows about 7% volume contraction. The nano-meter sized grains grow up generally after 1200 °C. Keeping on hot pressing a period of time at this temperature firstly can make the crystal structure of zirconium dioxide be transformed from monoclinic to tetragonal; The second, it also can accelerate the sintering densification when the grain growth is not obvious which is benefit to obtain a more ideal microstructure and mechanical property of the nano-composite ceramic material.

As shown in Table 3, composite 2 has the same components and sintering process with composite 1 except this 120 min is sintering at 1100 °C, but all the mechanical properties are noticeably higher than that of composite 1.

Investigation on the Vickers indentation is one of the effective methods to characterize the change of hardness and toughness. Fig. 12 shows the morphologies of Vickers indentation of both composite 1 and 2. As shown in Fig. 12 (a) and (b), the Vickers indentation of composite 2 is smaller than that of the composite 1 and the cracks of composite1 are obviously. It indicates that this sinter process can enhance the hardness of ZrO_2 ceramic materials. As shown in Fig. 12 (c) and (d), the crack of composite 2 is shorter and thinner than that of composite 1, which suggests that the fracture toughness of composite 2 is higher than that of composite 1 obviously.

As shown in Fig. 12 (c) and (d), although the mechanical property of two materials has changed, the grain size and the distribution of Al_2O_3 and TiB_2 are similar. It indicates that the enhancement of mechanical properties is mainly because the addition of ZrO_2 and its phase transformation. These results suggest that keeping sintering of 120 min at 1100 °C can make all of the ZrO_2 phase be transformed from monoclinic to tetragonal. It is a volume expansion process when the ZrO_2 grains transforms from tetragonal symmetry to monoclinic symmetry. It is very difficult to make the tetragonal be transformed to monoclinic resulted from the high hot pressing pressure. Thus, most of the tetragonal ZrO_2 grains can be kept after finish sintering until the room temperature.

When a crack appears and extends through the ZrO_2 grains, the tetragonal grains will be transformed to monoclinic under the stress of crack tip. On the one hand, this process can absorb the fracture energy and reduce crack tip stress; on the other hand, phase transformation often follows by the volume expansion which can press the crack tip and cause the crack thinning or even stop extending. This process is the typical stress-induced transformation toughening.

(a) composite 2 (200×)

(b) composite 1 (200×)

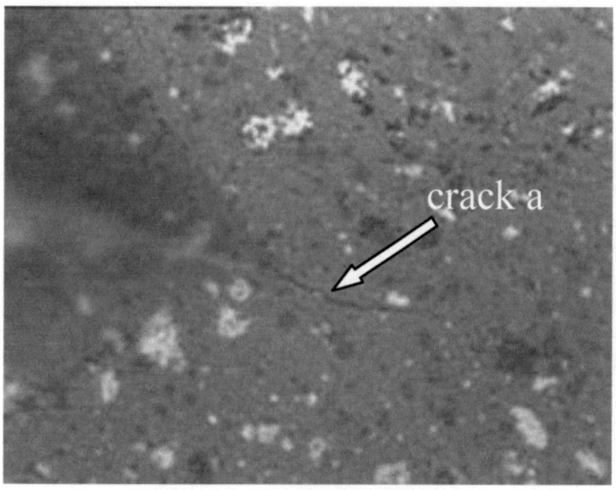

(c) composite 2 (400×)

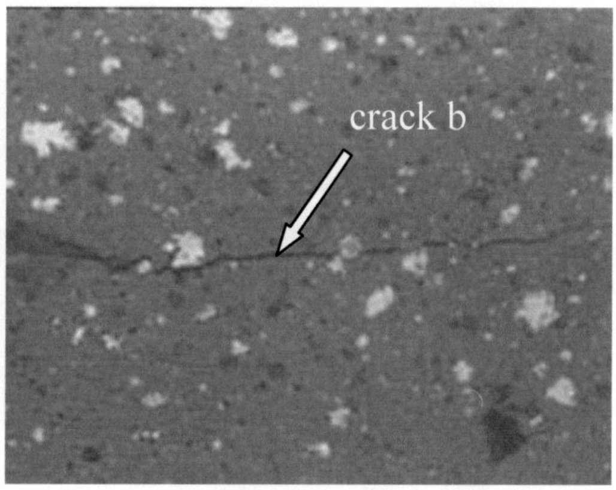

(d) composite 1 (400×)

Fig. 12. Morphologies of Vickers indentation

In order to study the change of phase transformation, the material surface is analyzed by XRD. As shown in Fig. 13(a) and (b), two diffraction peaks appears in nearby 30° after the common sintering process, which is the monoclinic ZrO_2. After the optimal sintering process, the monoclinic ZrO_2 disappeared, and all of the ZrO_2 grains are tetragonal symmetry.

Microstructure of both composite 1 and 2 materials under SEM are shown in Fig. 14. It can be seen that grains in composite 2 are finer than that in composite 1. Most of the composite 1 material grains are about 200 nm while the composite 1 material only has few of fine grains distributed in the grain boundary. The finer the grain is, the higher the mechanical property of nano-meter composite ceramic material is. After 120 min sintering at 1100 °C, materials have reached nearly the fall densification which can limit the grain growth space.

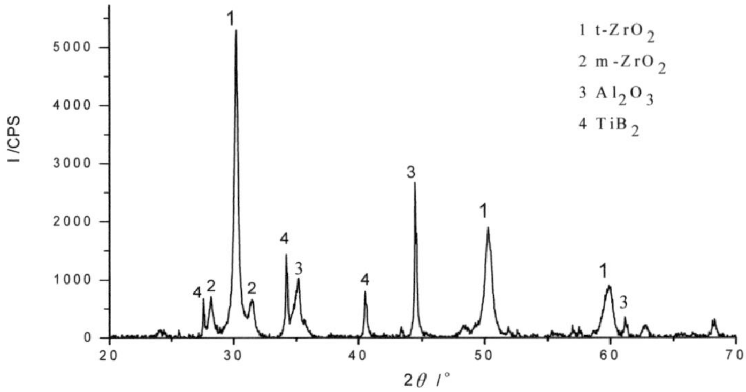

(a) mixed powder

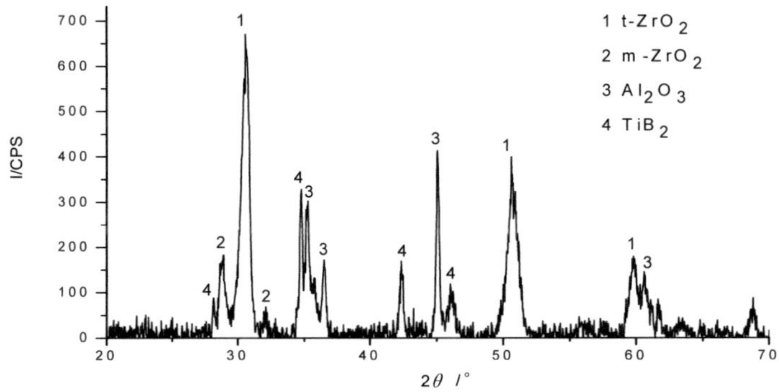

(b) common sintering process

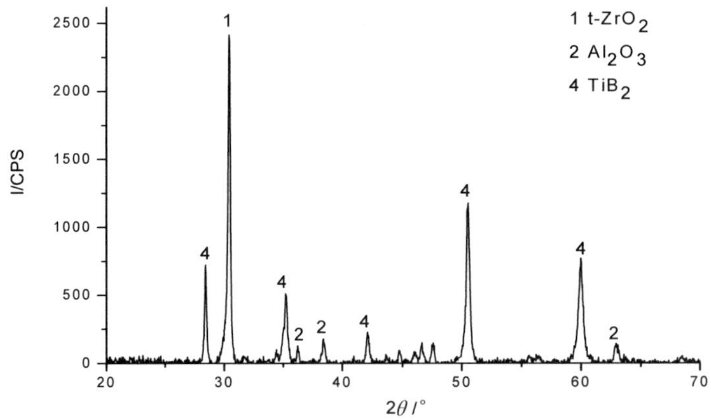

(c) optimal sintering process

Fig. 13. XRD analysis of nano-composite ceramic tool and die material

(a) composite1 (b) composite 2

Fig. 14. SEM morphology of the composites with different sinter process

4. Tribological design of ZrO_2 nano-composite ceramic tool and die material

In this experiment, the mechanical properties of ZrO_2 nano-composites can be seen from Table 4.

Material	TiB_2 /Vol. %	Al_2O_3 /Vol. %	Fracture toughness /MPa·m$^{1/2}$	Hardness /GPa	Flexural strength /MPa
a	5	10	10.59	10.20	501
b	10	5	10.51	10.37	617
c	10	10	11.37	10.71	612
d	10	15	12.20	10.19	565
e	15	10	7.91	10.22	524

Table 4. Compositions and Mechanical properties of ZrO_2 nano-composites

Fig.15 shows the friction coefficient and the wear rate of different TiB_2 contents under the 200r/min rotational speed and the load of 160N. The result shows that the friction coefficient of the composites decreases with the increase of TiB_2 content, reaches the minimum when TiB_2 content amounts to be 15 vol. %. The wear rate of the composites decreases first and then increases with the increase of TiB_2 content, reaches the minimum of 1.29×10^{-6} mm^3/N·m when TiB_2 content amounts to be 10 vol. %.

Contrast Fig.15 with Table 4, the changing trends of the friction and wear properties and the mechanical properties of the composites are roughly the same. The results indicated that the proper additive of TiB_2 could improve both the friction and wear properties and the mechanical properties. Different TiB_2 contents will significantly affect density of the material, and then the density directly affects the mechanical properties of materials. Although the additive of TiB_2 could improve the friction and wear properties, the high TiB_2 content also affect the microstructure of materials and the surface of materials was easy to be broken, and is bad for the friction and wear properties.

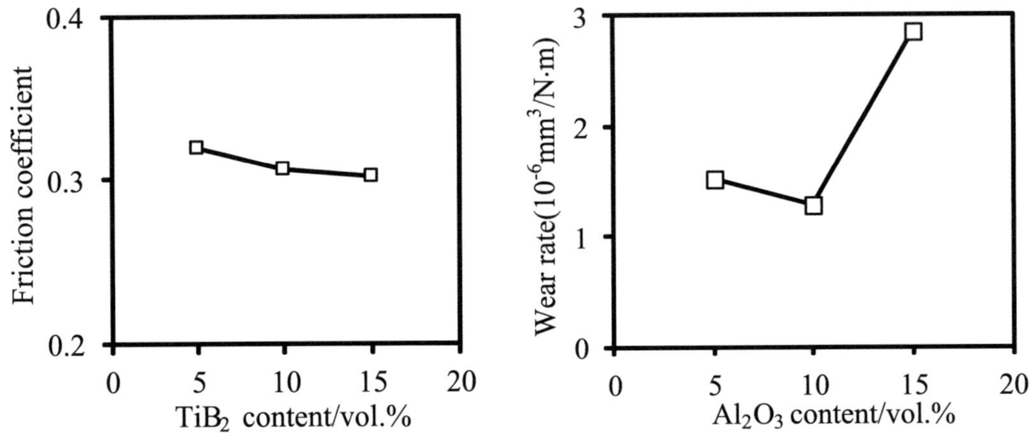

(a) friction coefficien (b) wear rate

Fig. 15. Effect of different TiB_2 contents on friction coefficient and wear rate

Fig.16 shows the friction coefficient and the wear rate of different Al_2O_3 contents under the 200r/min rotational speed and the load of 160N. The result shows that the friction coefficient of the composites decreases first and then increases with the increase of Al_2O_3 content, reaches the minimum when Al_2O_3 content amounts to be 10 vol. %. As the content of Al_2O_3 increases, the change of the wear rate is nearly the same in trend as that of the friction coefficient but not obvious, reaching the minimum of 1.286×10^{-6} $mm^3/N \cdot m$ when Al_2O_3 content amounts to be 10 vol. %. Al_2O_3 is proved a strengthening material of ZrO_2, the microstructure and the density of materials were obtained by the addition of Al_2O_3 particles, thus the mechanical properties of materials are increased, and reaches the maximum when Al_2O_3 content amounts to be 10 vol. % (Table 4),and with increasing Al_2O_3 content (from 5 vol. % to 10 vol. % in composites), the friction and wear properties of ZrO_2 composites are continuely increased, but increasing Al_2O_3 content further (up to 10 vol. %), it is decreased.

The effect of material content on the friction and wear properties of the composites is obviously, proper material content not only increase the mechanical properties, but also improve the friction and wear behaviors. Compared with friction coefficient, the changing trends of wear rate with mechanical properties is obviously, the reason is that both of the good wear rate and well mechanical properties needed the finer microstructure, and the friction coefficient are mainly dependent on the TiB_2 content [Mazaheri et al, 2008].

Fig.17 shows that the surface of the composite is smooth and a small amount of defect area can also be observed. The smooth surface is mainly because of the friction between the ceramic material and the steel ring. The friction chip has high surface activity and easy adheres to the material surface, forming the continual surface layer on the surface with the increase of wear time. The smooth and continue surface layer can effectively reduce the coefficient and slow the wear. With the temperature of friction area increase, wear increase because of the plastic deformation and material transfer take place, form the adhesive wear.

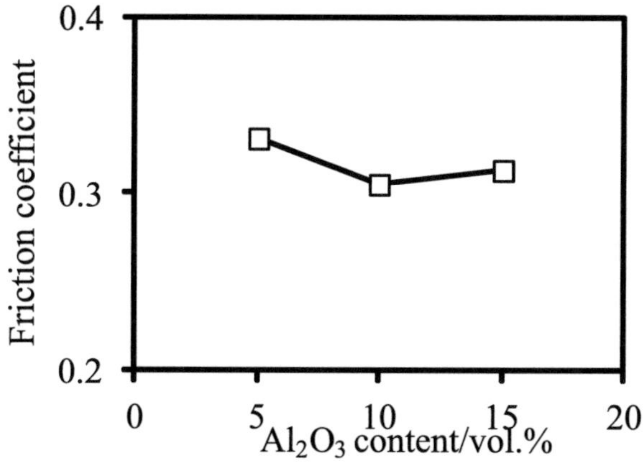

(a) Friction coefficient

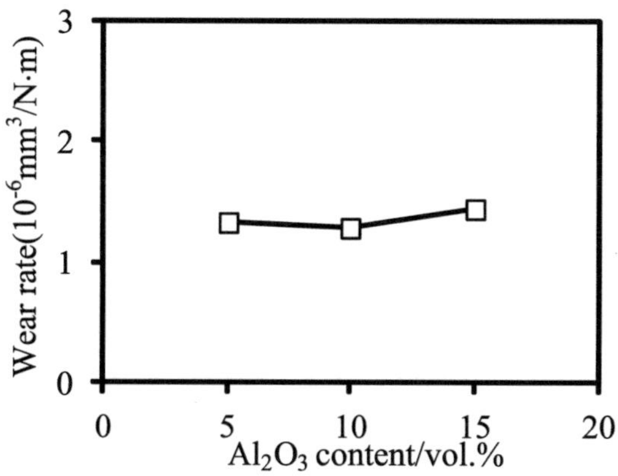

(b) Wear rate

Fig. 16. Effects of different TiB_2 content on friction coefficient and wear rate

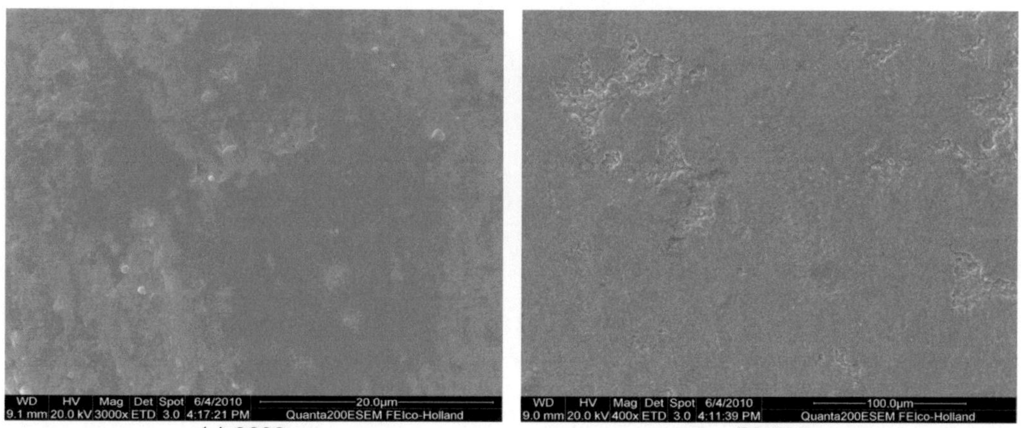

(a) 3000× (b) 400×

Fig. 17. SEM morphologies of ZrO_2 nano-composite ceramic tool and die material

XRD was used to analyze the physics of surface film(Fig.18). More TiB_2 and Al_2O_3 can be seen from Fig.18(b), the main reason is that the hardness of TiB_2 and Al_2O_3 is higher than ZrO_2 and this can reduce the wear in friction. Secondly, TiB_2 is proved a self-lubricating material which can low friction coefficient. Besides, FeO was found on the material surface after the test (Fig.18(b)), this was come from the reaction between the steel chip and the atmosphere under the friction heat and moved by friction. The softer FeO chips and the ceramic chips were mixed by the mechanical pressure, formed the surface layer which can protect the ceramic material and reduced the wear.

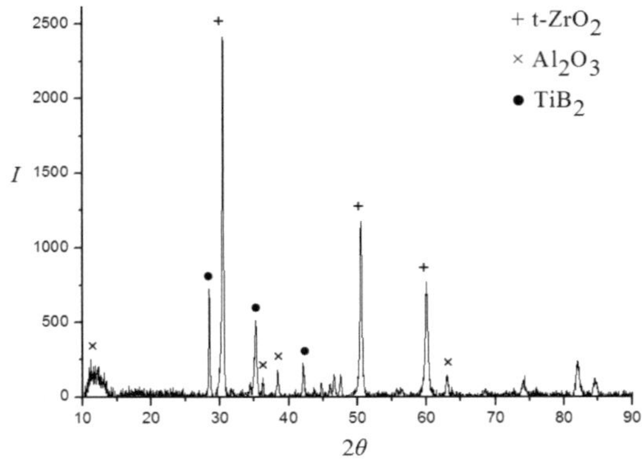

(a)before the test

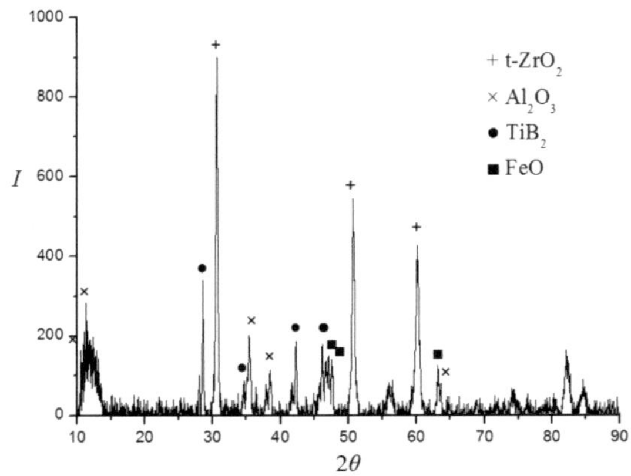

(b) after the test

Fig. 18. XRD analysis of ZrO_2 nano-composite ceramic tool and die material

5. Friction and wear behaviour

Fig.19 shows the effects of load and rotational speed on the friction coefficient of composites. The friction coefficient of the composites first and then decreases with the increases of load under the 200r/min rotational speed, and the friction coefficient of the composites decreases with the increases of rotational speed under the 160N load. The friction coefficient reaches the minimum of 0.3 and 0.29 when the load is 240N and the rotational speed is 200r/min, respectively.

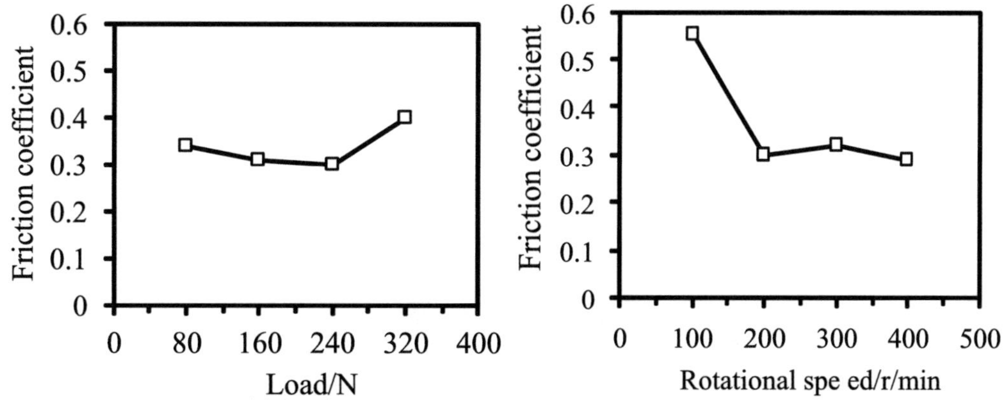

(a) load (b) rotational speed

Fig. 19. Effects of load and rotational speed on the friction coefficient

Fig.20 shows the effects of load and rotational speed on the wear rate of composites. The change of wear rate is near the same with the friction coefficient. The wear rate decreases with the load from 80N to 240N under the 200r/min rotational speed, but when the load increases to 320N, the wear rate reach the maximum, $5.44\times10^{-6}mm^3/N{\cdot}m$. The wear rate decreases with the speed increase as can be seen in Fig.20(b).

Fig.21 shows the SEM of the wear surface with 160N and 200r/min. The surface is smooth and most of the surface is covered by a nearly continuous layer in Fig. 21(a). The wear appearance of ceramic surface mainly includes two parts, part 1 is a smooth and grey area (point 1 in Fig.21), part 2 is the saddle(point 2 in Fig.21). In order to attribute the phase differences of the two parts, the EDAX electron spectrum analysis was carried out. Fig.21 (a) and (b) shows the electron spectrum analytic curve of point 1 and point 2.

As can be seen from Fig. 22, main elements come from the composite ceramic material. A few Fe exist obviously in point 1, but no Fe element can be found in point 2. Fe comes from the 45# steel work-piece. The chips which produced in the friction process has the high activeness, easily adheres in the friction surface and some shifted to the ceramic surface along with the friction process. After the progression rolling, the mix chips forms a soft and continuous film on the ceramic surface.

XRD was used to analyze the phase change in the friction. Fig.23 shows the XRD of the material surface before and after the wear test. FeO can be found on the ceramic surface after the friction(in Fig. 23(b)), and it came from the reaction between Fe and oxygen of air. The results indicated that the layer united by the work piece materials and ceramic materials. In addition, more TiB_2 were found on the ceramic surface. TiB_2 is a self-lubricating material and harder than the matrix material. Along with the friction, the TiB_2 content was become more and slow down the wear aggravation.

Defects also can be found from the film (point 3 in Fig. 21(b)). The colour of defects is similar to the film but obviously different from the original material. Moreover, defect is smaller and shallower than the original material. The results indicated that the material composition of the defect and film is roughly the same. Zirconium dioxide is ionic crystal, the adsorption affinity to the chip is higher. The ceramic chips mixed with workpiece chips and form the continuous film, the defect is the damage of the mixed film.

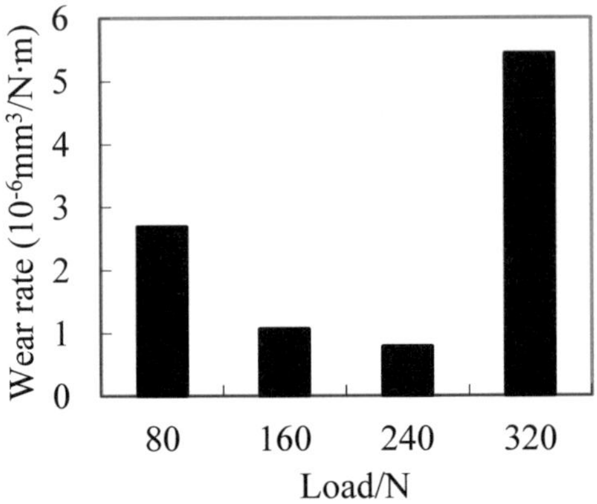

(a) load

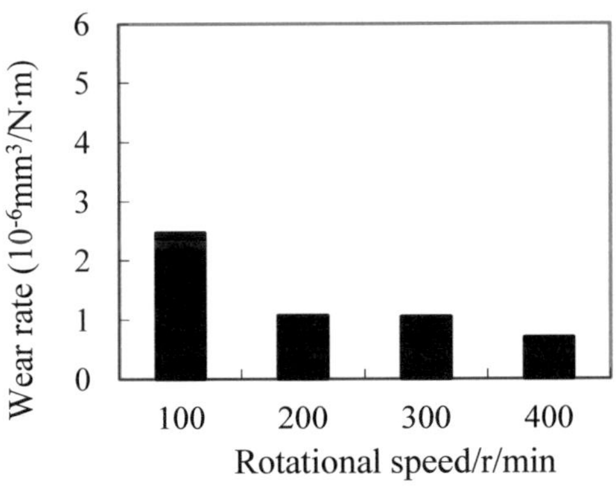

(b) rotational speed

Fig. 20. Effects of load and rotational speed on the wear rate

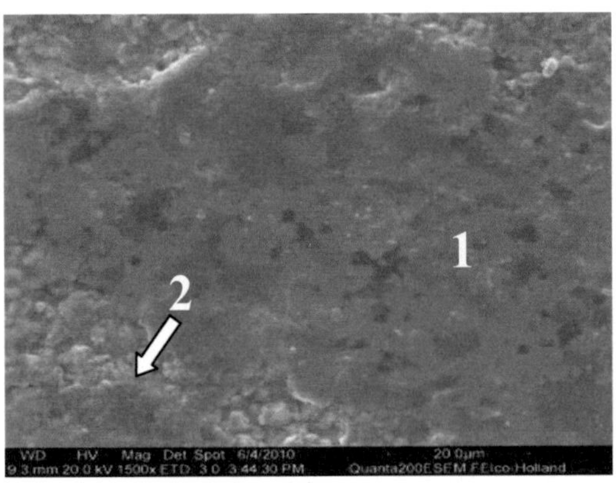

(a)

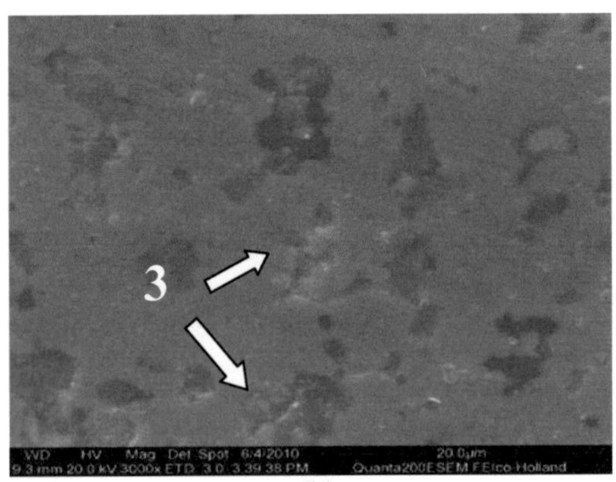

(b)

Fig. 21. SEM morphologies of the wear surface with the load of 160N and speed of 200r/min

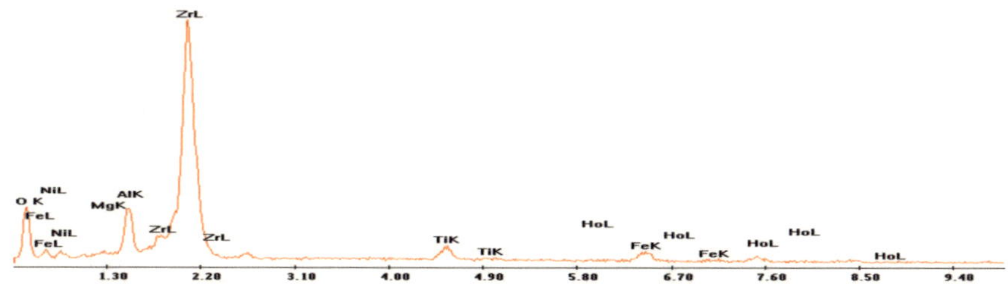

(a) point 1

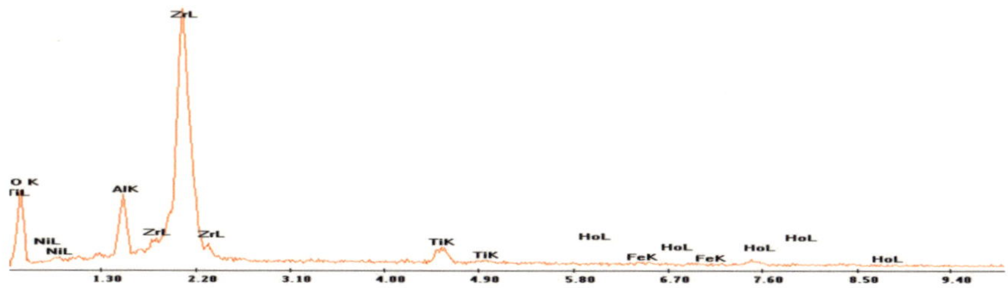

(b) point 2

Fig. 22. EDAX electron spectrum of the wear surface

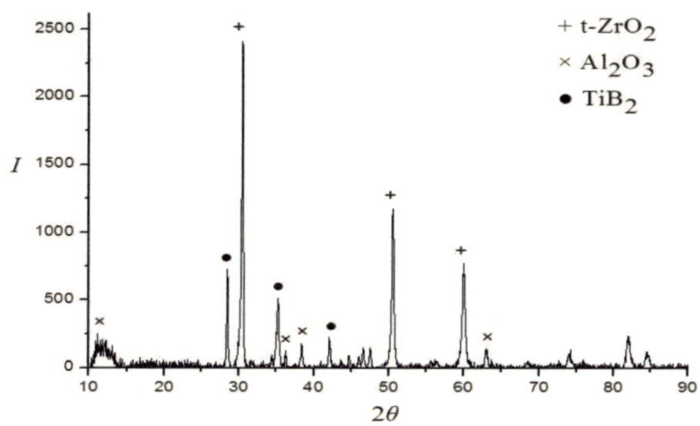

(a) before the wear test

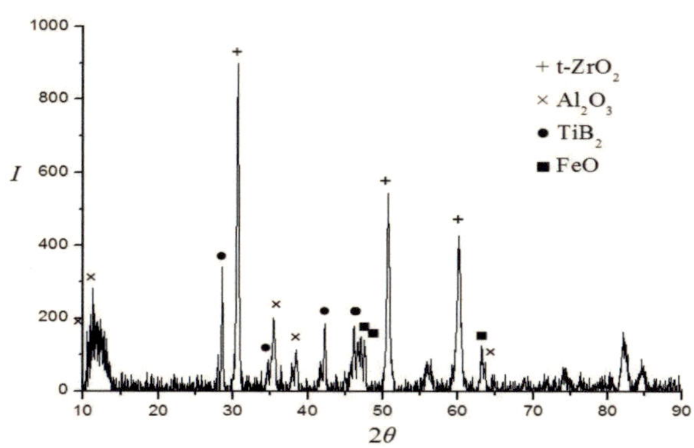

(b) after the wear test

Fig. 23. XRD analysis of nano-composite ceramic tool and die material surface

The result shows that the soft layer was produced in the stage of adhesive wear process, the soft work-piece material chips were mixed the ceramic chips and coated on the hard composite ceramic surface under the friction. Next, the friction mainly between the soft layer and work-piece, this is the main reason of the low friction coefficient and wear rate. First, the soft layer form a continual smooth rubbing surface on the composite materials surface, and increases the actual friction contacted area, reduces the friction moment; Second, the soft layer reduce the direct contact between the work-piece and ceramic, thus slowed down the wear of composite ceramic.

Fig.24 shows the SEM morphology of the wear section of the composite when the rotational speed is 200r/min and the normal load is 160N and 320N, respectively.

(a) 160N (b) 320N

Fig. 24. SEM of the wear section with 200r/min and different load

As shown in Fig. 24(a) and (b), when the rotational speed is 200r/min and the normal load is 160N, the wear of composite ceramic is light, there is only a few flaw in the wearing course and mainly distributes in about 10 microns wear courses. The flaw is exist independently and without jointed, so it is not easy to form the serious wear as seen in Fig. 24(b).

As shown in Fig. 24(b), some transverse cracks were found in the wear layer when the load increases to 320N. The cracks could cause the wear layer broken and finally form the spalling wear. When the load is low, the wear layer does not have the obvious change, wear is mainly by the slight scuffing of ceramic, and because of the hardness of ceramic is far high than the work-piece, the wear rate is low. When the load is 320N, the wear layer is easy broken and can't effectively protect the matrix material, wear rate is higher than that in low load. The test result of wear rate can be seen from Fig.20. When the load is 160N the wear rate is $1.06\times10^{-6}mm^3/Nm$. While when the load is 320N the rate of wear increases rapidly to be $5.44\times10^{-6}mm^3/Nm$.

6. Conclusion

A new ZrO_2 nano-composite ceramic tool and die material was prepared by vacuum hot pressing technique. ZrO_2-TiB_2-Al_2O_3 nano-composite ceramic tool and die material with 10 vol. % TiB_2 and 10 vol. % Al_2O_3 micrometer sized powders reaches the maximum mechanical property. ZrO_2-TiB_2-Al_2O_3 nano-composite ceramic tool and die material with optimum mechanical properties can be achieved when the hot pressing temperature is

1430°C, and the holding time is 60min. The flexural strength, hardness and fracture toughness reaches 1055 MPa, 13.59GPa and 10.57MPa·$m^{1/2}$, respectively. In the ZrO_2 nano-composite ceramic tool and die materials, the optimum sinter parameters could improve the microstructure, and the optimum sinter process could nearly completely stabilize the t-ZrO_2 to the room temperature condition that can enhance the toughening effect of ZrO_2. The additive of self-lubricating material TiB_2 could reduce the friction coefficient and improve the abrasion resistance. Moreover, the TiB_2 content was become more under the continuous friction condition which is able to slow down the wear aggravation. The continuous friction film was formed by the ceramic and work-piece chips under the friction. The film could reduce the friction and protect the matrix material. The wear rate of ZrO_2 nano-composite ceramic tool and die material is $1.06\times10^{-6}mm^3/Nm$ when the rotational speed is 200r/min and the normal load is 160N. It indicated that this new ceramic composite have good friction and wear properties. Therefore, it can be expected that the developed ZrO_2 nano-composite ceramic material will get further application in the field of cutting tools, dies and other wear resistant parts, etc. with high wear resistance and performance.

7. Acknowledgement

This work was supported by Shandong Provincial Natural Science Foundation, China (Grant No. ZR2009FZ005), the Program for New Century Excellent Talents in University of China (Grant No. NCET-10-0866), the National Natural Science foundation of China (Grant No. 51075248), Shandong Provincial Natural Science Foundation for Distinguished Young Scientists, China (Grant No. JQ201014).

8. References

Liu J. & Zhou F. (2003). Properties and Applications of Ceramic Materials for Hot Extrusion Dies. *Rare Metal Materials and Engineering*, Vol. 32, No. 3, Mar, 2003, pp. 232-235, ISSN 1002-185X

Kar A.; Tobyn M. J. & Ron S. (2004). An Application for Zirconia as a Pharmaceutical Die Set. *Journal of the European Ceramic Society*, Vol. 24, No. 10-11, Sep, 2004, pp. 3091-3101, ISSN 09552219

Basu B; Lee J & Kim D. (2004). Development of Nanocrystalline Wear-resistant Y-TZP Ceramics. *Journal of the American Ceramic Society*, Vol. 87, No. 9, Sep, 2004, pp. 1771-1774, ISSN 00027820

Hirvonen A.; Nowak R. & Yamamoto Y. (2006). Fabrication, Structure, Mechanical and Thermal Properties of Zirconia-based Ceramic Nanocomposites. *Journal of the European Ceramic Society*, Vol. 26, No. 8, May, pp. 1497-1505, ISSN 09552219

Guicciardi S.; Shimozono T.& Pezzotti G. (2006). Nanoindentation Characterization of Sub-micrometric Y-TZP Ceramics. *Advanced Engineering Materials*, Vol. 8, No. 10, Oct, 2006, pp. 994-997, ISSN 14381656

Zhang Y. S.; Hu L. T.; Chen J. M. & Liu W. M. (2009) . Fabrication of Complex-shaped Y-TZP/Al_2O_3 Nanocomposites. *Journal of Materials Processing Technology*, Vol. 209, No. 3, Feb, 2009, pp. 1533-1537, ISSN 09240136

Liu H. W. & Xue Q. J.(1996). The Tribological Properties of TZP-graphite Self-lubricating Ceramics. *Wear*, Vol. 198, No.1-2, Oct, 1996, pp.143-149, ISSN 00431648

Yang C. T. & Wei C. J.(2000). Effects of Material Properties and Testing Parameters on Wear Properties of Fne-grain Zirconia (TZP). *Wear*, Vol. 242, No. 1-2, Jul, 2000, pp.97-104, ISSN 00431648

Baharvandi H.; Hadian A. & Alizadeh A.(2006). Processing and Mechanical Properties of Boron Carbide-titanium Diboride Ceramic Matrix Composites. *Applied Composite Materials*, Vol. 13, No. 3, May, 2006, pp. 191-198, ISSN 0929189X

Basu B.; Vleugels J. & Biest O. (2005). Processing and Mechanical Properties of ZrO_2-TiB_2 Composites. *Journal of the European Ceramic Society*, Vol. 25 , No.16, May,2005, pp.3629- 3637 , ISSN 09552219

Hirvonen A.; Nowak R. & Yamamoto Y. (2006). Fabrication, Structure, Mechanical and Thermal Properties of Zirconia-based Ceramic Nanocomposites. *Journal of the European Ceramic Society*, Vol. 26, No. 8, May, 2006, pp. 1497-1505 , ISSN 09552219

Gupta T. K.; Bechtold J. H. & Kuznicki R. C.(1977). Stabilization of Tetragonal Phase in Polycrystalline Zirconia. *Journal of Materials Science*, Vol.12, No.12. Dec, 1977,pp.2421-2426 , ISSN 00222461

Hannink R.; Kelly P. & Muddle B.(2000). Transformation toughening in zirconia-containing ceramics. *Journal of the American Ceramic Society*. Vol.83, No. 3, Mar, 2000, pp. 461–487. ISSN 00027820

Tu J.& Li J. Effect of phase transformation induced plasticity on the erosion of TZP ceramics. *Materials Letters*, Vol. 31, No. 3-6, Jun, 1997, pp. 267-270 , ISSN 0167577X

Guicciardi S.; Shimozono T. & Pezzotti G. (2006). Nanoindentation characterization of sub-micrometric Y-TZP ceramics. *Advanced Engineering Materials*, Vol. 8, No. 10, Oct, 2006, pp. 994-997, ISSN 14381656

Elshazly E.; Ali M. & Hout S.(2008). Alumina effect on the phase transformation of 3Y-TZP ceramics. *Journal of Materials Science and Technology*, Vol. 24, No.6, Nov, 2008, pp. 873-877,ISSN 10050302

Mazaheri M.; Simchi A. & Golestani F.(2008). Densification and grain growth of nanocrystalline 3Y-TZP during two-step sintering. *Journal of the European Ceramic Society*, Vol. 28, No. 15, Nov, 2008, pp. 2933-2939 , ISSN 09552219

Part 3

Structural Ceramics

6

Synthesis, Microstructure and Properties of High-Strength Porous Ceramics

Changqing Hong, Xinghong Zhang,
Jiecai Han, Songhe Meng and Shanyi Du
Center for Composite Materials and Structure, Harbin Institute of Technology, Harbin, PR China

1. Introduction

Porous ceramics containing tailored porosity exhibit special properties and features that usually cannot be achieved by their conventional dense counterparts. Thus, porous ceramics find nowadays many applications as final products and in several technological processes. Porous ceramics are of significant interest due to their wide applications in high-temperature filters, thermal gas separation, lightweight structural components and thermal structural materials (Peng, H.X et al, 2000; Corbin, S. F and Apte, P. S 1999; Fukasawa, T et al; 2001).

To meet these requirements, various fabricating methods are adopted to produce highly porous ceramics. These processing methods include the replication of polymer foams by ceramic dip coating, the foaming of aqueous ceramic powder suspensions, the pyrolysis of preceramic precursors, partial sintering by pressureless sintering, and the firing of ceramic powder compacts with pore-forming fugitive phases (Kamal, M. M et al 2006; Amir, M. and Rohwer, K, 2007; Hong, C. Q et al 2007; Dale, H &David, J. G, 1995; Oh, S.-T et al, 2001; Deng, Z.-Y et al,2002; Deng, Z. Y et al,2001)

Often for porous ceramics used in high temperature environment, they require high temperature melting point, high mechanical strength, low thermal conductivity, open pore structure and special pore distribution. For example, with the development of modern spaceflight technology, active cooling is becoming a very efficient mode of thermal protection for high heat load structures. In active cooling, coolant is usually injected or infiltrated into a porous ceramic, and then transpirated to reduce the bulk temperature for thermal protection (Hong, C.Q et al 2007). The components used for active cooling are always subjected to severe environments with ultra-high temperature, high-pressure, and usually supersonic velocity. Therefore, the porous ceramic in active cooling must have high melting point, moderate framework strength, and reasonable pore distribution regularity.

The properties for porous ceramics can be tailored for specific environmental application by controlling the composition and microstructure of the porous ceramic. Changes in open and closed porosity, pore size distribution and pore morphology can have a major effect on a material's properties. All of these microstructural features are in turn highly influenced by the processing route used for the production of the porous material. For the mechanical properties, porous ceramics are determined by their structure parameters, such as porosity, pore size, and pore structure (Deville, S et al, 2006; Gough, J. E et al, 2004). Additionally, the microstructure of the solid phase related to neck growth and solid phase continuity strongly affect the mechanical properties. Several important issues regarding the neck growth between touching particles by surface diffusion and volume diffusion can significantly increase the mechanical properties with minimal increase in density. The microstructure in porous ceramics can be controlled not only by adjusting the particle size and shape of the initial powders, but also by the sintering process (Studart, A. R et al, 2006; Koh, Y. H et al,2006; Schmidt, H et al, 2001)

The objective of this chapter is to adopt partial/pressureless sintering and freeze-casting routes for the preparation of high-strength porous titanium diboride (TiB_2) and porous ZrO_2 ceramics, with particular emphasis on the processing-microstructure-property relations inherent to each process.

2. Production of porous diboride titanium

2.1 Porous TiB_2

TiB_2 has been regarded as promising candidate materials for structural applications for its unique combination of higher melting point, good strength, good thermal stability, and corrosion resistance.

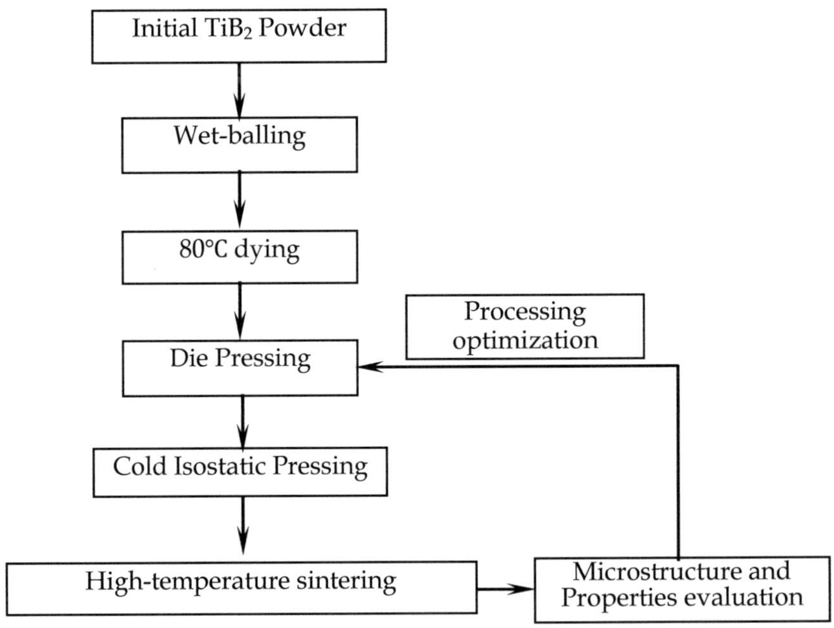

Fig. 1. Technologic flow for preparing porous TiB_2 ceramic

The constituent materials used in this study were TiB_2 powder (produced by Northwest Institute for Non-Ferrous Metal Research, Xi'an, People's Republic of China) with particle mean size of 2–4μm and purity of 98 per cent. In order to reduce the agglomerated TiB_2 powder to single particles, TiB_2 starting powder was milled by the wet-milling method for 4 h in a plastic bottle with agate balls and acetone as media. The resultant slurry was dried in a vacuum evaporator. The dried powder was screened through an 80-mesh screen. After that, the powder was prepressed in the cylindrical type mould with the L/D ratio of 1:2 and 30mm diameter under a uniaxial pressure of 10MPa. Then they were compacted by a cold isostatic press (CIP) machine at room temperature. The CIP pressures were 50, 100, and 200MPa, and they were maintained for 200s. The green-compacted billets were sintered in a furnace, in a vacuum atmosphere, at a heating rate of 10 °C/min. All specimens were held at the sintering temperature for 30 min and then cooled to room temperature at the same rate of 10 °C/min. Different sintering temperatures in a range of 1650–2000 °C were used in the present experiment to produce specimens of different porosities.

2.2 Microstructure

Fig.2 shows the dependence of relative density on sintering temperature of TiB_2 porous ceramic. It can be seen that the relative density increases as the sintering temperature increases. For the sake of the lower density of the formed green compacts, the billets with low initial pressure required a higher temperature to reach the same density when compared with the high initial pressure.

Fig 3(a) and (b) show the macro-morphology of sintered and machined porous TiB_2 samples. The sintered sample fundamentally kept its shape and no macroscopic defects were found after machining.

Fig.4 shows the microstructures of porous TiB_2 specimens sintered from different compacts at 2000 °C. Because of the low sintering temperature, the original morphology of TiB_2 grains produced by the compaction can be seen clearly. A number of voids and small flaws existed in the specimen, due to a loose connection between the TiB_2 grains. However, these voids and small flaws were reduced greatly with the increasing temperature, indicating a better connection between the TiB_2 grains. The weak interface strength in porous TiB_2 ceramic sintered from lower pressured compacts originated from the poor connections of TiB_2 grains (Fig. 4(a)). However, the highly packed regions densified faster than the less-dense regions, as shown in Fig. 4(b) and (c), suggesting that the surface diffusion is enhanced and promotes the preferential neck growth with certain increase in density.

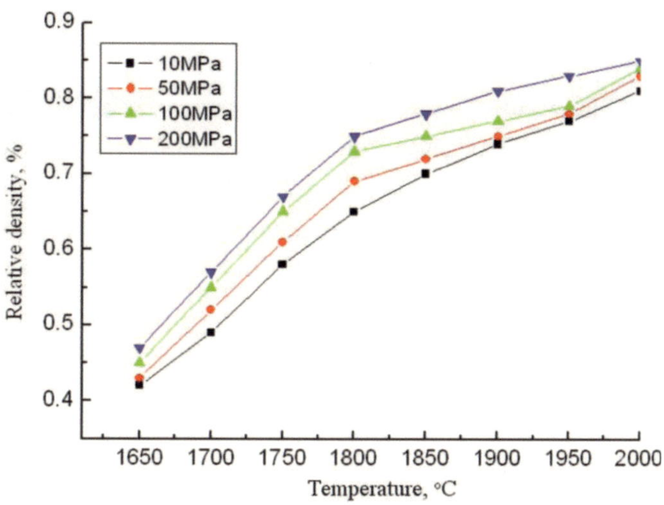

Fig. 2. Dependence of relative density on sintering temperature of TiB2 porous ceramic

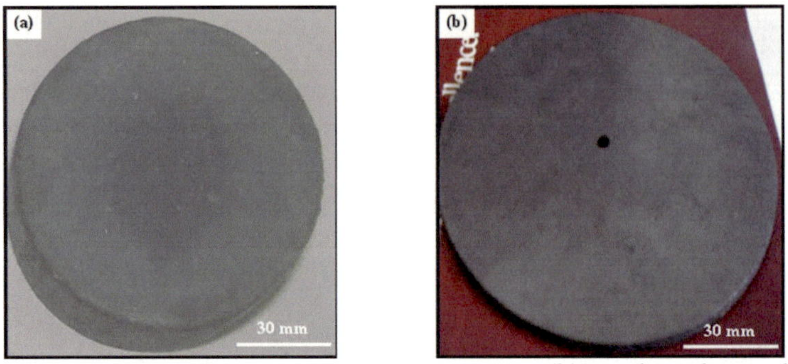

(a) sintered sample (b) machined sample

Fig. 3. Macro-morphology of porous TiB2 sintered at 2000 °C,

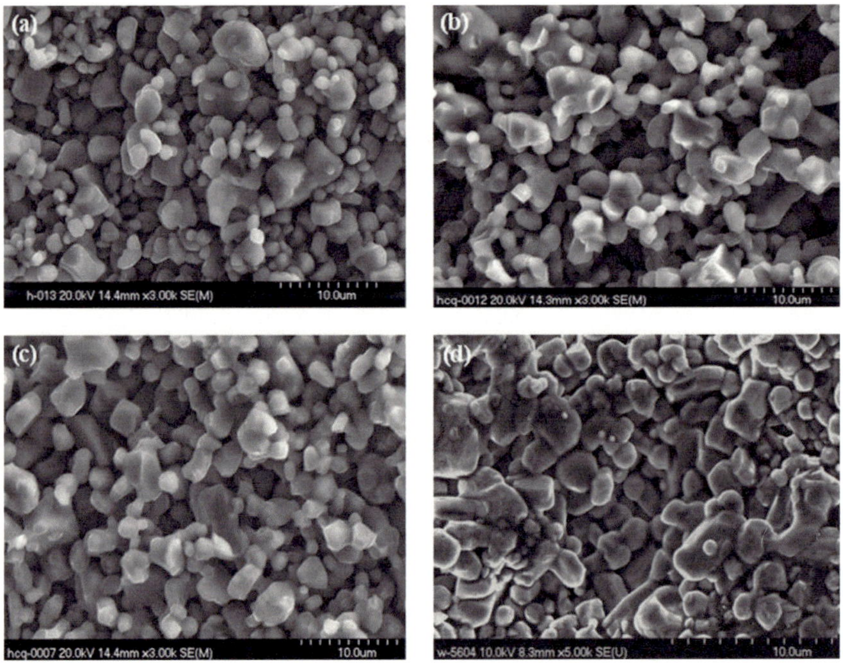

Fig. 4. SEM micrographs of TiB2 porous ceramics sintered at 2000 °C with different initial compacted pressures: (a) 10MPa; (b) 50MPa; (c) 100MPa; (d) 200MPa

Fig. 5 shows the microstructure of the specimens (initial pressures of 100MPa) sintered at 1750, 1800, 1850, and 1950 °C. Clearly, the size of the pores between grains decreases with the increase of sintering temperature, which is consistent with the variation of relative density in Fig. 2. Moreover, it can be observed (Fig. 5) that the grain size does not increase dramatically with sintering temperature. Another noteworthy observation in Fig.5(d) is the formation of well developed necks between grains, which should be responsible for the improved mechanical strength.

Because TiB_2 belongs to the covalently bonded solids, the intrinsic diffusivity is very low and therefore the Peierl's stress is high for the movement of dislocations. The preferential neck growth in mass transfer was affected by evaporation-condensation and surface

Fig. 5. SEM micrographs of TiB2 porous ceramics sintered with initial compacted pressure of 100MPa at different sintering temperatures: (a) 1750 °C; (b) 1800 °C; (c) 1850 °C; (d) 1950°C

diffusion. According to Bhaumik (Bhaumik, S. K et al, 2001) and Wang (Wang and Fu, 2002), the dominant sintering mechanism forTiB_2 is surface diffusion and subsidiary volume diffusion at such an experimental sintering condition. The surface diffusion does not result in shrinkage but in the formation of solid bonds between adjacent particles. As the temperature increases, the enhanced neck size of TiB_2 grains was enhanced and bridges between particles were well developed (Fig. 5(d)). The promoted neck formation between particles by surface diffusion gives rise to such microstructural feature.

2.3 Mechanical properties

Figures 6 and 7 show the dependence of bending strength and fracture toughness on the relative density of porousTiB_2 ceramic sintered from the compacts under different pressures. In general, the mechanical properties of porous TiB_2 followed the expected trend, i.e. fracture strength and toughness increased with increasing relative density. This indicated that high compaction pressures improve the mechanical properties of porous TiB_2 ceramics. As shown in Fig. 5, e.g. the strengths (with initial pressure of 10 and 100MPa, respectively) having 80 per cent relative density retain high strength of about 175 and 215MPa, respectively. The high strength must be related to the growth of the interparticle contacts (neck growth) by surface diffusion, which will be discussed in next section.

The fracture toughness of the porous TiB_2 ceramic shows a similar trend as the flexural strength, as demonstrated in Fig. 7. The fracture toughness of porous TiB_2 ranged from 0.5 to 2.4MPa$m^{0.5}$ for initial pressure of 10MPa; while the fracture toughness increased from 0.8 to 3.5MPa$m^{0.5}$for initial pressure of 100MPa, indicating initial compacted pressure and the relative density (induced by sintering temperature) have an important influence on the mechanical properties.

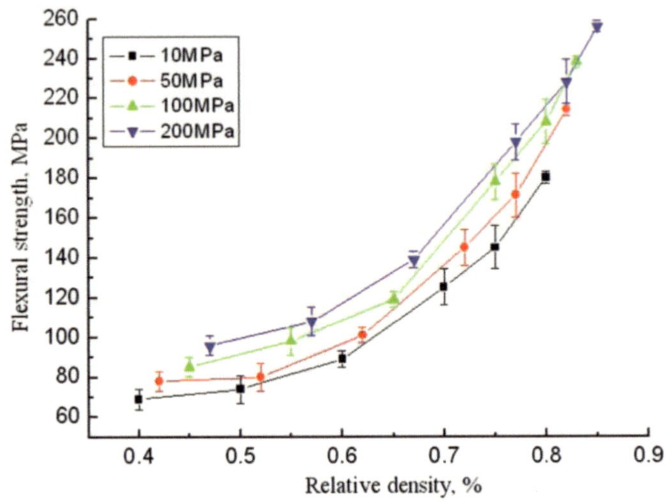

Fig. 6. Dependence of flexural strength on relative density of TiB2 porous ceramic

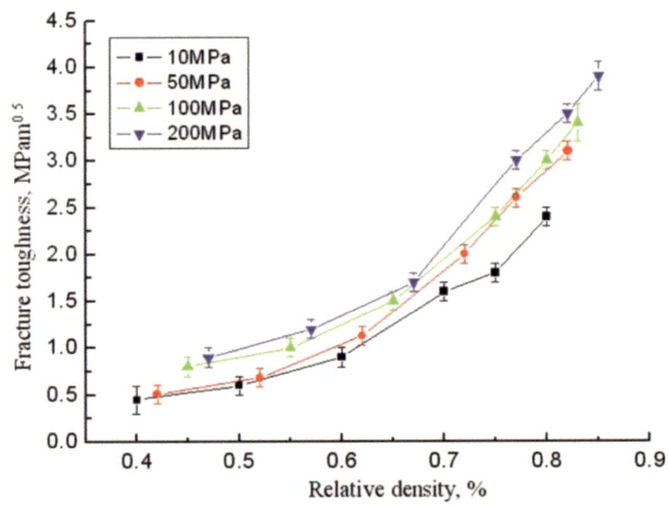

Fig. 7. Dependence of fracture toughness on relative density of TiB2 porous ceramic

2.4 Mechanical mechanism

Generally, high mechanical strength is desirable simultaneously with high porosity for almost all applications of porous ceramics. The relationship between strength and porosity has been investigated by many researchers, and a number of models have been proposed to explain the strength-porosity behaviors. Based on the assumption that the fracture strength of porous ceramics is determined by the minimum solid area, one of the simplest expressions was given by Rice (Rice, R. W et al, 1993)

$$\sigma = \sigma_0 \exp(-bP) \quad (1)$$

where σ_0 is the strength of a non-porous structure, σ is the strength of the porous structure at a porosity P, and b is a constant that is dependent on the pore characteristics. This expression indicates that strength will increase with decreasing of porosity, and this seems to agree with our results. However, from Fig. 6, it can be seen that the fracture strength increased with the increasing of initial pressure for a certain relative density. This reveals that the strength for porous ceramics is not only controlled by pore volume content but also influenced by special microstructure, sintered neck bonding or neck growth. Actually for highly porous materials, the stress concentration associated with pores no longer defines the mechanical behavior, as the effects due to surface defects are negligible. Hence, the strength values obtained in this study may have no dependency on effective pore volume. It is also noteworthy to mention that the enhanced strength of porous

TiB_2 must be related to the growth of the interparticle contact (neck growth) by surface diffusion, in the initial stage of sintering. This increase in strength with minimum densification is a significant factor for the mechanical behaviour of porous ceramics. The strength improvements are also afforded by control of the sintering mechanism and the microstructural homogeneity.

Extensive investigations revealed that mechanical properties of porous ceramics are related to the theory of minimum solid contact area (MSA) [Rice, R. W et al, 1993] when no special reinforcing mechanisms exist. If the particle stacking is a cubic array and the particles are spherical, MSA is the necking area between the particles, i.e. the grain bonding area. If the particle stacking is not the cubic array, the grain bonding area is a representation of MSA. For this study, as the fracture mode of porous TiB_2 ceramics is intergranular (Fig. 8), the bonding interface is the place of greatest stress concentration. Moreover, the shape of TiB_2 grains is equiaxed and no special reinforcing mechanisms can be found in this material. The weak interface strength in porous TiB_2 ceramics sintered from an initial pressure of 10 and 50MPa originates from the TiB_2 fragments and their poor connections (Figs 4(a) and (b)). Because highly packed regions (under higher compacted pressure) densified faster than the less-dense regions (under lower compacted pressure), as shown in Figs 4(c) and (d), where the TiB_2 grains on neck regions show poor connections. Comparably the TiB_2 grains from an initial pressure of 100 and 200MPa reveal good necking bonding and connection.

At the initial stage of densification, the sintering shrinkage is small and the difference in boundary defects for different compacts is small. The resultant interface bonding and mechanical properties of porous TiB_2 display limited differences between the compact with relatively lowpressures. As densification progresses, the sintering shrinkage increases and the difference in boundary defects and the resultant interface bonding increase. Therefore, mechanical property differences increase with increasing relative density according to the MSA analysis.

Fracture toughness is a property that enables the materials resist to the propagating of cracks. However, for a porous ceramic, the crack tip field distribution differs from the dense ceramic material. On one hand, the crack tip becomes blunt as a crack meets an open pore in a porous ceramic. Deng *et al.*(Deng, Z.-Y et al, 2002) had demonstrated the reinforcing effect of crack-tip blunting in porous SiC ceramics. This decreases the stress-concentration at the crack tip and increases the external load to propagate the crack so that the fracture

Fig. 8. Magnified fracture morphology of porous TiB2 sintered at 2000°C with initial pressure of 100 MPa

toughness increases (as shown in Fig.9). On the other hand, as discussed above, the higher initial compacted pressure and higher sintering temperature will enhance the neck growth and grain-boundary strength, which contributes to the increase of fracture toughness.

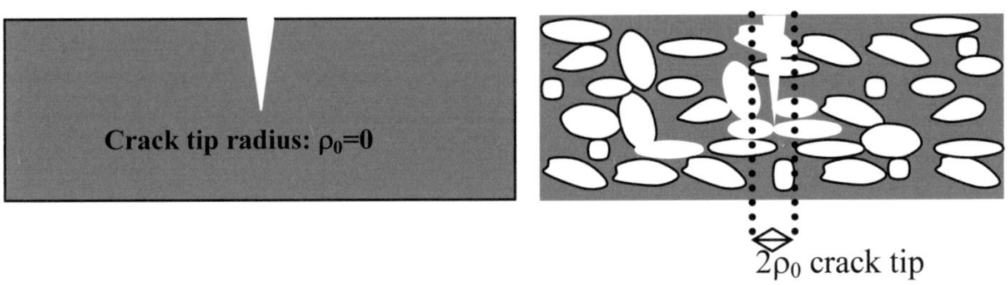

Fig. 9. Sharp and blunt crack tip of two types of materials

3. Production of porous zirconia

3.1 Porous ZrO_2

Commercially available yttria-stablized zirconia doped with 5 mol%Y_2O_3 (0.4μm, Fanmeiya powders Co. Ltd., Jiangxi, China) was used as the ceramic framework. Camphene ($C_{10}H_{16}$) (95% purity, Guangzhou Huangpu Chemical Factory, Guangzhou, China) was used as the freezing vehicle without any further purification. In addition, Texaphor 963 (Guangzhou Haichuan Co. Ltd., Guangzhou, China) was used as dispersant (density at 25 °C of 0.89-0.91 gcm^{-3}). The dispersant concentration was 3 wt.% of the ZrO_2 powder for all 10, 15 and 20 vol.% of solid loadings.

The first part of the manufacturing process involved the slurry preparation. This was achieved by melting the camphene at a temperature of 60 °C on a heating plate to create a clear and fluid vehicle. The dispersant concentration was 5 wt.% of ZrO_2 powder for all solid loadings. The ZrO_2 powder was then added in quantities of 10, 15and 20 vol.%. The slurry was stirred via the use of a motor and stirrer with a cap on the top to prevent any camphene vapour escaping. Zirconia/camphene/dispersant slurries with various Zirconia contents (quantities of 10, 15 and 20 vol%) were prepared by ball-milling at 60oC for 20h, before pouring into silicone rubber die for freezing; the moulds were 42 mm in internal diameter and 80 mm in height.

The samples were left to cool at ice-water environment (0 °C) for 30 min. The detailed above sketch of the test setup is shown in Fig.10 (a). Meanwhile some samples were cooled at liquid nitrogen environment (-196°C) for the same length of time to study the effect of cooling rate on the solidification characteristics, which can be seen from Fig.10 (b). After solidification, the green body was removed from the moulds and left to sublime (optimized to 24 h) at room temperature in order to remove the camphene entirely and achieve a highly porous structure. Following sublimation, sintering of the green body at 1400-1550 °C enabled the densification of the samples and concomitant improvements in mechanical

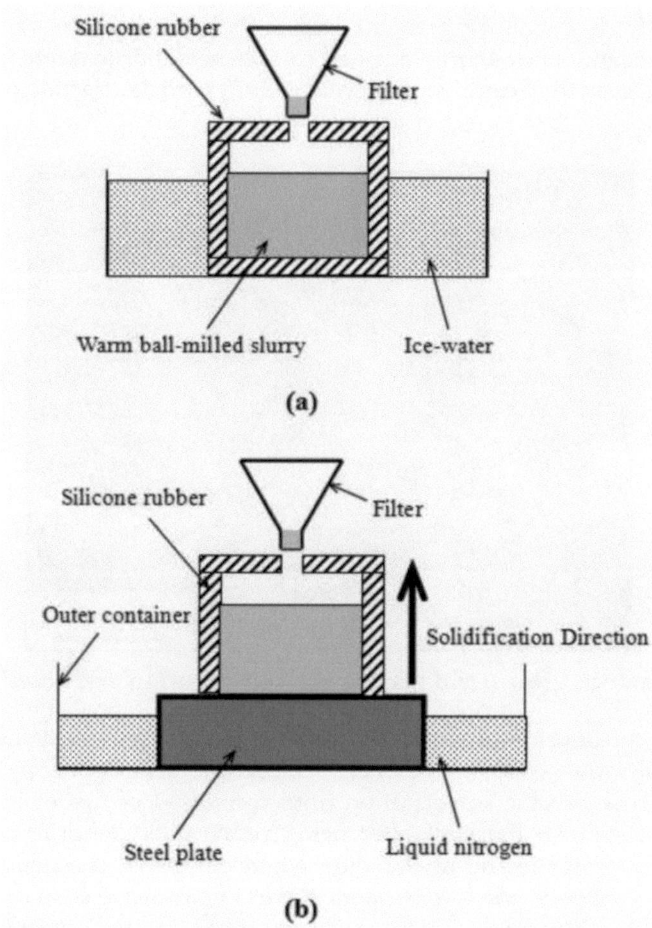

Fig. 10. A sketch of freezing assembly for fabricating cast body (a) in ice-water environment; (b) in liquid nitrogen cooling environment

strength. The sintering regime entailed heating the samples at 0.5 °C/min up to 600 °C followed by 1 h of dwell time. They were then heated at 1 oC/min up to final sintering temperature and held at this temperature for 2 h, prior to cool down to room temperature at the same rate of 20 °C /min.

3.2 Camphene solidification observations

The solidification phenomenon of the mixed slurry was investigated so as to acquire an understanding of the characteristics of pore and dendrite formation. Fig. 12(a) shows the development of long and straight dendritic branches near the marginal corner along the freezing plane, which runs towards the left corner (as shown by the arrow). Since the temperature gradient at the marginal region decreases fast, the camphene crystals grow dendritically in certain crystallographic directions. When the solidification was completed, a unique phase separated structure was produced, in which aligned camphene dendrites with a well-defined morphology surrounded by ZrO_2 particle networks were formed.

Fig. 12(b) shows the interconnectivity of the camphene branches near the central region of the slide where the temperature gradient was sufficiently high to stimulate secondary dendritic formation, and many short elongated camphene dendrites were formed randomly. A typical dendritic growth of camphene as well as ZrO_2 particle rejection of the warm slurry is shown in Fig.11.

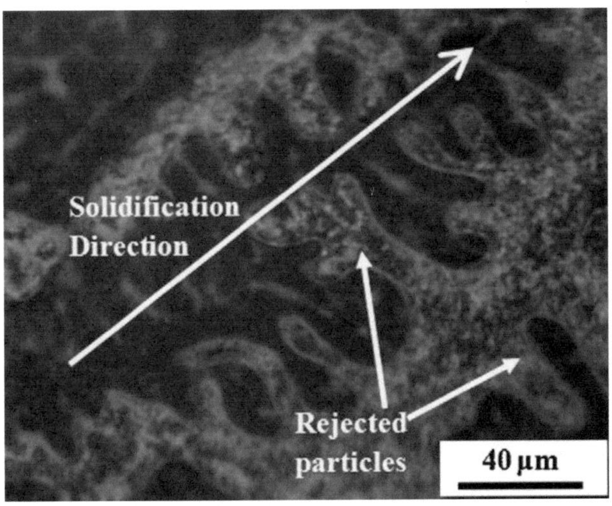

Fig. 11. Typical dendritic growth and particle rejection for camphene-based slurry

It can be seen that the development of dendritic branches along the freezing plane, which, in the image, runs towards the top right corner. The particles are not only pushed along ahead of the advancing macroscopic solidification front composed of tips of growing dendrites, but are also connected on the spot after being rejected by dendrite arms. The macro-morphology in the center region of the slide where no camphene dendrites were found (shown in Fig.12(c)). Due to the low temperature gradient in the center region (since both top and bottom slide glass were covered by silicone rubber insulation), some of the dendrite side arms might be melted off and then act as seeds for new dendrites, resulting in the formation of equiaxed pore structures. The above observations reveal the morphology of the dendrites that are produced during freezing of the camphene and the important role of the heat transfer gradient on determining the final shape and orientation.

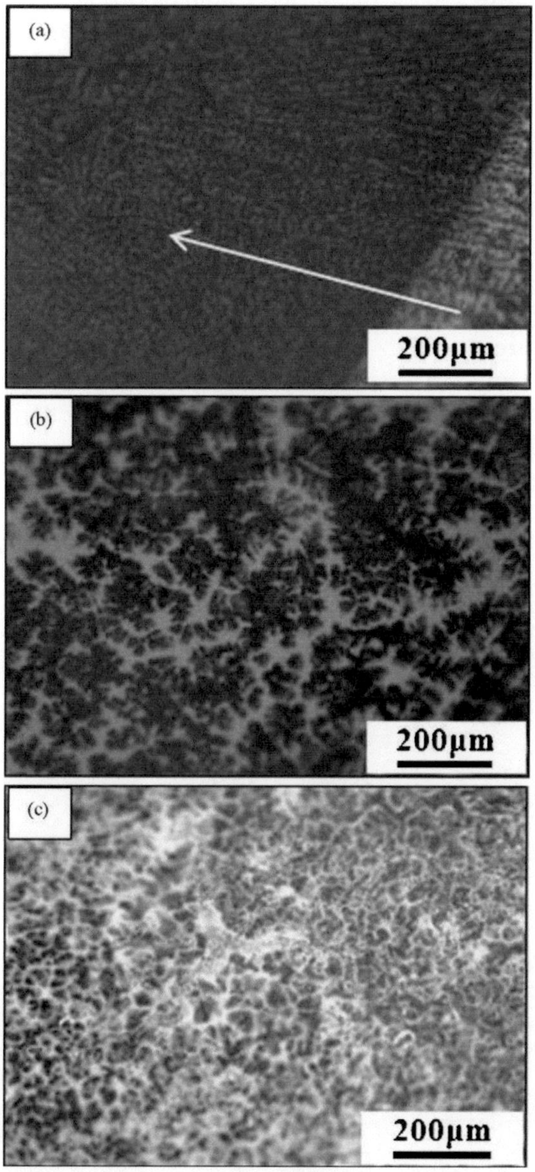

Fig. 12. Optical micrographs of the camphene dendritic growth during solidification (a) aligned camphene growth (shown by the arrow); (b) typical dendritic growth; (c) equiaxed camphene morphology.

3.3 Unidirectional solidification

Because the sintered pore structure form as replicas of camphene dendrites, unidirectional solidification was tried to control the growth direction of the camphene dendrites. A special mold composed of a steel bottom plate and a silicone rubber die on it was prepared to create a cylindrical cavity. Warm ball-milled ZrO_2 slurry of 60 °C with 15 vol.% solid content was poured at room temperature into the prewarmed mold having almost the same temperature as the slurry. Just after casting, liquid nitrogen was poured into the outer container to cool only the steel bottom plate so that the solidification of the camphene-based slurry would occur unidirectionally from the bottom towards the top in the mold. The solidification was completed in about 3 min, and then the cast body was dried in an ambient atmosphere for 30h.

Sintered samples generated with solid loadings of 15 vol.% were analyzed by SEM. The SEM micrographs (Fig. 13(a) and (b)) showed that the pores tend to align in the direction of freezing, indicating the possibility of controlling pore orientation by controlling parameters such as the heat transfer gradient and direction of freezing. Since the chilled mold was suddenly cooled with liquid nitrogen about −160 °C, the camphene rapidly cooled below its solidification temperature, and thus many nuclei of the camphene then form on the mold wall and begin to grow into the warm slurry. Under these conditions, most of the nuclei do not have a preferential orientation that corresponds to the direction of the heat conduction. Therefore, these camphene crystals cannot overgrow dendritically, and these results in the formation of long straight channels in the sintered body (see Fig. 11).

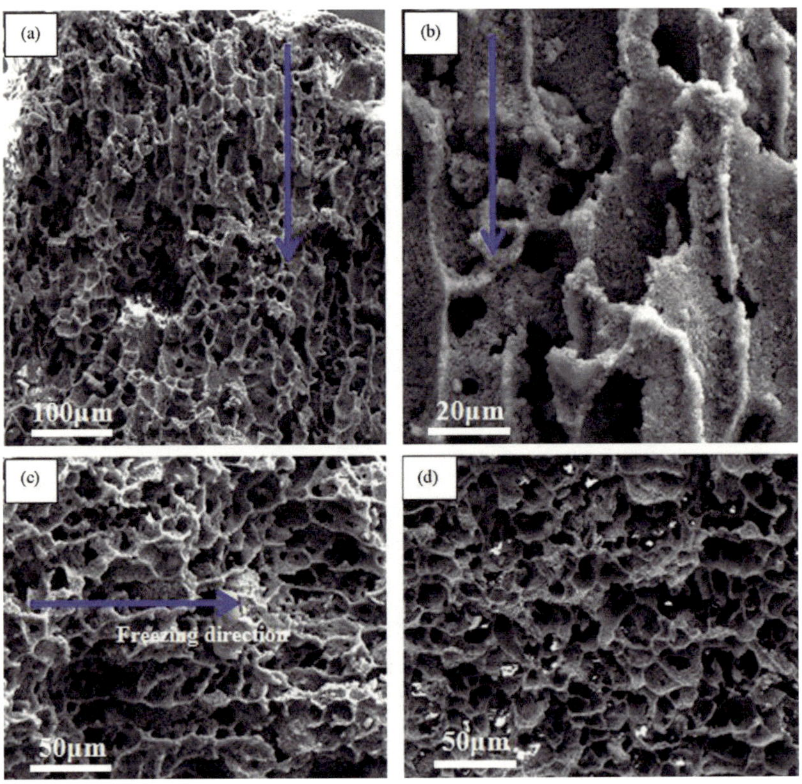

Fig. 13. (a) Pore alignment in the direction of freezing; (b) magnified microstructure in Fig. 2(a); (c) dendritic pore distribution beyond the die wall; (d) equiaxed pore morphology in the center of sintered sample

Beyond the above region, the temperature gradient near the die wall decreases and the camphene crystals began to grow dendritically in certain crystallographic directions. Those crystals with a preferential orientation close to the direction of heat flow, i.e., parallel to the mold wall, grow faster and can lead to their secondary dendritic formation. It is accordingly results in the elongated aligned pore channels and short arms channel in the sintered body, as shown in Fig. 13(c).

In the center of the cast body, some of the dendrite side arms might be melted off and then act as seeds for new dendrites, resulting in the formation of equiaxed pore structures (shown in Fig. 13(d)). This unique pore structure in the center of the cast body might be related to the breakaway of the side arms from the primary dendrite of the camphene.

3.4 Effect of initial solid loading

Solid loading plays an important role in determining the porosity and mechanical strength characteristics of the sintered samples. Fig. 14(a)-(c) shows the pore structures after sintering at 1500 °C with the solid loading varying from 10 to 20 vol.%. It is clear that lower solid loading results in higher porosity and larger pore sizes, while the sintered zirconia walls became thinner. During freezing, the camphene dendrites can grow until the force created by the particle concentration exceeds the capillary drag force pushing the particles with the solid/liquid interface (Yoon, B. H et al, 2008). Therefore, it is reasonable to suppose that a lower solid loading will lead to the formation of larger camphene dendrites and thinner concentrated ZrO_2 ceramic walls.

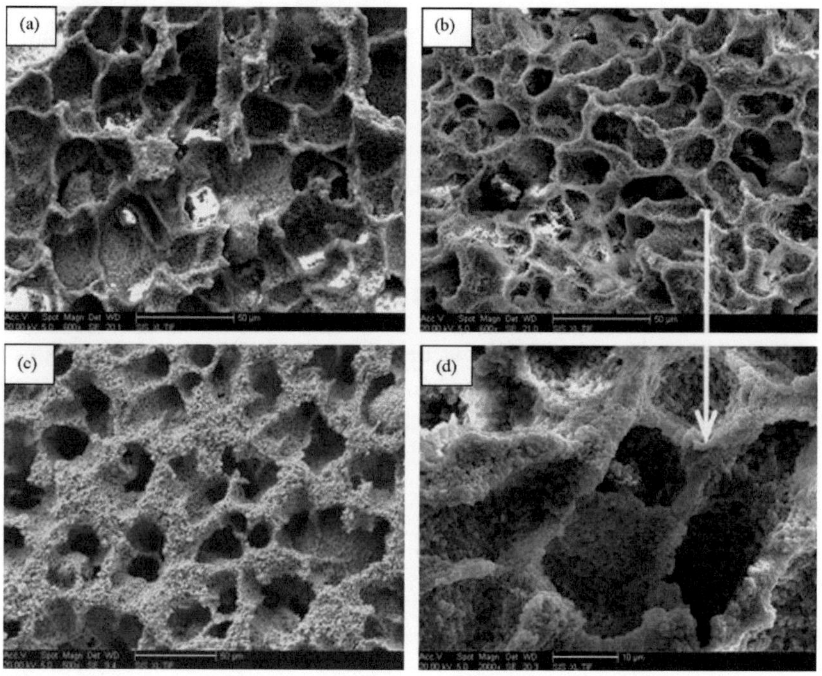

Fig. 14. SEM micrographs of porous structures sintered at 1450 °C for (a) 10 vol.%, (b) 15 vol.%, (c) 20 vol.%, and (d) sintered ceramic wall in solid loading of 15 vol.%.

Fig. 15 shows the relationship between the porosity and the initial solid loading of the ceramic slurry. As the solid loading is increased, the porosity was found to decrease proportionately. The porosity was reduced from 81.5 to 65.5% by increasing the solid loading from 10 to 20 vol.%. The linear relationship between the porosity and the initial solid content can be expressed as follows:

$$P = 97.5 - 1.6x \tag{2}$$

where P is the porosity (vol.%) and x is the solid loading (vol.%).

This result suggests that the porosity can be manipulated by empirically controlling the initial solid loading used in the ceramic/camphene slurry. It should be noted that the possible sublimation of the molten camphene during ball-milling the freeze-casting is not considered in the above equation. From these observations, it is obvious that the camphene-based freeze casting is very useful for producing high porous ceramics with porosities and completely interconnected pore channels, as well as well sintered ZrO_2 walls.

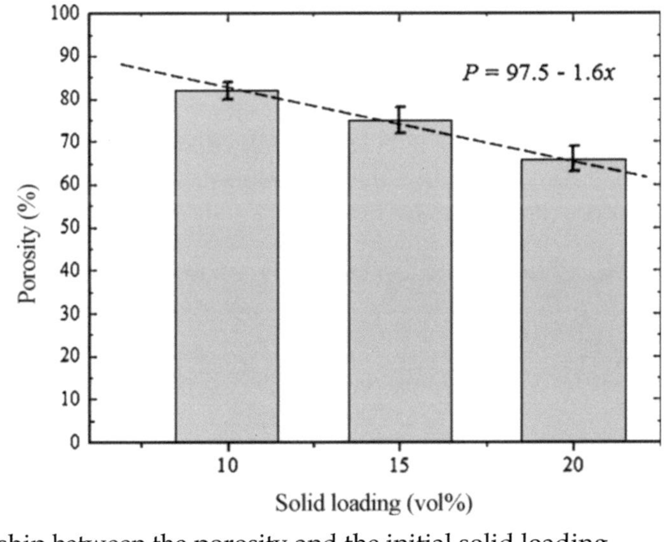

Fig. 15. Relationship between the porosity and the initial solid loading

The main physical and mechanical properties of the sintered samples are summarized in Table 1. The results show that after being sintered with a lower solid content, the sample acquired from 10 vol.% slurry was so light that bulk densities were even lower than water (1.0 g/cm^3). As the initial solid loading was increased, the compression strength increased from 16.2 to 53.4MPa.

Slurry solid loading (vol.%)	Bulk density after sintering(g/cm^3)	Compressive strength(MPa)	Porosity content
10	0.93-0.96	16.2	81.5%
15	1.20-1.29	32.9	74.4%
20	1.68-1.69	53.4	65.5%

Table 1. Properties of ZrO_2 with different initial densities after sintered at 1500 °C for 2 h

Fig. 16 shows the stress-strain curves for 10, 15 and 20 vol.% ZrO_2 solid content sintered at 1500 °C. All curves shown almost elastic deformation followed by a transition or plateau stage. No sudden drops or catastrophic fracture are observed, although load drops are found after showing a peak load. Generally, the strength of a porous ceramic is strongly affected not only by the porosity, but also by the formation of sintering neck on the ceramic wall as well as the smaller pore size (several tens of microns) compared conventional processing methods (>100μm). In this observation, pore size was determined by measuring the average size of pores from the SEM micrographs taken at several points on the polished surface. For example, for the 10 and 15 vol.% ZrO_2 solid content sintered at 1500 °C, the pore sizes of the prepared sampled approximately ranged from 15 to 30 μm(Fig. 17(a) and (b)). Furthermore remarkable sintering necks were formed (shown in Fig. 17(d)), resulting in high mechanical properties.

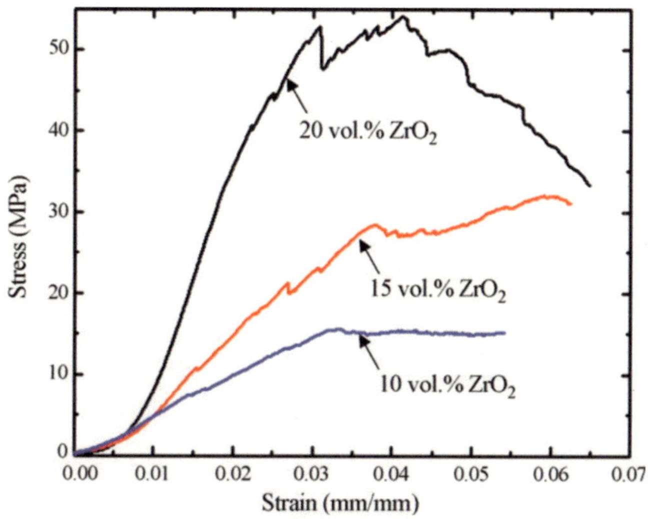

Fig. 16. The stress-strain curves for 10, 15 and 20 vol.% ZrO_2 solid content sintered at 1500 °C.

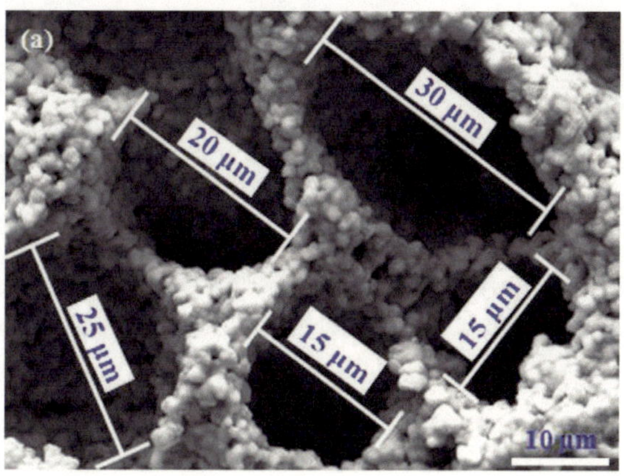

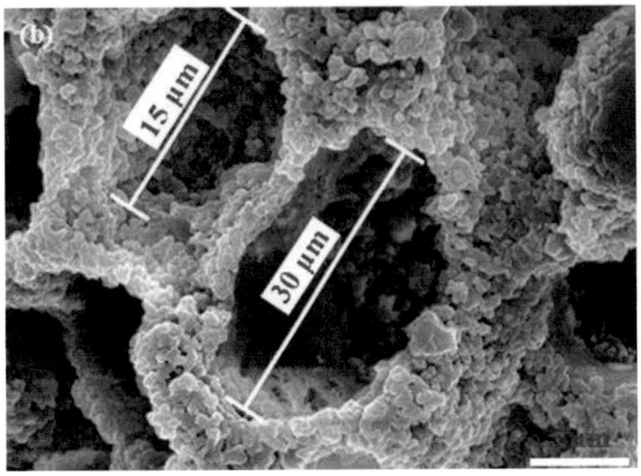

Fig. 17. The pore sizes of the prepared sampled approximately ranged from 15 to 30 μm for the 10 and 15 vol.% ZrO2 solid content sintered at 1500 °C

3.5 Effect of sintering temperature

The evolution of microstructure sintered at elevated temperatures ranging from 1400 to 1550 °C is shown in Fig. 18(a)-(d). It can be concluded that with a given initial solid loading (15 vol.%), the sintering temperature significantly contributes to the change of ceramic struts and porosity. As the sintering temperature is increased, the ZrO_2 ceramic struts are greatly densified (i.e., loosely bonded particles are observed from Fig. 18(a), while the sintering struts are greatly enhanced in view of Fig. 18(d)).

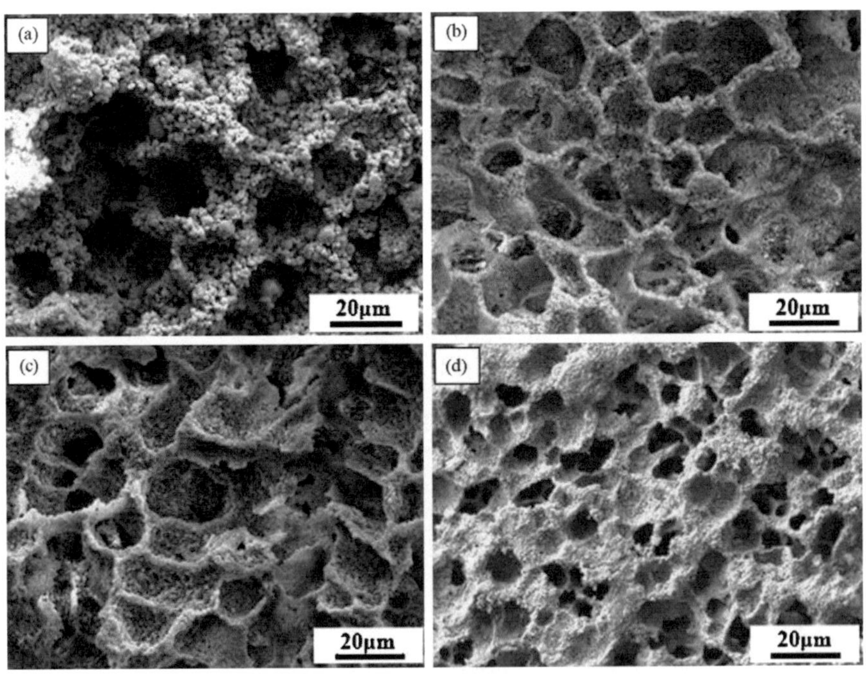

Fig. 18. Scanning electron micrographs of the samples sintered for 2 h in air at (a) 1400 ◦C, (b) 1450 °C, (c) 1500 °C and (d) 1550 °C in solid loading of 15 vol.%.

The effect of sintering temperature on the porosity was also investigated, and the results are shown in Fig. 19. The experimental examinations (as shownin Fig. 19) revealed that the porosity decreases and compressive strength goes up when sintering temperature increases from 1400 to 1550 °C (i.e., 18–59MPa for compressive strength). Both then reach a comparable plateau value, which means that densification is achieved at 1500 or 1550 °C. Further increase in sintering temperature will deteriorate the pore structure. Therefore, the optimum sintering temperature was determined to be 1500 °C in this work. Here, it is worth noting the increasing of ZrO_2 grain size with temperature variation is very limited from the experimental observation (see Fig. 18(a)–(d)).

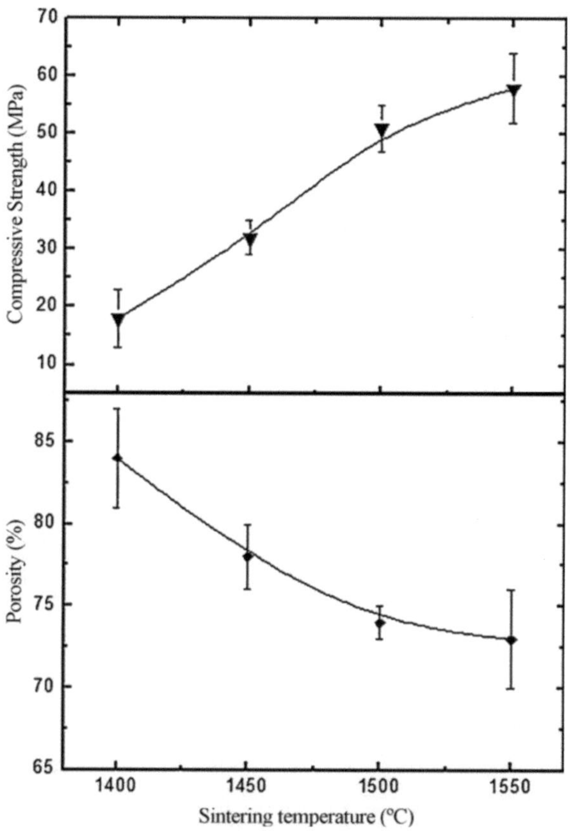

Fig. 19. Influence of sintering temperature (2 h at dwell temperature) on compressive strength and total porosity

Although the proper reasons of this particular phenomenon are still to be explored, it is probably partly related to the pinning of grain boundary at the surface during sintering of thin films, when the grain size lies in the same range of order as the lamellate thickness. Highly porous ceramics fabricated by conventional methods often contain defects, such as cracking and surface flaws. For example, reticulated porous ceramics produced using the polymer replication method have longitudinal cracks and surface flaws on the sintered ceramics struts during pyrolysis of the polymeric sponge. Cellular structures prepared by direct foaming usually exhibit mechanical strengths higher than that of replica techniques due mainly to the absence of flaws in the cell struts, the sintered foams show moderate compressive strengths of upto 16MPa in alumina foams with porosities of 88% produced from particle-stabilized foams.

Using camphene as the solvent, dilute ZrO_2 slurries were produced and large shrinkage during drying was avoided. By adjusting the initial solid loading and sintering temperatures, light-weight ZrO_2 ceramics with controlled microstructures and properties were prepared. It was possible to increase the porosity from 65.5 to 84.5%, while still obtaining high mechanical strength (over 16MPa). In addition, the prepared samples produced using camphene as solvent showed well-constructed zirconia walls without any noticeable defects, which might enable them to have good mechanical properties.

4. Summary and future perspectives

This chapter introduces pressureless sintering and novel freeze-casting methods are nowadays available for the production of porous ceramics.

Although the two techniques differ greatly in terms of processing features and final microstructures/properties achieved, they have their own merits for producing high-strength ceramics for actual application with special purposes. The pressureless sintering technique is an easy and well-established method to prepare high-strength structures with porosity between 40%and 80%. The freeze-casting based on the novel camphene-based freezing route is a good method for preparing high-porosity ceramic (60-85%) with special microstructure.

For porous TiB_2 discussed here, the microstructure of the compacts prepared at lower pressures and lower sintering temperature appeared to decrease the relative density and degrade the interface bonding strength of the TiB_2 grain boundaries. The high mechanical properties are related to the enhanced neck growth by surface diffusion at high temperature. Initial solid loading played an important role in the resulting porosity of the materials.

For freeze-casting porous ZrO_2 ceramics, the porosity was reduced from 81.5 to 65.5% by increasing the solid loading from 10 to 20 vol.%. As a result of this, the compressive strength was affected, increasing from 16.2 to 53.4MPa for the respective increase in solid loading. The rate of heat transfer affected the final morphology of the dendrites; camphene dendrites were found to orient according to the direction of freezing under liquid nitrogen environment; the sintering temperature (from 1400 to 1550 °C) also affected the porosity and mechanical properties. The porosity (for 15 vol.% solid content) ranged from 73.5 to 84.5 vol.%, while the compressive strength increased from 18 to 59MPa, respectively.

In conclusion, this manufacturing technique shows great potential for generating defect-free porous ceramics with controlled porosity and pore size and appropriate compressive properties for use in modern engineering applications.

5. Conclusion

This chapter provides a comprehensive introduction of the synthesis, structure, mechanical properties and characterization of high-strength porous ceramics. It introduces pressureless sintering and novel freeze-casting methods for the production of porous TiB_2 and ZrO_2 ceramics. Taking into account the decisive influence of the processing method on the material's microstructure and properties, the selection of the processing route for the production of porous ceramics depends primarily on the final properties and application aimed. For freeze-casting method, this manufacturing technique shows great potential for producing highly porous ceramics with controlled porosity and pore size for use in special engineering applications.

6. References

Amir, M. and Rohwer, K(2007). Proc. IMechE, Part G: J. Aerospace Engineering, Vol. 221, 831–845.

Araki, K. and Halloran, J. W(2005). J. Am. Ceram. Soc.Vol. 88, 1108–1114.

Bhaumik, S. K., Divakar, C., Singh, A. K., and Upadhyaya, G. S(2000). Mater. Sci. Eng. A. Vol.279, 275–281.

Cadirli, E., Marasli, N., Bayender, B. and Gunduz, M(2000). Mater. Res. Bull. Vol.35, 985-995.

Chen, R. F., Huang, Y., Wang, C. A. and Qi, J. Q.(2007). J. Am. Ceram.Soc. Vol.90, 3424–3429.

Corbin, S. F. and Apte, P. S. (1999). J. Am. Ceram. Soc., Vol. 82, 1693–1701.

Dale, H. and David, J. G(1995). J. Eur. Ceram. Soc. Vol. 15, 769–775.

Deng, Z. Y., Fukasawa, T., Ando, M., Zhang, G. J., and Ohji, T(2001). Acta Mater. Vol. 49, 1939–1946.

Deng, Z.-Y., Yang, J.-F., and Beppu, Y(2002). J. Am. Ceram. Soc. Vol. 85, 1961–1965.

Deville, S.(2008) Adv. Eng. Mater. Vol.10, 155–169.

Deville, S., Saiz, E. and Tomsia, A. P.(2006) Biomaterials. Vol.27, 5480–5489.

Deville, S., Saiz, E., Nalla, R. K. and Tomsia, A(2006). Science,Vol.311, 515–518.

Fukasawa, T., Ando, M., Ohji, T., and Kanzaki, S(2001). J. Am. Ceram. Soc. Vol.84, 230–232.

Gonzenbach, U. T., Studart, A. R., Tervoort, E. and Gauckler, L. J(2006). J. Am. Ceram. Soc. Vol.90, 16–22.

Gough, J. E., Clupper, D. C. and Hench, L. L(2004)J. Biomed. Mater. Res.Vol.69, 621–628.

Hong, C. Q., Han, J. C., Zhang, X. H., and Meng, S. H.(2007). Mater. Sci. Eng. A, Vol.447, 95-98.

Kamal, M. M. and Monhamad, A. A(2006). Proc. IMechE, Part A: J. Power and Energy, Vol. 220, 487–508.

Kiyoshi, A. and John, W. H.(2004). J. Am. Ceram. Soc.Vol. 87, 1859–1863.

Koh, Y. H., Lee, E. J., Yoon, B. H., Song, J. H. and Kim, H. E(2006). J. Am. Ceram. Soc. Vol. 89, 3646–3653.

Oh, S. T.,Tajima, K. I., Ando, M., and Ohji,T(2000). J. Am. Ceram. Soc., 2Vol. 83, 1314–1316.

Oh, S.-T., Tajima, K.-I., Ando, M., and Ohji, T(2001). Mater. Lett. Vol. 48, 215–218.

Peng, H. X., Fan, Z., Evans, J. R. G., and Busfield, J. J. C(2000). J. Eur. Ceram. Soc., Vol.20, 807–813.

Rice, R. W(1993) .Mater. Sci. Vol. 28, 2178–2190.

Schmidt, H., Koch, D. and Grathwohl, G.(2001) J. Am. Ceram. Soc.Vol. 84, 2252–2255.

She, J. H., Yang, J.-F., and Kondo, N. (2002) J. Am. Ceram. Soc. Vol. 85, 2852–2854.

Studart, A. R., Gonzenbach, U. T., Tervoort, E. and Gauckler, L. J(2006). J. Am. Ceram. Soc. Vol. 89, 1771–1789

Wang, J. H., Gan, M., and Shi, J. X(2007). Mater. Charact. Vol.58, 8–12.

Wang,W. M. and Fu, Z. Y. (2002)J. Eur. Ceram. Soc. Vol.22, 1045-1049.
Waschkies, T., Oberacker, R. and Hoffmann, M. J(2009). J. Am. Ceram. Soc.Vol.92, S79-S84.
Yoon, B. H., Choi, W. Y., Kim, H. E., Kim, J. H. and Koh, Y. H.(2008) Scripta Mater.Vol.58, 537-540.

7

Composites Hydroxyapatite with Addition of Zirconium Phase

Agata Dudek[1] and Renata Wlodarczyk[2]
[1]*Institute of Materials Engineering, Czestochowa University of Technology,*
[2]*Department of Energy Engineering, Czestochowa University of Technology,*
Poland

1. Indroduction

Osteoporosis is a disease in which the density and quality of bone are reduced, leading to weakness of the skeleton and increased risk of fracture. Osteoporosis literally means "porous bone". When a bone has become osteoporotic or osteopenic (low bone mass), the risk of a fracture increases. The forearm, spine and hip are the most common fracture sites, accounting for more than 80 percent of all fractures. Osteoporosis and associated fractures are an important cause of mortality and morbidity. 20 percent of hip fractures lead to death within a year. Osteoporosis is a widespread public health problem. The costs to national healthcare systems from osteoporosis-related hospitalization are staggering. In the US, the cost to the health care system associated with osteoporotic fractures is approximately $17 billion annually. This converts to more than $45 million a day! Each hip fracture represents an estimated $40,000 in total medical costs.

Human skeleton makes up a load-bearing structure for motor organs and it constitutes a location for attachment of tendons and ligaments. Mature bone has a lamellar structure with layers and comprises compact (cortical) bone and trabecular (spongy) bone. Cortical layer consists of cylindrically arranged lamellae, while spongy bone is composed of threedimensional, irregular trabeculae in the form of a network. Space between trabeculae is filled with hematopoietic system cells and adipose tissue (Fig. 1). Matrix of the bone consists of organic compounds called osseine, which is responsible for bone elasticity as well as of mineral compounds: magnesium phosphate, calcium carbonate and calcium hydroxyapatite, which provide bones a particular hardness. Deviation from proper level of minerals in relation to organic compounds might cause lack of bone elasticity, thus its brittleness. Main mineral compound of bones is *calcium hydroxyapatite* with crystal size that ranges from 4-50 nm. Hydroxyapatite is a 'warehouse' for the most of calcium (99%) and phosphorus (85%) contained within the body. Hydroxyapatite crystals account to as much as 77% of organic stroma the bones are composed of (Fig. 1).

Moreover, hydroxyapatite is the main mineral component of dentine (Fig. 2). Dentine comprises essential part of the mass of hard tissues, which human teeth are made of. In the area of the crown, the dentine is covered with enamel, while in the area of the root - with cementum. Teeth (singular: tooth) are dense structures found in the jaws of many vertebrates. They have various structures to allow them to fulfill their different purposes. The primary function of teeth is to tear, smell and chew food, while for carnivores it is also a weapon. Therefore, teeth have to withstand a range of physical and chemical processes, including compressive forces (up to ~ 700 N), abrasion and chemical attack due to acidic foods or products of bacterial metabolism.

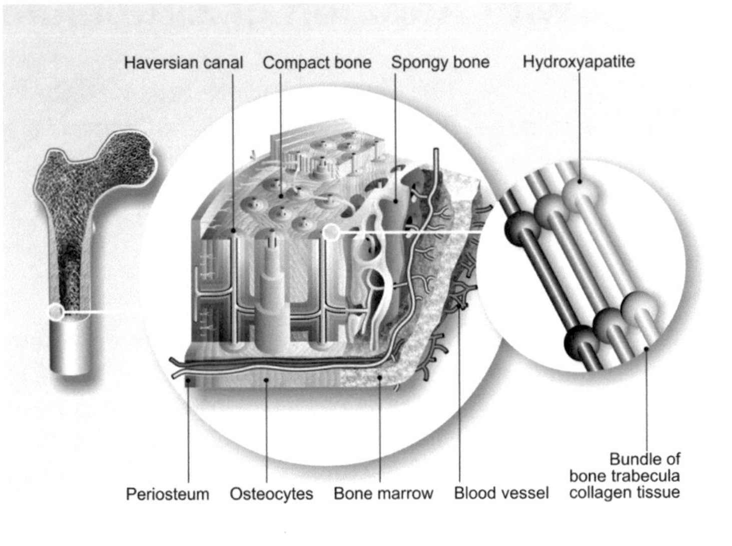

Fig. 1. Section of a bone

Nowadays dentists, in their practice, often encounter problems of bone defects which occur as a result of removal of bone cyst or through alveolar process artophy. Bone graft is a standard procedure during treatment of such defects. The locations from which transplants are taken include: mentum (area next to canines), cranial vault and iliac ala. This methodology has some advantages, e.g. no graft rejection reaction, however, its fundamental drawback is the fact that this requires additional surgical intervention, which might lead to some disorders in the location the grafts are taken from.

Due to the abovementioned facts, a great emphasis is on searching for bone-replacing materials which would allow for filling of bone defects resulting from a variety of reasons. Such materials would eliminate many complications which occur during use of materials of autogenic [1], allogenic[2] or xenogenic [3] origins.

Development of materials for medicine applications was first recorded in 1860, when doctor J. Lister developed aseptic techniques used in surgery. Previous attempts with use of biomaterials frequently ended up with infections spreading throughout the patients' bodies, thus causing demise. Since then, further developments started to spread rapidly. Some discoveries were made almost 'by the way', e.g. during the Second World War lack of

[1]Graft from the same body

[2]Material for the graft taken from a specimen of the same species, genetically different than recipient

[3]Graft taken from a specimen of different species

chronic adverse reactions was observed in wounded pilots as some parts of aircraft canopies, made of polymethyl methacrylate, were left lodged in their bodies for a longer time. Nowadays, this material is widely used, e.g. it replaces parts of skull bones.

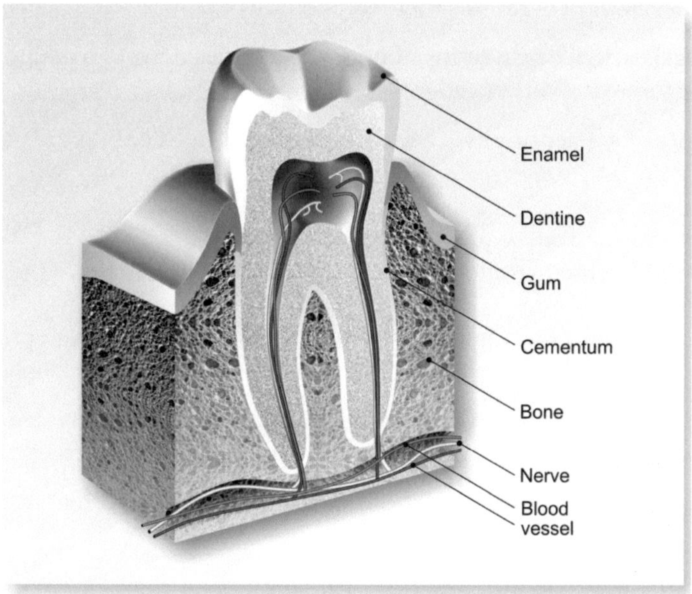

Fig. 2. Section of a tooth

2. Material for Implants – Biomaterials

According to the definition adopted by the European Society for Biomaterials, biomaterials include the compounds which are not medicines or combination of natural and synthetic substances, and, which might replace a part or the whole tissue or organ.

Materials that implants are made of [4] must not be hazardous for human body, i.e. carcinogenic, toxic or radioactive, they have to be resistant to corrosion (depending on the environment they work within), biocompatible (the materials which show tissue compatibility and do not trigger allergic reactions), well tolerated by living tissues. Implant acquire their fundamental properties and biocompatibility through specific chemical constitution of a material it is made of. It is characterized by biotolerance, .i.e. biological compatibility and harmony of interaction with living matter. Biotolerance causes that an implant, having been implemented into the body, does not trigger acute or chronic reactions or inflammatory condition in adjacent tissues. The biggest importance for implant acceptance by a tissue and for the process of osseointegration is the composition of its surface layer. Another important issue is osteointegration[5], i.e. ingrowth of bone tissue into the implant surface, and, in consequence, integration of the graft with the bone.

---

[4]Medical devices made of one or more biomaterials which might be located in the body, partly or entirely under the skin

[5]Osteointegration - ingrowth of living bone tissue into the titanium implant surface

Biomaterials are used, among other things, in orthopaedics, cardiology, nanosurgery and dentistry (Fig. 3).

There are several groups of materials which have been used and which replace steels and temporarily damaged or ill organs or their parts (Fig. 4).

Analysis of clinical experience in terms of human body reactions to metallic implants allows to emphasise the following complications:

- immunological oversensitivity, which contributes to bone necrosis or soft tissue necrosis,
- thrombuses,
- risk of immediate and subsequent infections,
- reduction in cells' resistance to bacteria, resulting from reduction of pH factor near the graft,
- weakening of neutrophiles and macrophages that facilitate development of bacterial flora, caused by electrical potential gradient at the interface of metal - body fluid.

Lack of metallic biomaterial which are entirely neutral to human body causes that research works toward searching, improvement and modification of such materials are a necessity.

One of the suggestions for improvement of biotolerance in applied metallic materials is the use of ceramic coatings.

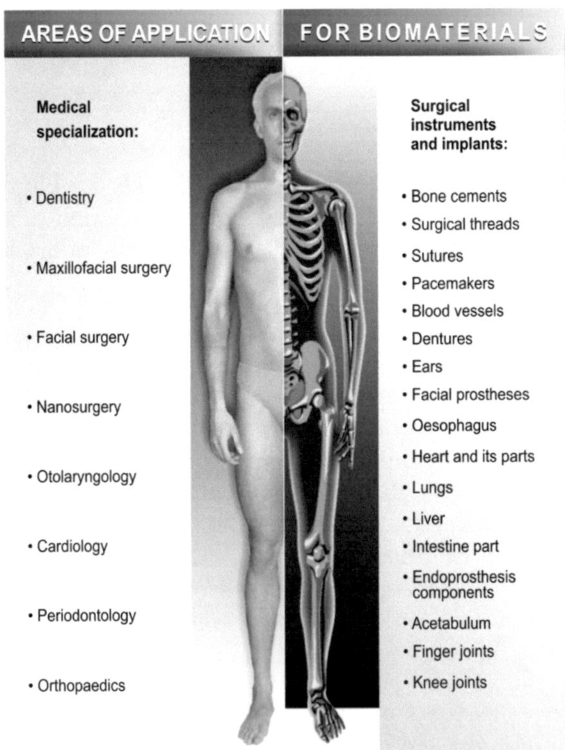

Fig. 3. Areas of application for biomaterials

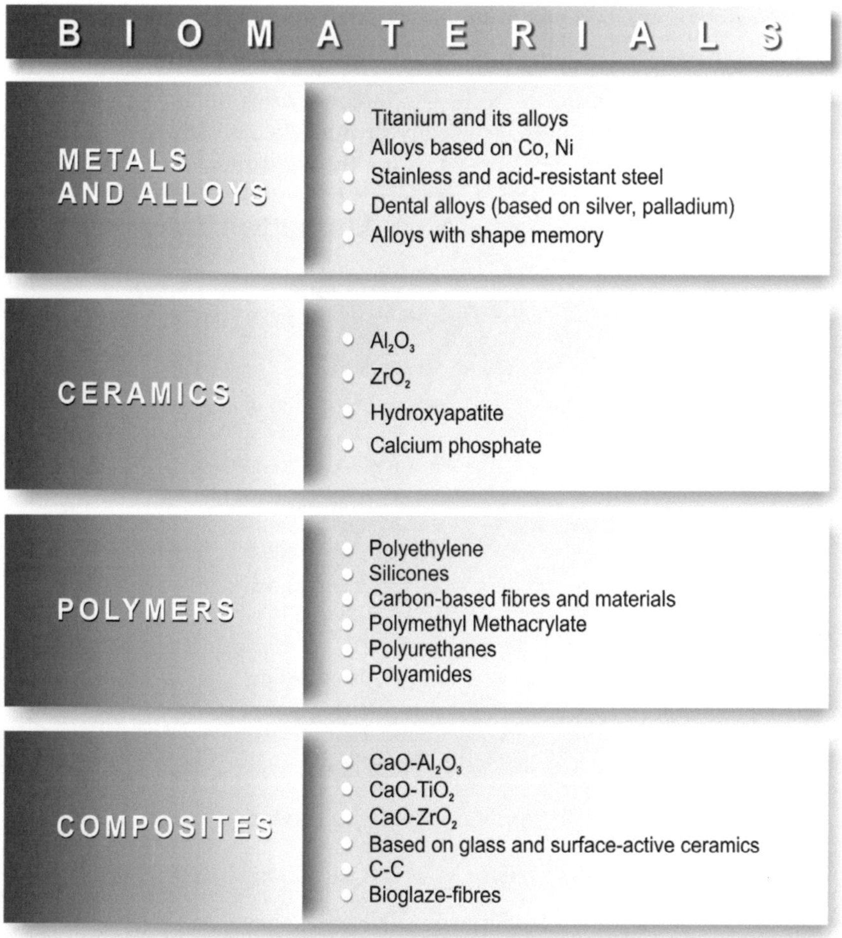

Fig. 4. Comparison of biomaterials and areas of their application

Bioceramics, despite their high hardness and, in consequence, higher brittleness, have an essential advantage i.e. their porosity, which ensures integration of vascularised soft tissue, which, in effect, becomes a permanent connection. Good adhesion of the material grafted into the bone is a very important problem of implantology. There have been attempts to solve this problem through application of bioactive ceramics which 'spontaneously' connects with the bone.

Ensuring effective biological connection of ceramics (composite) with the bone depends on pores size and their spatial distribution, i.e.:

- pores over 5μm - no tissue ingrowing occurs,
- over 25μm - ingrowing of fibrous tissue and vessels,
- over 50μm - mineralization of ingrown pores,
- over 75μm - mineralization occurs at the depth of 500μm,
- większe od 100μm - mineralization occurs even at the depth of over 1000μm, ensuring proper bone formation.

According to many authors, the most optimal pore size maintains at the level of 100-500μm.

Phase and chemical composition of bioactive materials is selected so as to ensure that implant surface adjacent to tissue or body fluids constitutes an intermediate layer which connects an implant with bone tissue. A variety of materials which fulfil such criteria have been developed in recent years. The most frequently used bioceramics, due to their high mechanical, corrosion and wear resistance as well as their non-toxicity and biocompatibility, include oxides: Al_2O_3 (whose use for medicine is dated back to the thirties of the past century), ZrO_2 and calcium phosphates (Tab. 1, 2).

Symbol	Chemical formula	Name	Ca/P - - Atoms relation
DCPD	$CaHPO_4$ $2H_2O$	Dicalcium phosphate dihydrate	1.00
DCPA	$CaHPO_4$	Dicalcium phosphate anhydrous	1.00
OCP	$CaH(PO_4)_3$ 2,5 H_2O	Octacalcium phosphate	1.33
TCP	$Ca_3(PO_4)_2$-α $Ca_3(PO_4)_2$-β	α - Tricalcium phosphate β - Tricalcium phosphate	1.50
HA	$Ca_{10}(PO_4)_6(OH)_2$	Hydroxyapatite	1.67
TTCP	$Ca_4(PO_4)_2O$	Tetracalcium phosphate	2.00

Table 1. Comparison of calcium phosphates existing at the temperature of 25^{0}C

	Human bone	Ti-6Al-4V	Al_2O_3	HAp	ZrO_2
ρ [g cm^{-3}]	1.99	4.5	>3.9	3.16-3.23	6.0
YS [MPa]	200	884	4000	509-917	-
TS [MPa]	130	940	-	-	>650
E [GPa]	18-19	113	380	4.0-117	210
Porosity [%]	80	-	-	-	-
Microhardness HV [MPa]	-	260	2300	343	1200

Table 2. Properties of human bone and comparison with selected materials

Calcium orthophosphates are chemical compounds of special interest in many interdisciplinary fields of science, including geology, chemistry, biology and medicine. The main driving force behind the use of calcium orthophosphates as bone substitute materials is their chemical similarity to the mineral component of mammalian bones and teeth calcium orthophosphates are also known to be osteoconductive (able to provide a scaffold or template for new bone formation).the major limitations to use calcium orthophosphates as load-bearing bioceramics are their mechanical properties; namely, they are brittle with a poor fatigue resistance. In general, calcium orthophosphate bioceramics should be characterized from many viewpoints such as the chemical composition (stoichiometry and purity), homogeneity, phase distribution, morphology, grain sizes and shape, grain boundaries, crystallite size, crystallinity, pores, cracks, surface, *etc*. From the chemical point of view, the vast majority of calcium orthophosphate bioceramics is based on HA, β-TCP, α-TCP.

A sintering procedure appears to be of a great importance to manufacture bulk bioceramics with the required properties. Usually, this stage is carried out according to controlled temperature programs of electric furnaces in adjusted ambience of air with necessary additional gasses; however, always at temperatures below the melting points of the materials. The heating rate, sintering temperature and holding time depend on the starting materials. For example, in the case of HA, these values are in the ranges of 0.5–3 °C/min, 1000–1250 °C and 2–5 h, respectively. In the majority cases, sintering allows a structure to retain its shape. However, this process might be accompanied by a considerable degree of shrinkage, which must be accommodated in the fabrication process.

Hydroxyapatite (HA, HAp), with formula of $Ca_{10}(PO_4)_6(OH)_2$, is a compound which is chemically and mineralogically similar to inorganic substances from which human bone tissue, including teeth, is made of. Hydroxyapatite crystallizes stechiometrically in monoclinic system, while synthetic, mineralogical and biological one shows hexagonal structure (Fig. 5). In its chemical constitution, HA contains 1.8% of H_2O. While heated, HA is stabile until 1703K, then it irreversibly looses hydroxyl groups, gradually changing into oxyapatite $Ca_{10}(PO_4)_6O$. At 1503K it might lose as much as 75% of water, maintaining its apatite structure.

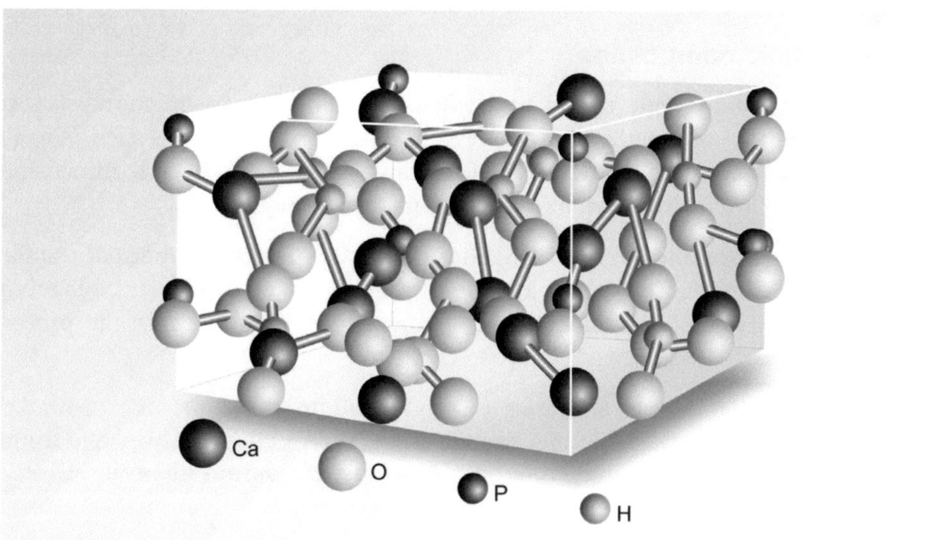

Fig. 5. Crystallographic structure of hydroxyapatite

Hydroxyapatite material is osteoconductive[6] and, on smaller scale, osteoinductive, thus implants made of this material can be directly integrated with the bone. After implantation of the material, osteogenesis[7] occurs and the layer which connects the surface of the hydroxyapatite graft with bone tissue appears. HA is a material which is characterized with small biodegradation rate - resorption in the body amounts to from 5 to 15% yearly. The process of connection of implant with the bone takes ca. a year.

Application of phosphate-calcium ceramics for implantology depend mainly, in terms of their bioactivity, on their:

1. porosity,
2. chemical and phase composition,
3. crystallinity degree.

Hydroxyapatite ceramics have been successfully used in dentistry, facial surgery, orthopaedics and otolaryngology in the form of shaped pieces and porous granules for replenishment of bone defects in the locations which do not bear mechanical load (e.g. malleus). Wider use of this ceramics, despite their best bioactivity and biocompatibility, is limited due to their poor mechanical properties.

The researchers have frequently attempted to improve mechanical strength and resistance to cracking through introduction of ZrO_2, Y_2O_3 and CaO oxides into HA. Zirconium oxide, commonly used for production of femoral joint prostheses, while its impact on composite reinforcement is determined, among other things, by stress distribution in material during polymorphous t↔m (tetragonal to monoclinic) transition of ZrO_2 during martensite transition.

Thus it seems to be necessary to strive for development of a composite with HAp matrix, while reinforcing phase should consist of other materials well-tolerated by living tissues.

3. HA+ZrO_2 ceramic composites

From the standpoint of common use of hydroxyapatite and HAp-based composites (with addition of ZrO_2 phase) for medicine, it is very important to determine such percentage content of ZrO_2 phase addition in the mixture that invariable or predictable dimensions of an implant or coating are maintained after the process of sintering.

The goal of the investigations: The investigations aimed to determine thermal stability in hydroxyapatite and HAp + ZrO_2 composites and the impact of 8%wt. and 20%wt. Y_2O_3 additions of ZrO_2 on phase composition in the composites obtained after the process of sintering.

Thesis: Increase in the amount of addition of ZrO_2 zirconium phase, modified with 8% wt. and 20% wt. Y_2O_3, to hydroxyapatite bioceramics leads to reduction of shrinkage triggered by sintering of the mixture prepared from both powders. Zirconium phase in HA + ZrO_2 mixture stabilizes the dimensions of final, sintered composite.

---

[6] Being a source of substances that induct bone genesis in the tissues surrounding the bone defect
[7] The process of bone creation on connective-tissue or cartilaginous base

In order to prepare sinters based on HAp, powder metallurgy method was employed in this study, which allowed to obtain porous materials with beneficial biofunctional properties.

The obtained materials, as a group of biomaterials which are widely used in bone surgeries, were then subjected to structural and phase tests. The *mathematical description that allows for assessment of volume shrinkage in HAp + ZrO_2 sinters* was developed after the process of sintering, which is a significant result of this work.

3.1 Procedure for the experiment

During the investigations the following powders were used:

- HAp ($Ca_{10}(PO_4)_6(OH)_2$) with average particle size of 50 micrometers, characterized by high purity level (over 99%wt.): Pb = 0.8ppm, As<1.0 ppm, Cd, Hg<0.1 ppm; Ca/P = 1.67,
- ZrO_2 oxide with addition of 8% (wt.) Y_2O_3 (YSZ - Yttria Stabilized Zirconia) (Fig. 6),
- ZrO_2 oxide with addition of 20% wt. Y_2O_3.

Both zirconium oxide powders were 100%crystalline, while their grains were of regular, spherical shape of grains.

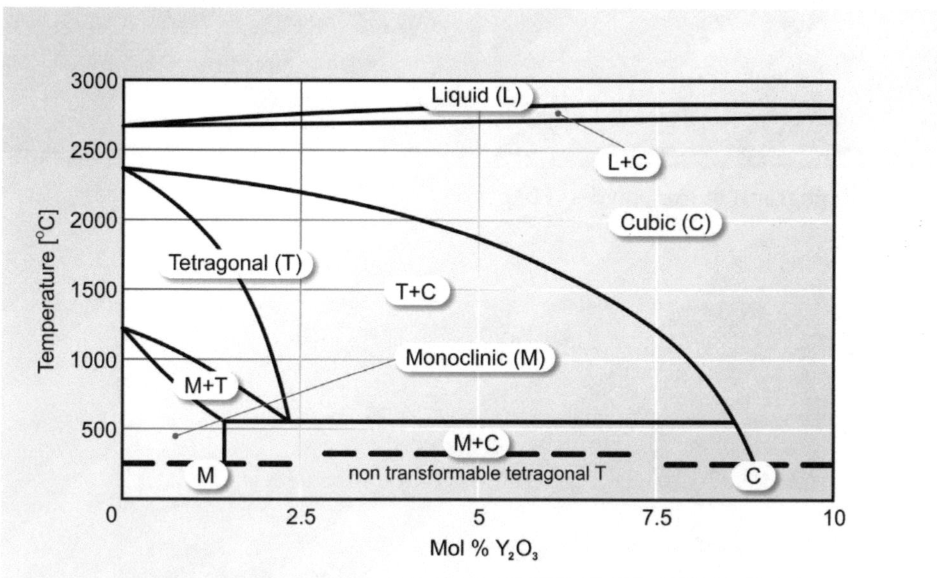

Fig. 6. Types of zirconium ceramics depending on the content of Y_2O_3 modifying phase

The composites based on hydroxyapatite with addition of different content of the both zirconium ceramic powders modified with ytrrium oxide were prepared:

- 100% HAp,
- (90%÷30%) HAp + (10%÷70%) YSZ; where: (YSZ-Ytrria Stabilized Zirconium, means ZrO_2+8%wt.Y_2O_3)
- (80%÷30%) HAp + (20%÷70%) ZrO_2 + 20%wt. Y_2O_3

3.2 Stereological tests of the used powders

Histograms of distribution of stereological parameters in the investigated powders, whose morphology is shown in Fig. 7, are presented in Fig. 10-12 (for hydroxyapatite) and in Fig. 8, 13-15 (for zirconium ceramics with addition of 8%wt. Y_2O_3) and Fig. 9, 16-18 (for zirconium ceramics with addition of 20%wt. Y_2O_3).

Shape factor R was calculated using the following formula:

$$R=L^2/(4\pi A) \quad (1)$$

where: L-particle circumference, A-particle surface area.

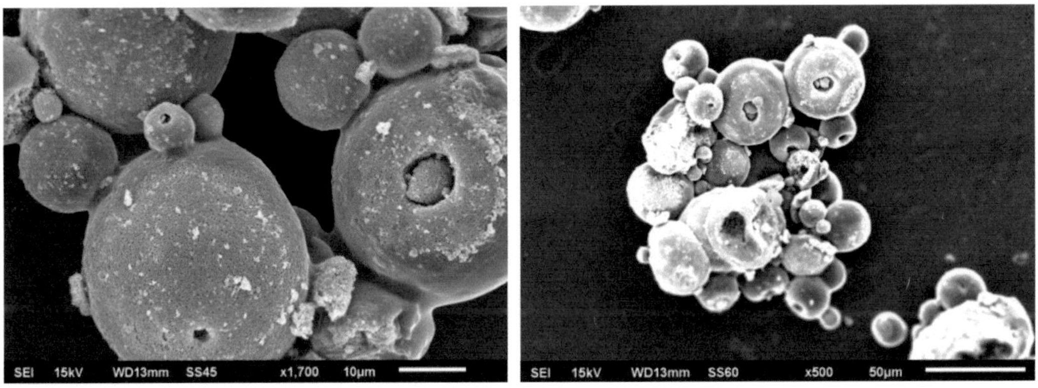

Fig. 7. Microphotograph of the powder: HAp

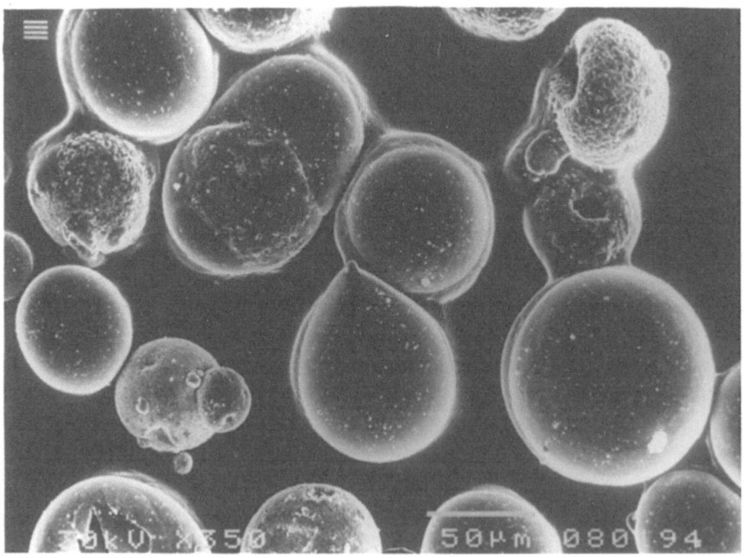

Fig. 8. Microphotograph of the powder: YSZ

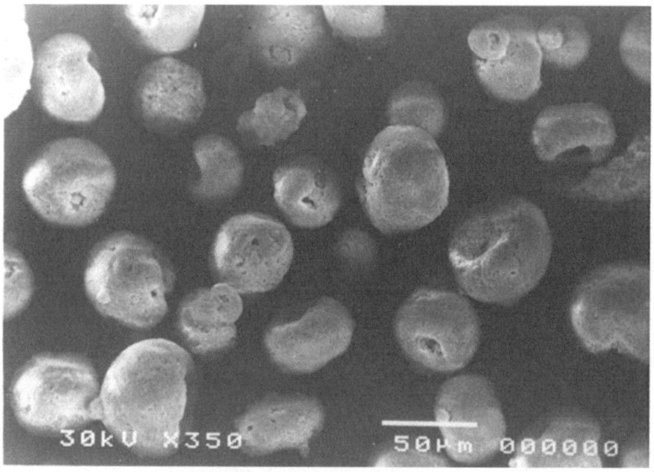

Fig. 9. Microphotograph of the powder: ZrO_2+20%Y_2O_3

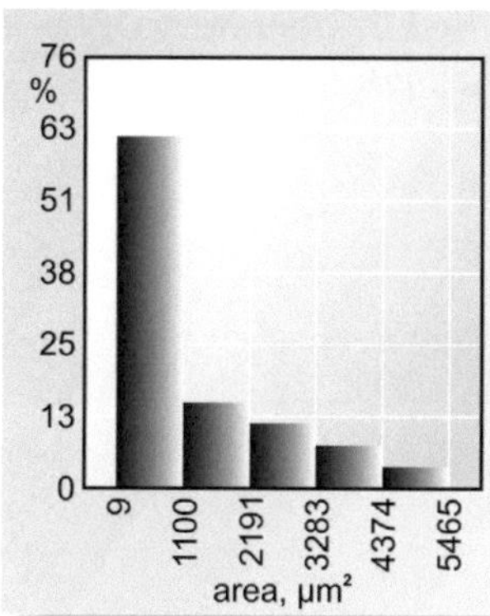

Fig. 10. Surface area histogram in particle of HAp powder

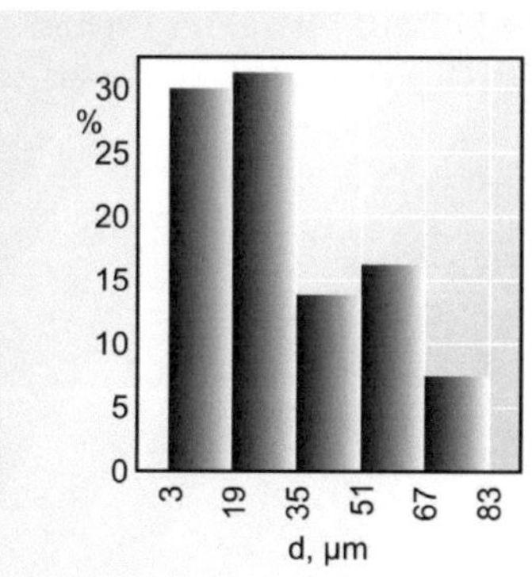

Fig. 11. Chord histogram in particles of HAp powder

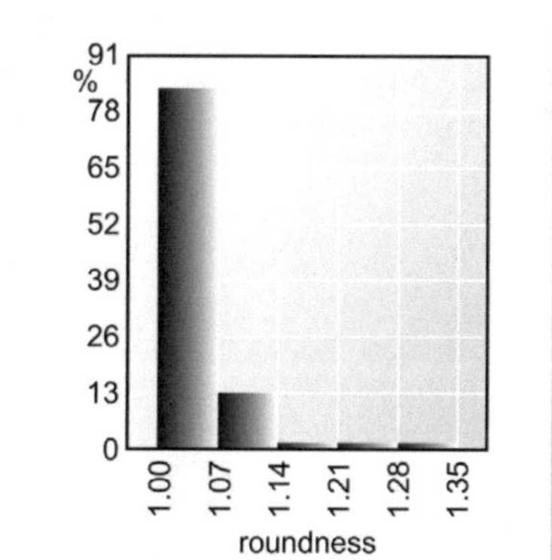

Fig. 12. Shape factor histogram in particles of HAp powder

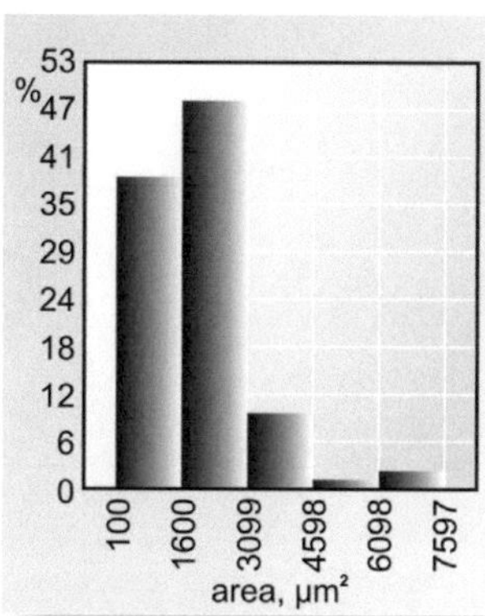

Fig. 13. Surface area histogram in particle of ZrO_2 + 8wt.% Y_2O_3 powder

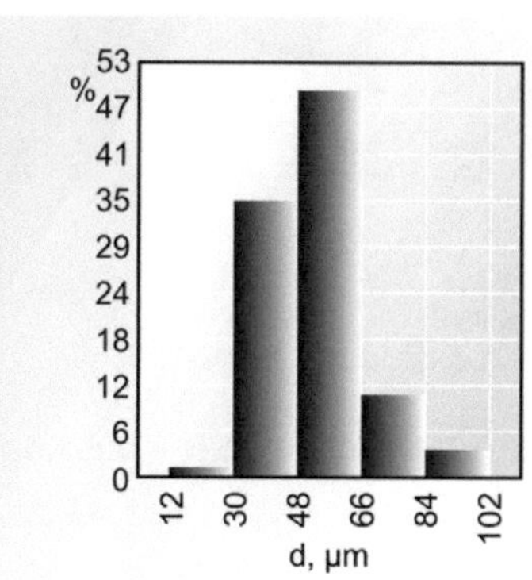

Fig. 14. Chord histogram in particles of ZrO_2 + 8wt.% Y_2O_3 powder

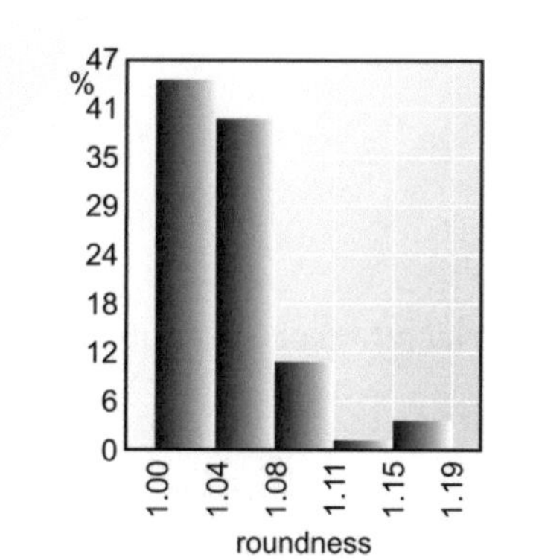

Fig. 15. Shape factor histogram in particles of ZrO_2 + 8wt.% Y_2O_3 powder

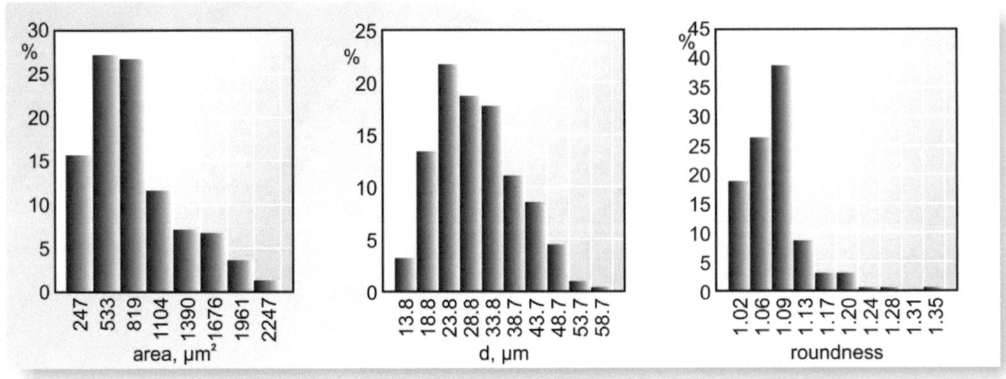

Fig. 16. Surface area histogram in particle of ZrO_2 + 20wt.% Y_2O_3 powder

Fig. 17. Chord histogram in particles of ZrO_2 + 20wt.% Y_2O_3 powder

Fig. 18. Shape factor histogram in particles of ZrO_2 + 20wt.% Y_2O_3 powder

After homogenization of the mixtures of selected particular compositions, powders were then axially compressed with the load of 110 MPa (Fig. 19) and dried in laboratory dryer. As a result of the procedure the moulded pieces were obtained with nominal dimensions of: ϕ=30mm, h=5mm.

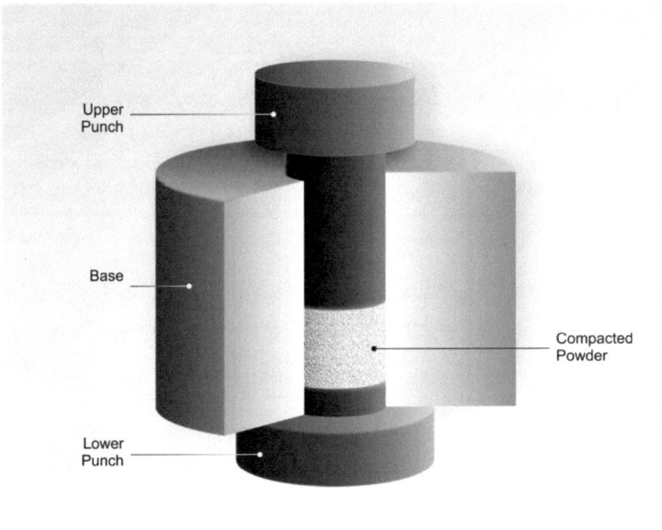

Fig. 19. Diagram of the mould used for powder compaction

Compaction of the powders was carried out in the following stages:

- movement of particles in relation to each other,
- elastic deformations in particles,
- crushing of particles.

The obtained moulded pieces where then subjected to the process of sintering at the temperature of 1100-1300^{0}C for two hours.

As a result of physical and chemical processes that accompany the process of sintering, a change in properties and dimensions of the moulded pieces occurred.

3.3 Microstructure tests

Macroscopic changes in the product during sintering are a result of numerous physical and chemical processes that occur in the material. The moulded pieces were subjected to microstructure tests using JEOL JSM 5400 scanning microscope.

Analysis of one-phase system (100 % HA) (Fig.20) and sinters of HA+ZrO_2 before and after sintering (1100-1300°C) (Fig.21) reveals that grains after the process of sintering were partially crushed and they were adjacent to each other. After the process of sintering a reduction in porosity (i.e. in number and size of pores) occurred.

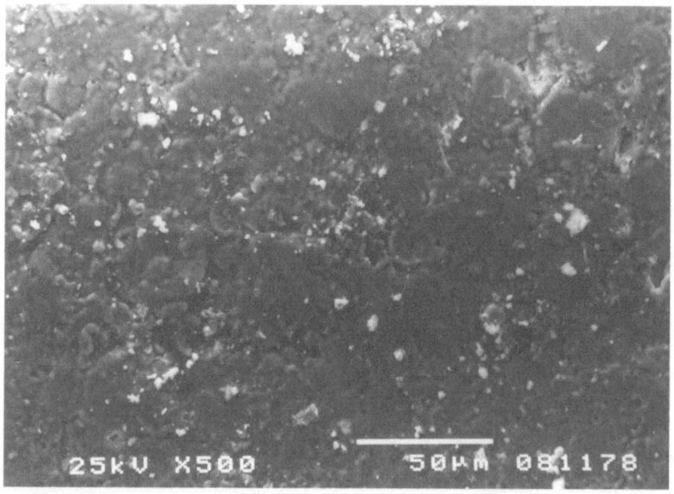

Fig. 20. 100% HAp powder after sintering

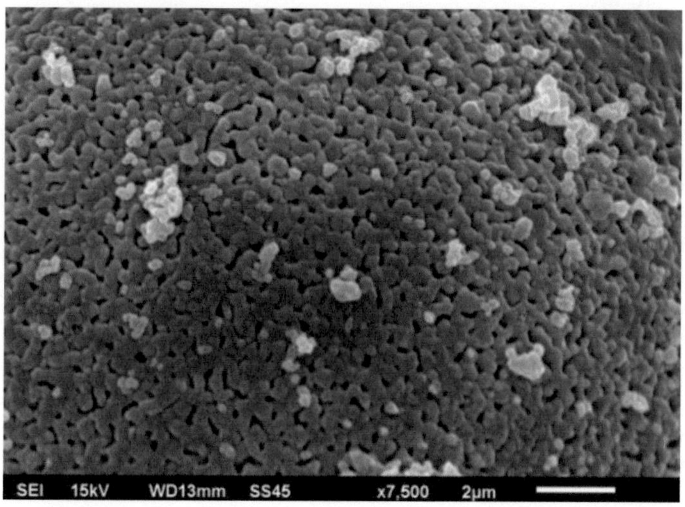

Fig. 21. Powder of: 50%HA + 50% ZrO_2 (8%wt.Y_2O_3) after sintering

Analysis of Hap+ZrO_2 sinters reveals that decreases largeness of grain and increase largeness of pore.

In the case of calcium orthophosphates, several specific processes occur during sintering. Firstly, moisture, carbonates and all other volatile chemicals remaining from the synthesis stage, such as ammonia, nitrates and any organic compounds, are removed as gaseous products. Secondly, unless powders are sintered, the removal of these gases facilitates production of denser ceramics with subsequent shrinkage of the samples. Thirdly, all chemical changes are accompanied by a concurrent increase in crystal size and a decrease in the specific surface area. Fourthly, a chemical decomposition of all acidic orthophosphates and their transformation into other phosphates takes place.

3.4 X-Ray structural analysis

It seems to be necessary, from the standpoint of wider use of composite materials, to determine phase stability in the obtained composites. After the sintering process, phase analysis was carried out using Seifert 3003 T-T X-ray diffractometer with radiation of wavelength of $\lambda_{K\alpha Co}$=0,17902 nm.

First stage encompassed phase analysis of the powders. X-ray quality analysis of hydroxyapatite powder revealed its 100% crystallinity and presence of hexagonal phase of HA with the following parameters of the cell: a = b = 9,418 nm, c= 6,884 nm, space group P63/m.

As results from the analysis of the obtained diffractogram, ZrO_2 modified with 8% wt. of Y_2O contains two polymorphous modifications of zirconium dioxide: tetragonal phase and small amount of monoclinic phase.

ZrO_2 powder with addition of 20% wt. of Y_2O_3 mainly consisted of monoclinic and small amount of regular phase. Lack of tetragonal phase, which indicates stabilization of zirconium phase proves undoubtedly that the powder is only a mixture of the two oxides rather than their solution.

Analysis of phase composition for sintered samples of 100% HA revealed that they are composed of hexagonal phase - HAp and TCP_α, CaO phases.

Diffractogram of sinters with analysed percentage contents of both powders are presented in: Fig. 22 for HA+ ZrO_2 (modified with 8wt.% Y_2O_3) and Fig. 23 for HA + ZrO_2 sinters (with addition of 20wt.% Y_2O_3).

Analysis of diffractograms of sinters that contain from 20-60% of YSZ phase revealed presence of HA phase, ZrO_2 with tetragonal modification, insignificant amount of TCP_α and CaO phase. The amount of TCP_α and CaO decreases as zirconium phase addition rises for all sintering temperatures.

The results of analysis of sinter diffractograms with addition of ZrO_2+20%Y_2O_3 phase were comparable - as HA, TCP_α and insignificant amount of CaO phase (that gradually disappears as zirconium phase addition rises) occur in it. The peaks from tetragonal ZrO_2 phase were also observed; its presence seems to be of much interest since monoclinic phase predominated in initial powder.

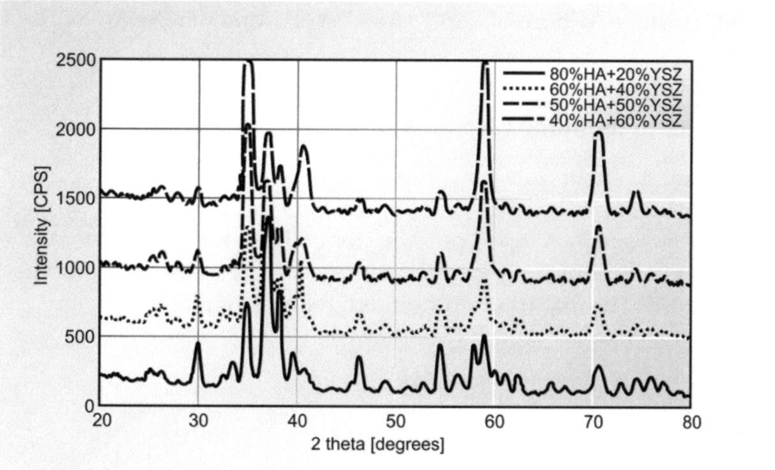

Fig. 22. Collective diffractogram for sinters of HA+ZrO_2 modified with 8% wt. Y_2O_3

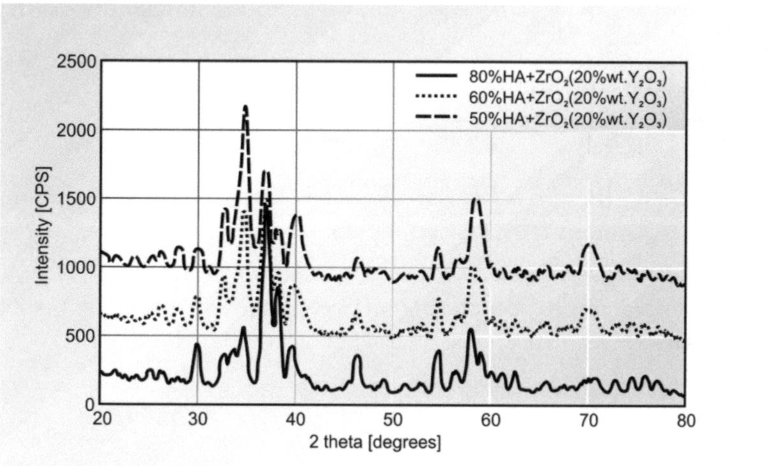

Fig. 23. Collective diffractogram for sinters of HA + ZrO_2 with addition of 20% wt. Y_2O_3

The process of sintering of HA + ZrO_2 (+20%Y_2O_3) mixtures caused, according to the investigations, decline in monoclinic modification and its transition into tetragonal modification. Thus, the applied treatment resulted in m↔t transition, i.e. transformation of monoclinic form into tetragonal one took place. This fact can be explained by impact of CaO phase that appeared during hydroxyapatite decomposition on stability of tetragonal phase. Moreover, presence of monoclinic modification of ZrO_2 in ZrO_2+20%wt.Y_2O_3 powder proves undoubtedly the absence, in the case of the analysed composition, of solid solution of Y^{+3} ions in crystallographic network of zirconium oxide. The process of sintering could have led to appearance of such solutions and also to stabilization of tetragonal phase.

The conducted analysis allows to observe that addition of zirconium phase impacts on rise in temperature of hydroxyapatite decomposition, which manifests in decline in CaO and

TCP_α phase and rising addition of ZrO_2+8wt.%Y_2O_3 and ZrO_2+20%wt.Y_2O_3 phases in powder mixtures.

3.5 Apparent density of moulded pieces and sinters

During development of materials for implants one must consider proper density of HA+ZrO_2 composites. Proper level of density must ensure that specific strength properties are obtained as well as necessary open porosity that allows for ingrowing of the implant. In order to achieve this, apparent density measurements in the obtained moulded pieces and sinters were carried out. The results are presented in Fig. 24 and 25.

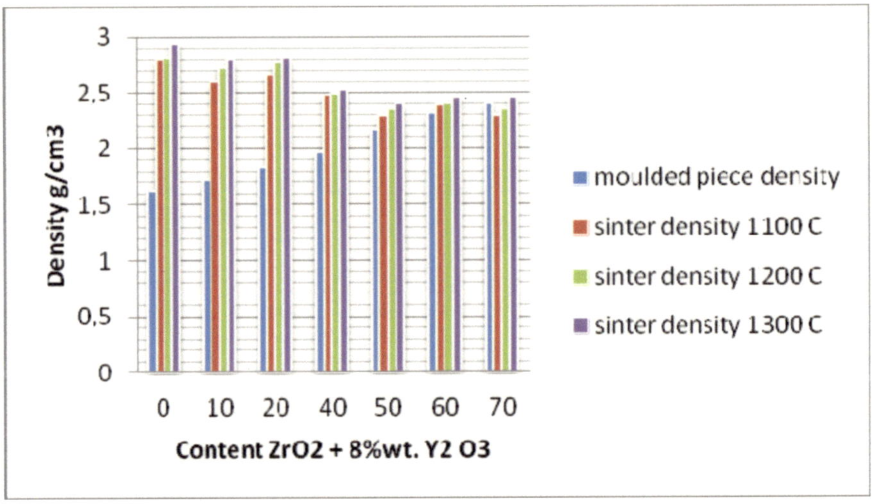

Fig. 24. Change in sample density depending on percentage content of ZrO_2 + 8 % wt. Y_2O_3 before and after sintering

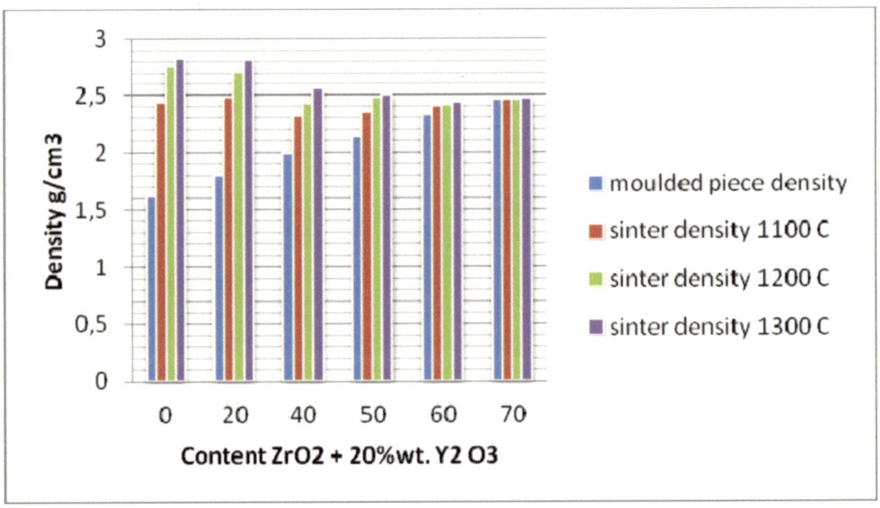

Fig. 25. Change in sample density depending on percentage content of ZrO_2 + 20% wt. Y_2O_3 before and after sintering

Densities of moulded pieces and sinters assessed in the case of addition of the same amounts of ZrO_2 modified with 8% wt. as well as 20% wt.Y_2O_3 to HA were comparable. Density of moulded pieces (HA+ ZrO_2) increases with addition of zirconium oxide, which is connected with difference in the value of density in both powders (ρ_{HA}=3gcm^{-3}, ρ_{ZrO2}=6gcm^{-3}). In consideration of the density in each sample before and after sintering, it is remarkable that the difference between density of a moulded piece and the sinter decreases with addition of zirconium phase. A density of sintered pieces increases with sintering temperatures.

This fact proves that addition of zirconium ceramics (regardless of ittria content used as modifier) stabilizes dimensions of the sintered composite based on HA.

Analysis of test results reveals that, in the case of composite samples with addition of ZrO_2 modified with both 8% wt. and 20% wt.Y_2O_3, density of the moulded piece equals density of the sinter at the content of ca. 70% of zirconium phase. This means that the obtained sinter is characterized by zero shrinkage after the treatment.

3.6 Mathematical description of the shrinkage in HA+ZrO_2 composites

An essential issue is to determine exact percentage content of addition of ZrO_2 phase to the mixture in terms of their impact on dimensional stability.

In order to formulate generalized dependencies in the obtained sinters, the analysis of dependencies between content of zirconium phase and the shrinkage was carried out using correlation and regression of two variables.

For the obtained experimental points linear regression equations were matched by means of least squares method, correlation coefficient was calculated and the tests were made with the significance level of α=0.05 using the following dependencies:

- significance test for correlation coefficient

$$t = \frac{r}{\sqrt{1-r^2}}\sqrt{n-2} \tag{2}$$

where: n – number of degrees of freedom, r – correlation coefficient

- significance test for coefficients of regression equation

$$t = \frac{A - A_0}{S_r}\sqrt{\sum_{i=1}^{n}(x_i - \bar{x})^2}\ ; S_r = \sqrt{\frac{1}{n-2}\sum_{i=1}^{n}(y_i - \hat{y})^2}\ ; t = \frac{B - B_0}{S_r}\sqrt{\frac{n\sum_{i=1}^{n}(x_i - \bar{x})^2}{\sum_{i=1}^{n}x_i^2}} \tag{3}$$

where: S_r – remaining deviation, A, A_0 – values of direction component, B, B_0 – absolute term.

- confidence interval for direction component of simple regression

$$P\left[A - t_\alpha \frac{S_r}{\sqrt{\sum_{i=1}^{n}(x_i - \bar{x})^2}} < A' < A + t_\alpha \frac{S_r}{\sqrt{\sum_{i=1}^{n}(x_i - \bar{x})^2}}\right] = 1 - \alpha \tag{4}$$

where: value of T-Student statistics for n-2 degrees of freedom and the accepted confidence level 1-α.

- asymptotes equation for confidence corridor

$$\hat{y}_1 = B_1 + A_1 \ \hat{y}_2 = B_2 + A_2 x \tag{5}$$

- confidence corridor for simple regression

$$P\left[\hat{y} - t_\alpha S_{\hat{y}_i} < \tilde{y} < \hat{y} + t_\alpha S_{\hat{y}_i}\right] = 1 - \alpha \quad S_{\hat{y}_i} = S_r \sqrt{\frac{1}{n} + \frac{(x_i - \bar{x})^2}{\sum_{i=1}^{n}(x_i - \bar{x})^2}} \tag{6}$$

The results of the performed calculations are presented in Fig. 26-27. These calculations show strong dependence between the contents of zirconium phase and sample shrinkage after sintering process. It was also proved by high values of correlation coefficient $r^2 > 0.98$.

On the basis of regression curves one can determine the content of zirconium phase at which sinters based on hydroxyapatite show zero shrinkage. In the case of HAp + ZrO_2 (8%wt. Y_2O_3) ceramics, a simple extrapolation method enabled determination of ZrO_2+8%wt. Y_2O_3 powder content which ensured zero shrinkage at the level of 70.6% wt. Content interval determined from confidence corridor amounts to from 65% to 76% wt.

Similar considerations were made for HAp + ZrO_2 (20 %wt. Y_2O_3) ceramics and they enabled determination of the amount of addition ZrO_2+20 %wt. Y_2O_3 powder that corresponded with zero shrinkage in sinter at the level of 76% wt.. Confidence corridor range for these samples ranged from 70% to 82% wt.

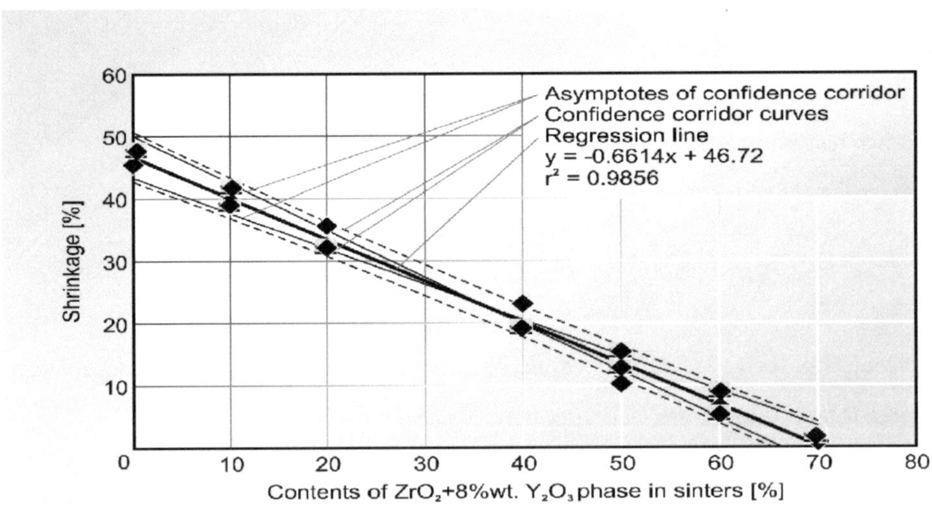

Fig. 26. Correlation chart with confidence corridor for HAp + ZrO_2 (8 %wt. Y_2O_3) sinters in 1100-1300°C

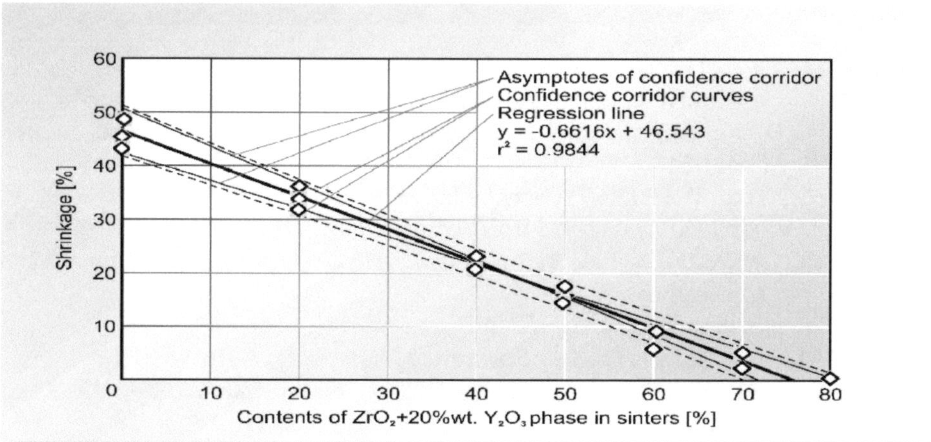

Fig. 27. Correlation chart with confidence corridor for HAp + ZrO_2 (20 %wt. Y_2O_3) sinters

4. Summary and conclusions

- A sintering procedure appears to be of a great importance to manufacture bulk bioceramics with the required mechanical properties.
- Hydroxyapatite ceramics have been found to be one of the best implantation materials successfully used in bone surgeries and dentistry. This ceramics allows for their easy connection with other types of materials, which creates opportunity for development of composites for medical applications.
- For the purposes of this work and creation of ceramic composites (HA+ZrO_2), powder metallurgy method was employed, which allowed to receive porous materials based on hydroxyapatite.
- Phase analysis of the sinters obtained from mixtures of powders (based on hydroxyapatite with addition of zirconium oxide modified with 8% wt. Y_2O_3 and 20% wt. Y_2O_3 revealed presence of HA phase as well as TCP_α and CaO phases that prove decomposition of hydroxyapatite. The investigations also revealed that addition of zirconium phase impacts on rise in temperature of hydroxyapatite decomposition, which manifests in decline of CaO and TCP_α phases with addition of ZrO_2+8wt.%Y_2O_3 ZrO_2+20%wt.Y_2O_3 phases in powder mixtures.
- In consideration of the density in each sample before and after sintering, it is remarkable that the difference between density of a moulded piece and the sinter decreases with addition of zirconium phase. This fact proves that addition of zirconium ceramics (regardless of ittria content used as modifier) stabilizes dimensions of the sintered composite based on HA.
- From the standpoint of common use of hydroxyapatite and HAp-based composites (with addition of ZrO_2 phase) for medicine, it is of key importance to determine such percentage content of ZrO_2 phase addition in the mixture that invariable or predictable dimensions of the implant or coating are maintained after the process of sintering. The presented mathematical description enables assessment of the amount of zirconium oxide (with different addition of stabilizing Y_2O_3 phase) in the mixture of powders that ensures zero shrinkage in sinters.

5. References

Ashok M. (2003) *Materials Letters*, 57, 2066-2070

Chevalier J., Deville S., Munch E., Jullian R., Lair F. (2005) *Biomaterials*, 25, 5539-5545

Cheng G., Pirzada D., Cai M., Mohanty P., Bandyopadhyay A. (2005) *Materials Science and Engineering C*, 541-547

Chiu C.Y., Hsu H.C, Tuan W.H. (2007) *Ceramics International*, 33, 715-718

Dorozhkin Sergey V. *BIO* (2011), 1, 1-51 http://ccaasmag.org/BIO

Dudek A. (2009) *Collective monograph, Materials and exploitation problems in modern Materials Engineering,* Czestochowa

Evis Z. (2007), *Ceramics International*, 33, 987-991

Fu L., Khor K.A., Lim J.P. (2001) *Materials Science and Engineering,* A316, 46-51

Gu Y.W., Loh N.H., Khor K.A., Tor S.B., Chrang P. (2002), *Biomaterials*, 23, 37- 43

Hartmann P., Jager C. (2001) *Journal of Solid State Chemistry*, 160, 460-468

Heimann R.B. (2006) *Surface and Coatings Technology*, 201, 2012-2019

Inuzuka M., Nakamura S., Kishi S. (2004) *Solid State Ionic's,* 172, 509-513

Jurczyk M., Jakubowicz J. (2008) *Bionanomaterials,* Wyd. Politechniki Poznańskiej

Kalkura S.N.: *Materials Letters* 57, (2003), 2066-2070

Khalil K.A., Kim S., Kim H.Y. (2007) *Materials Science and Engineering*, 456, 368-372

Li J., Liao H., Hermansson L. (1996) *Biomaterials*, 17, 1787-1790

Marciniak J. (2002) *Biomaterials*, Gliwice

Prado M.H., Silva Da, Lima J.H. (2001) *Surface and Coatings Technology*, 137, 270-276

Piconi C., Maccauro G. (1991) *Biomaterials,* 20, 1-5

Rapacz-Kmita A., Paluszkiewicz C., Ślósarczyk A., Paszkiewicz Z (2005) *Journal of Molecular Structure*, 744-747, 653-656

Rhee S.H. (2002) *Biomaterials*, 23, 1147-1152

Schulz U., Lin H-Tay (2007) *Andvanced Ceramic Coatings and Interfaces II*, Wiley-Interscience

Shackelfortd J.F., Doremus R.H. (2008) *Ceramic and Glass Materials*, Springer

Silva V., Lameiras, F.S., Dominguez R.Z. (2001) *Compos. Science Technology*, 1-2, 133-136.

Slosarczyk A. (1997) *Hydroxyapatite ceramics*, PAN, Ceramics 51, Kraków

Sung Y.M., Kim D.H. (2003) *Journal of Crystal Growth*, 254, 411-417

Subotowicz K. (2008) *Ceramics for each*, Elamed, Katowice

Vaccaro A.R. (2002) *The role of the osteoconductive scaffold in synthetic bone graft ortopedics*, 25, 571-578

Yoshida K., Hashimoto K., Toda Y., Udagawa S., Kanazawa T. (2006) *Journal of the European Ceramic Society*, 26, 515-518

Ziebowicz A. (2008) *Biomaterials in stomatology*, Wyd. Politechniki Śląskiej, Gliwice

Part 4

Simulation of Ceramics

8

Numerical Simulation of Fabrication for Ceramic Tool Materials

Bin Fang*, Chonghai Xu, Fang Yang, Jingjie Zhang and Mingdong Yi
School of Mechanical and Automotive Engineering, Shandong Polytechnic University, P. R. China

1. Introduction

Ceramic materials have good mechanical properties, such as high hardness, good wear resistance and elevated-temperature anti-oxidation. So, Ceramic materials are widely applied not only in the field of aeronautics and astronautics, building and mechanics in modern technology, but also in the field of the cutting tool materials. Ceramic tool materials are widely used in the dry cutting and high speed cutting. The preparation of ceramic tool materials includes powdering, forming, hot-press fabrication and machining process. A green compact before sintering is a porous packing of loose powder that is held together by weak surface bonds. The individual particles fabricated together to form a dense, strong monolithic part by sintering. The driving force for the sintering is the reduction in surface free energy of the particle. This reduction is performed by diffusion transport of material. Many factors affected the process of material transport and the exhalation of pores. Therefore, the hot-press fabrication is a very complicated process in which the compact formed of fine powder materials fabricated at the temperature below the melting point of the main constituent for the purpose of gaining the enough strength of the compact by bonding particles together. The hot-press fabrication is a key process, which governs the mechanical properties of the ceramic tool materials as well as the components and content.

Most sintering theory models are based on the simple assumption of particle shape and the mass-transport mechanisms. With the rapid development of the computer simulation technology and the computational material science, modeling and numerical simulation of ceramic sintering process becomes a new and promising approach to investigate the sintering process on the micro- and meso- scale or macro-scale (Jagota& Dawson, 1990; Bordère, 2002; Wakai et al, 2004, 2005; Mori, 2006; German, 1998). At the same time, it is very significant to simulate Microstructural evolution of sintering process on the micro- and meso-scale for the theory development and the evaluation of relationship between the mechanical properties and the microstructure of ceramic materials (Guo, 1998). The two-dimensional Monte Carlo grain-growth simulation methods, firstly developed by Srolovitz and co-workers (Anderson et al, 1984; Srolovitz et al, 1984), is lately extended into the three dimensional simulation. Monte Carlo Potts' model is largely applied to simulate the solid-phase sintering process of ceramics (Hassold et al, 1990; Chen et al, 1990; Tikare & Holm,

* Corresponding Author

1998; Tikare et al, 2003) . Because of no assumption of the particle shape, the model can be properly used to simulate the mass-transport mechanisms. Computer simulation of final-stage sintering process of ceramics is initially developed by Hassold G N with Monte Carlo Potts' model (Hassold et al, 1990; Chen et al, 1990; Tikare & Holm, 1998). Tikare V et al (Tikare et al, 2003) simulated the densification of ceramic sintering processes with Monte Carlo Potts' model, which is coupled with vacancy-annihilation method. However, the mentioned simulation methods are only applied to simulate the sintering processes for one-composition, single-phase ceramic system. To meet the need of the severe cutting conditions and the durable tool-life in high speed cutting process, ceramic tool materials should have good mechanical properties, such as high hardness, high flexural strength, high fracture toughness, good wear resistance and elevated-temperature anti-oxidation. In order to improve the mechanical properties further, the second-phase powders are combined on the base of matrix powder to develop the composite ceramic tool materials by hot-press fabrication (Lang, 1982; Guo, 1998, Mukhopadhyay et al, 2007; Belmonte et al, 2006).

In this paper, the new Monte Carlo Potts' model for simulating the sintering process of single- and two-phase ceramic materials is developed and the two-dimensional (2D) grain-growth process is successfully simulated with considering the presence of pores in the green compact. A detailed description of the new Monte Carlo Potts model and simulation procedure is presented. The simulation results are also discussed.

2. Porosity and mean pore size of green compact

2.1 Porosity of green compact

Although the fine powders of ceramic are tightly packed by the effect of the pre-pressure and the gravity, the pore is present among the powders. Assuming the fine powders are the sphere which is equal in the radius, the porosity can be calculated by the ball-packed model.

According to the crystal structure of metallurgy (Jin et al, 2002) , as Fig.1 was shown, it is a hypothesis that there are three types of the particle arrangement, the body-centered cubic (BCC), the faceted-centered cubic (FCC) and the hexagonal closed-packed (HCP).

The porosity of three types of the particle arrangement is given by Eq. (1), Eq. (2) and Eq. (3), respectively.

1. Porosity of BCC

$$f_{\mathrm{BCC}} = 1 - \frac{2\times\frac{4}{3}\pi R^{3}}{\left(\frac{4}{\sqrt{3}}R\right)^{3}} = \frac{\sqrt{3}}{8}\pi \approx 0.32 \tag{1}$$

2. Porosity of FCC

$$f_{\mathrm{FCC}} = 1 - \frac{4\times\frac{4}{3}\pi R^{3}}{\left(\frac{4}{\sqrt{2}}R\right)^{3}} = \frac{\sqrt{2}}{6}\pi \approx 0.26 \tag{2}$$

3. Porosity of HCP

$$f_{\text{HCP}} = 1 - \frac{2 \times \frac{4}{3}\pi R^{3}}{6 \times \frac{\sqrt{3}}{4} \times \sqrt{\frac{8}{3}} \times (2R)^{3}} = \frac{\sqrt{2}}{6}\pi \approx 0.26 \tag{3}$$

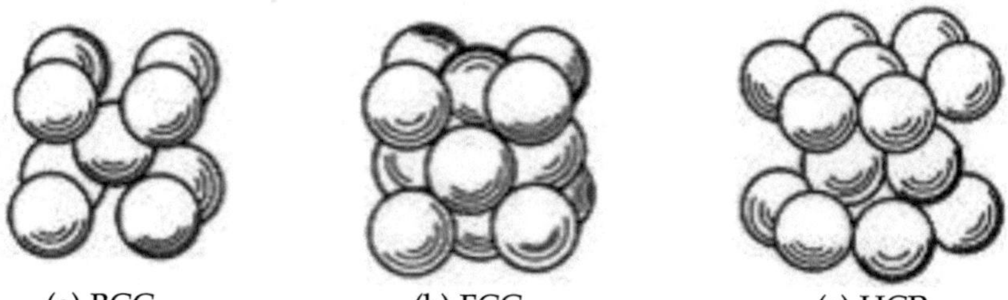

(a) BCC (b) FCC (c) HCP

Fig. 1. Schematic Diagram of Three Type of the Particle Arrangement (Jin et al, 2002)

The particle arrangement of green compact is not as tight as the atoms in the crystal structure of metallurgy. The particle size has some distribution, and the smaller particles in the size are filled into the bigger hole. So, it is proper that the porosity of green compact is 30% during the simulation.

2.2 Mean pore size of green compact

Pore size is a function of the particle size, the particle shape and the density of the green compact in the green compact. For the specific particle size and density, the mean pore size of green compact is given by Eq. (4) (Myers et al, 1989).

$$d = \frac{2D(1-f)}{3f} \tag{4}$$

where, d is mean pore diameter, D is powder particle diameter and f is the density of green compact.

As mentioned above, the porosity is 30% in the initial simulation system, that is to say, the value of f is 0.7. According to Eq. (4), the mean pore diameter (d) is 0.29D. Therefore, during the initialization of simulation system, the function of pore initialization is implemented when the particle diameter is the finial initialized particle size. Then, the initialized process is not finished until the required particle size is obtained.

3. Simulation method

The two-dimension Monte Carlo Potts simulation model utilizes two different approaches to simulate the grain growth and annihilation of pore (Braginsky et al, 2005) . It can model the following processes:

1. Simulation of grain growth by short range diffusion of atoms from one side of grain boundary to the other side;

2. Simulation of long range diffusion of pores by the surface diffusion and of vacancies/material by grain boundary diffusion;
3. Simulation of vacancy-annihilation at grain boundaries.

3.1 Monte Carlo Potts model of single-phase ceramic tool materials containing pores

In the Monte Carlo Potts' model for simulating the sintering process of single-phase ceramic tool materials with pores, the fine powders and pores are mapped onto a set of two-dimensional and discrete hexagonal sites. In order to describe the state of a site *i*, two parameters, S_i and P_i, are required. S_i represents the orientation of the grain that the site belongs to. P_i represents the phase state of the site. The fine powder sites are given one of Q distinct, degenerate states, where the individual state is designated by the symbol S_i. The value of S_i for the solid phase is from 1 to Q and Q is the maximum value for the grain orientation. At the same time, the P_i of fine powder sites is given 1, which is the symbol of solid. The pore sites can be assumed only one state, S_{pore}=Q+1 , and the P_i of pore sites is given 3, which is the symbol of pore. Continuous fine powder sites of the same state S_i forms the fine powders and the value of P_i for the continuous fine powder sites are 1. Continuous pore sites forms a pore and the value of P_i for the pore sites are 3. During evolution of simulation, grain boundaries exist between neighboring grain sites of different states, S_i, and pore-grain interfaces exist between neighboring pore and grain sites. There are two different boundaries in the simulation system. The boundary-one is formed between neighboring grain sites of different states, and its energy is J_1. The boundary-two is formed between neighboring pore and grain sites, and its energy is J_2.

The equation of state for these simulations is the sum of all the neighbor interaction energies of single-phase ceramic tool materials with pores can be written as:

$$E_{Tol} = E_1 + E_2 \tag{5}$$

$$E_1 = J_1 \sum_{i=1}^{N} \sum_{j=0}^{n_1} \left[1 - \delta(S_i, S_j) \right] \tag{6}$$

$$E_2 = J_2 \sum_{i=1}^{N} \sum_{j=0}^{n_2} \left[1 - \delta(S_i, S_j) \right] \tag{7}$$

where, E_{Tol} is the total energy (J). E_1 is the energy of the neighboring grain sites (J). E_2 is the energy of the neighboring pore and grain sites (J). N is the sites number of solid phase. n_1 is the solid-phase site number around one specific site of solid phase(≤6). n_2 is the pore-phase site number around one specific site of solid phase(≤6). And $\delta(q_i, q_j)$ is the Kronecker function defined as $\delta(S_i, S_j)$ =1 for $S_i = S_j$ and otherwise $\delta(S_i, S_j)$ =0, with S_i and S_j denoting the orientation parameters of the neighboring sites i and j.

3.2 Monte Carlo Potts model of two-phase ceramic tool materials containing pores

In the Monte Carlo Potts' model for simulating the sintering process of two-phase ceramic tool materials with pores, the fine powders and pores are mapped onto a set of two-dimensional and discrete hexagonal sites. In order to describe the state of a site *i*, two parameters, S_i and P_i, are required. S_i represents the orientation of the grain that the site belongs to. P_i represents the phase state of the site. The fine powder sites are given one of Q distinct, degenerate states, where the individual state is designated by the symbol S_i. The value of S_i for the solid phase is from 1 to Q and Q is the maximum value for the grain orientation. At the same time, the P_i of fine powder sites is assigned the value of 1 or 2, which is the symbol of solid phase-one or phase-two. The pore sites can assume only one state, S_{pore}=Q+1, and the P_i of pore sites is assigned the value of 3, which is the symbol of solid. Continuous fine powder sites of the same state S_i forms the fine powders and the value of P_i for the continuous fine powder sites is 1 or 2. Continuous pore sites forms a pore and the value of P_i for the pore sites is 3. During evolution of simulation, grain boundaries exist between neighboring grain sites of different states, S_i, and pore-grain interfaces exist between neighboring pore and grain sites. If two adjacent sites are phase-one and phase-two and their grain orientation is the same, then the composite grains are formed and the two phases exist diffusion each other. If two adjacent sites are single phase and their grain orientation is the same, then the grains for single phase are formed. Grain boundaries exist between neighboring grain sites of different states, S_i, and pore-grain interfaces exist between neighboring pore and grain sites. There are five different boundaries in the simulation system. The boundary-one is formed between the phase-one sites with different orientations, the boundary-two is formed between the phase-two sites with different orientations, the boundary-three is formed between the phase-one and phase-two sites, the boundary-four is formed between the phase-one and pore sites and the boundary-five is formed between the phase-two and pore sites. And their grain boundary energy is J_{11}, J_{22}, J_{12}, J_{13} and J_{23}, respectively. It is assumed that the grain boundary energy is dependent on the grain boundary and independent of the misorientation of the adjacent grains and of the grain orientation. Each of mass diffusion was performed by one site in the simulation.

The equation of state for these simulations is the sum of all the neighbor interaction energies of two-phase ceramic tool materials with pores can be written as:

$$E_{Tol} = E_{11} + E_{12} + E_{22} + E_{13} + E_{23} \tag{8}$$

$$E_{11} = J_{11}\sum_{i=1}^{N_1}\sum_{j=0}^{n_1}\left[1-\delta(S_i,S_j)\right] \tag{9}$$

$$E_{22} = J_{22}\sum_{i=1}^{N_2}\sum_{j=0}^{n_2}\left[1-\delta(S_i,S_j)\right] \tag{10}$$

$$E_{12} = J_{12}\sum_{i=1}^{N_3}\sum_{j=0}^{n_3}\left[1-\delta(S_i,S_j)\right] \tag{11}$$

$$E_{13} = J_{13}\sum_{i=1}^{N_1}\sum_{j=0}^{n_4}\left[1-\delta(S_i,S_j)\right] \tag{12}$$

$$E_{23} = J_{23}\sum_{i=1}^{N_2}\sum_{j=0}^{n_5}\left[1-\delta(S_i,S_j)\right] \tag{13}$$

where, E_{Tol} is the total energy (J). E_{11} is the energy of the neighboring phase-one grain sites (J). E_{22} is the energy of the neighboring phase-two grain sites (J). E_{12} is the energy of the neighboring phase-one and -two grain sites (J). E_{13} is the energy of the neighboring pore and phase-one grain sites (J). E_{23} is the energy of the neighboring pore and phase-two grain sites (J). N_1 is the sites number of phase-one. N_2 is the sites number of phase-two. N_3 is the smaller sites number of that of phase-one or phase-two. n_1 is the phase-one site number around one specific site of phase-one(≤6). n_2 is the phase-two site number around one specific site of phase-two(≤6). n_3 is the phase-two site number around one specific site of phase-one when N_3 is the phase-one site number(≤6), otherwise, n_3 is the phase-one site number around one specific site of phase-two when N_3 is the phase-two site number(≤6). n_4 is the pore-phase site number around one specific site of phase-one(≤6). n_5 is the pore-phase site number around one specific site of phase-two(≤6). And $\delta(S_i,S_j)$ is the Kronecker function defined as $\delta(S_i,S_j)$ =1 for $S_i = S_j$ and otherwise $\delta(S_i,S_j)$ =0, with S_i and S_j denoting the orientation parameters of the neighboring sites i and j.

3.3 Monte Carlo algorithm

3.3.1 Simulation algorithm of grain growth

The Monte Carlo simulation of the grain growth processes is implemented by the following steps:

1. The system energy is computed using Eq. (5) for the single phase material system or Eq. (8) for the two phase material system.
2. A site is randomly selected. If the site is solid-phase and located on the boundary, a new possible crystallographic orientation of the site is randomly chosen from one set of crystallographic orientations of neighboring solid-phase site.
3. The total energy of the simulated system with a new crystallographic orientation is computed using Eq. (5) for the single phase system or Eq. (8) for the two phase system and the difference (ΔE) between the new energy and the old is calculated.
4. As shown in Eq. (14), the energy difference (ΔE) is applied to compute the transition probability for the site with a new crystallographic orientation.

$$P(\Delta E) = \begin{cases} 1 & \Delta \mathrm{E} \leq 0 \\ \exp^{-\frac{\Delta \mathrm{E}}{k_B T}} & \Delta \mathrm{E} > 0 \end{cases} \quad (14)$$

   where, T is the temperature of the site which is considered and k_B is the Boltzmann's constant.

5. As the transition probability P(ΔE) is calculated, a random probability R is generated in a range of 0~1. The site with a new crystallographic orientation is accepted to change into the new crystallographic orientation if R≤P(ΔE), otherwise the old site orientation is kept.
6. The system energy for the current configuration is assigned to the step (1). The simulation procedure is repeated from the step (2). The attempted N (total sites number in the simulation system) times is regarded as one Monte Carlo Step (MCS).

During the simulation, the site leaps to the nearest-neighbor cell across the grain boundary if one which belongs to the cell has attempted a crystallographic orientation of that and been accepted during the reorientation attempt, which leads to the grain-boundary migration and the microstructural evolution. The leap of the site simulates the evaporation-deposition mechanism by the atoms during the sintering processes. The cell that loses the sites starts to shrink, on the other hand, the cell that gains the sites starts to grow. So, the growth or shrinkage of grains can be properly simulated during the fabrication of ceramic tool materials in the simulation system.

3.3.2 Simulation algorithm of pore migration

In order to keep the same total number of pore sites and grain sites throughout the simulation in the simulation system, pore migration is simulated using conserved dynamics (Hassold et al, 1990).

The Monte Carlo simulation of pore migration processes is implemented by the following steps:

1. The pore site neighboring grain boundary and the solid-phase site neighboring the pore site are chosen at random.
2. The two sites are temporarily exchanged with the solid-phase site assuming a new state S where S results in the minimum energy. This minimum-energy, pore-grain exchange simulates pore migration by surface diffusion.
3. The change in energy (ΔE) for this exchange is calculated using Eq. (5) for the single phase system or Eq. (8) for the two phase system.
4. The change in energy (ΔE) is applied to compute the transition probability for the pore migration attempt with the above mentioned Eq.(14).
5. As the transition probability $P(\Delta E)$ is calculated, a random probability R is generated in a range of 0~1. The temporarily exchange is accepted if $R \leq P(\Delta E)$, otherwise the old microstructure is kept. The attempted N (total sites number in the simulation system) times is regarded as one Monte Carlo Step (MCS).

At lower temperatures, the pores do not have sufficient energy to diffuse in the grain structure, and only can coalesce into small isolated pores and stagnate microstructural evolution. Therefore, the simulation temperature of the pore migration is $k_B T=0.7$. This higher temperature is necessary for accurate simulation of pore migration (Tikare & Holm, 1998).

3.3.3 Simulation algorithm of vacancy annihilation

During Monte Carlo simulation of ceramic tool material system with pores, vacancy is defined as a single pore site which is surrounded by grain sites. The densification of material in simulation is achieved by vacancy annihilation (Braginsky et al, 2005). In DeHoff's stereological theory of sintering densification (DeHoff et al, 1989), the densification mechanism comprises vacancy migration from pores to grain boundaries and vacancy annihilation at the grain boundaries. The densification is the process of vacancies being painted on the grain boundary, with an entire monolayer of vacancies annihilated, so that the mass centers of adjacent grains move towards the grain boundary. The rate of densification is governed by the time of vacancies that diffuse and cover the entire grain boundary. The densification is simulated with the algorithm of vacancy annihilation as shown in Fig.2. The schematic diagram of this algorithm is shown in Fig.3.

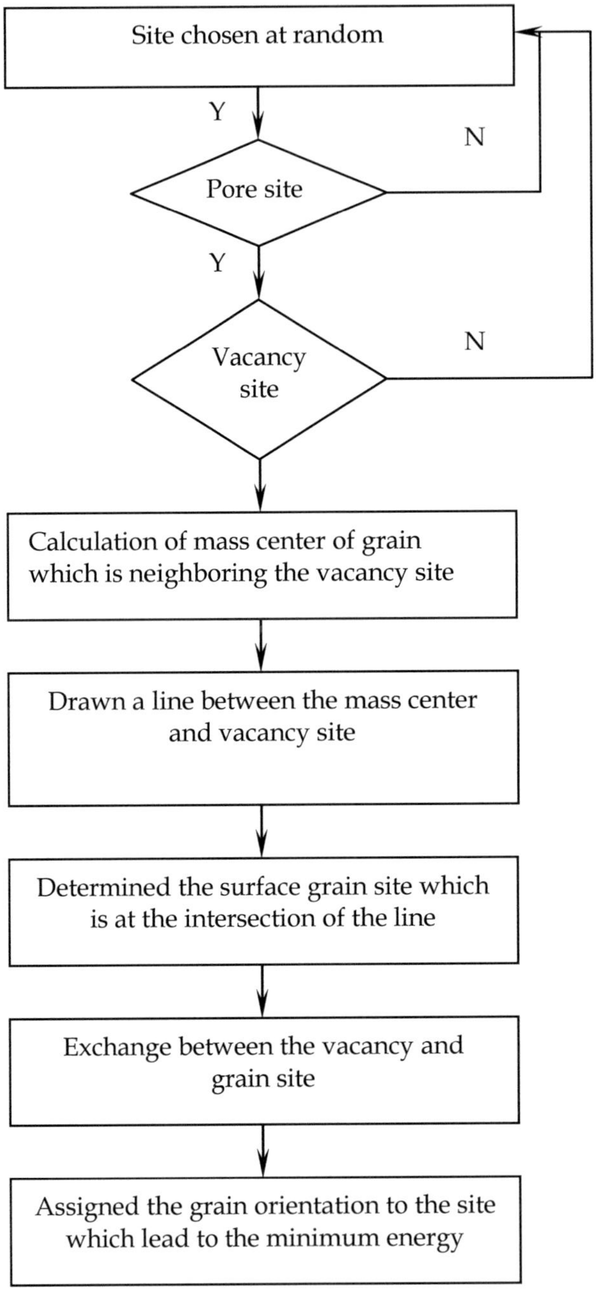

Fig. 2. Simulation Algorithm of Vacancy Annihilation

During pore annihilation, if the site chosen randomly is a site of vacancy, the grain site for the exchange is chosen at the intersection of a line which is drawn from the site of vacancy through the mass center of the adjacent grain and the outside boundary of the sintering compact. The grain site is assigned the state of the adjacent grain if the exchange is successful. The pore site, which jumps from inside to outside and coalesces with the others, can visualize the pore exhalation. The grain site which jumps from the outside to the inside can visualize the mass center of the grain migrated to the site of the vacancy. The whole process can conceptualize the densification of ceramic tool materials during fabrication. The site number and the contents of the solid phases are constant to keep the mass conserve during the simulation.

3.4 Simulation conditions

The software is developed with the Visual C++ compiler. The program was implemented in a Pentium-Pro 233MHz computer with 1 GB of RAM. The simulation domain is consisted of 500×500 hexagon sites. The edge length of hexagonal site is 0.6mm in simulation domain, so the area of each site is 0.9353mm^2.

The maximum value of orientation (Q) is 180. It is proved that the simulation results of different Q are nearly indistinguishable from each other when Q>50 is used (Hassold et al, 1990). The simulation time is expressed in term of the number of Monte Carlo Steps (MCS). Because the present simulations are performed on 500×500 sites, one MCS is equal to 250000 attempts in the computational domain. The ratio of the specific grain boundary energy is J_1:J_2=1:1.4 in the simulation of microstructural evolution of single-phase ceramic tool material. The ratio of the specific grain boundary energy is J_{11}:J_{22}:J_{12}:J_{13}:J_{23}=1:4:1:1.4:4.5 in the simulation of microstructural evolution of two-phase ceramic tool material. These values are comparable to the values of Al_2O_3 and SiC (Handwerker et al, 1990; Tanaka & Kohyama, 2003; Jiao et al, 1997). The simulation numerical data are the average results of the 10 independent simulations so as to eliminate the statistical fluctuations.

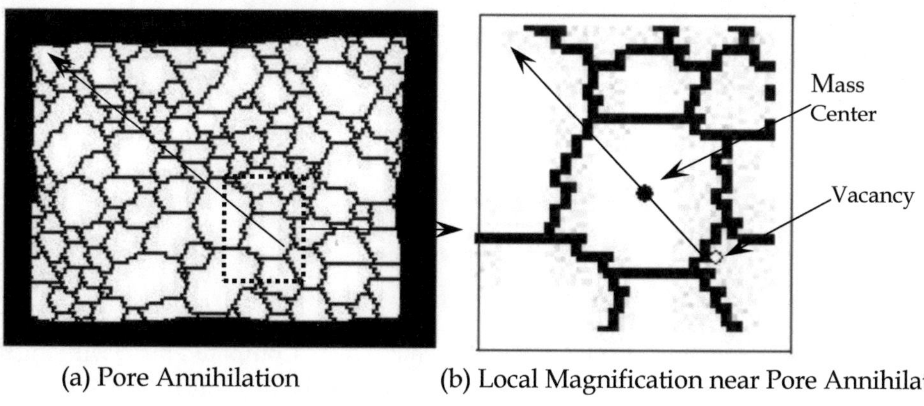

(a) Pore Annihilation (b) Local Magnification near Pore Annihilation

Fig. 3. Schematic of Pore Annihilation

4. Simulation results and discussion

4.1 Simulation of microstructural evolution of single-phase ceramic tool material

As mentioned above, there are 500×500 hexagon sites in the simulation domain, including the 75000 pore sites. That is to say, it is assumed that the porosity of green compact is 30%.

Initialization is needed to perfectly simulate the green compact. The initialization of pore is performed when the mean diameter is 0.29um. The initialized simulation microstructure is shown in Fig. 4. The white region denotes grains, and the black region denotes pores. Fig. 4 (a) is the part of the outside boundary, and the Fig. 4 (b) is the part of the inner. It is found from Fig. 4 (a) that pores exist in the inner of simulation domain and the density is lowest. Neighbor pore sites coalesce together. The bigger the pore is, the more the grain is around the pore. The state of the distribution of simulation microstructure is similar with the real green compact.

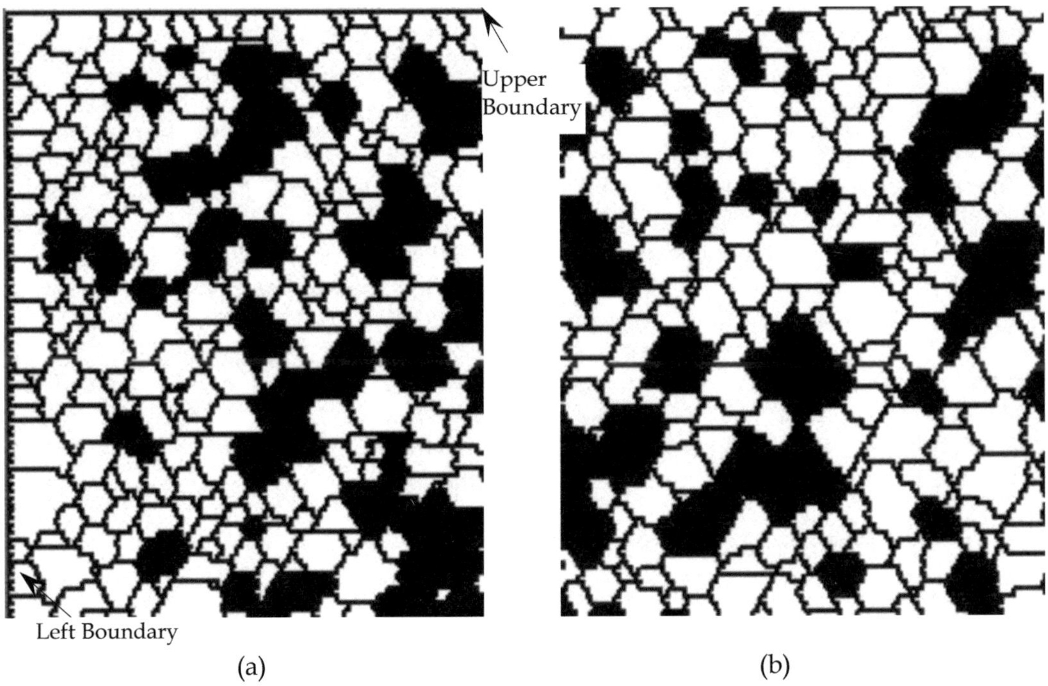

(White region denotes the grain, black region denotes pore)

Fig. 4. Initialized Simulation Microstructure of Single-phase

The resulting microstructure at 100 MCS is shown in Fig. 5. Compared Fig.5 with Fig.4, most of pore sites migrate from the inside to the outside. This is implied that the vacancy is annihilated and exhaled from the inside of simulation domain. Companied by the process, the densification and the grain size also increases. Kingery and Francois (Kingery & Francois, 1965) proposed that pores should exist at the grain boundary, especially, at the triple junction point of grain boundary when the porosity of green compact is fabricated. As shown in Fig.5, the simulation results accord with the theory. It is also seen in Fig.5 that the isolated pore site exists at the grain boundary, which means that the pores diffuse along the grain boundary. It is useful for the pore exhalation and the densification of material.

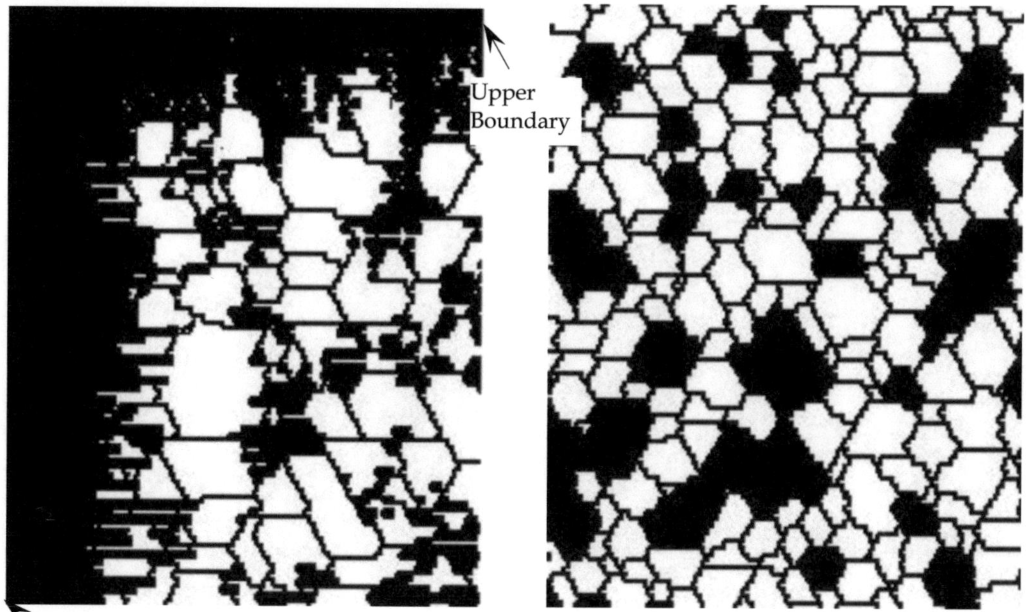

(a) (b)

(White region denotes the grain, black region denotes pore)

Fig. 5. Microstructural Evolution of Single-phase Ceramic Tool Materials

4.2 Simulation of microstructural evolution of two-phase ceramic tool material

As mentioned above, there are 500×500 sites in the simulation domain, including 75000 pore sites. It is assumed that the porosity is 30% in the initialization. The second-phase content is 5vol%. In order to perfectly simulate the real green compact, it is necessary to initialize the simulation domain. The initialized simulation microstructure of two-phase ceramic tool materials is shown in Fig.6. Fig.6 (a) is the part of simulation domain at the boundary, and Fig.6 (b) is the part of simulation domain at the inside. It is seen from Fig.6 (a) that the pore sites is in the inside of simulation domain. However, the second-phase sites are the boundary or in the inside of simulation domain. It is similar to the initialized single-phase microstructure where neighbor pore sites coalesce together. The second phase is randomly distributed in the composite and some of them agglomerate. Pore is surrounded by the composite and second-phase particle. It is accordance with the distribution of particle and pore in the green compact.

The simulated microstructure after 100 MCS is shown in Fig.7, where white region denotes the composite, black region denotes the pore and grey region denotes the second-phase particle. Fig.7 (a) is the boundary part of simulation domain. Fig.7 (b) is the inside part of simulation domain. Compared Fig.7 (a) with Fig.6 (a), it is seen that many pore sites jump from the inside of simulation domain to the outside and coalesce with the boundary pores. This implies the isolated pore is annihilated and exhaled from the inside. The density of simulation domain increases with the decrement of pores. It is similar to the simulation results of the single-phase microstructure. Grain size of simulation domain gradually grows. However, by comparison with Fig.5, it is seen from Fig.7 that the rate of grain growth is

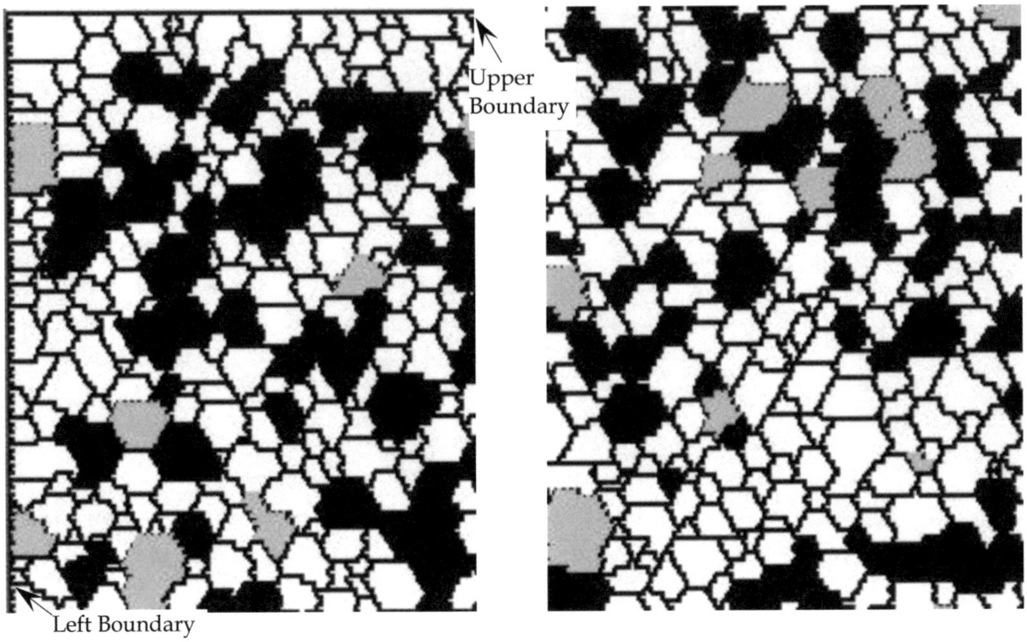

(a) (b)

(White region denotes composite phase, black region denotes pore and grey region denotes second phase)

Fig. 6. Two-phase Microstructure of Initialized Simulation Domain

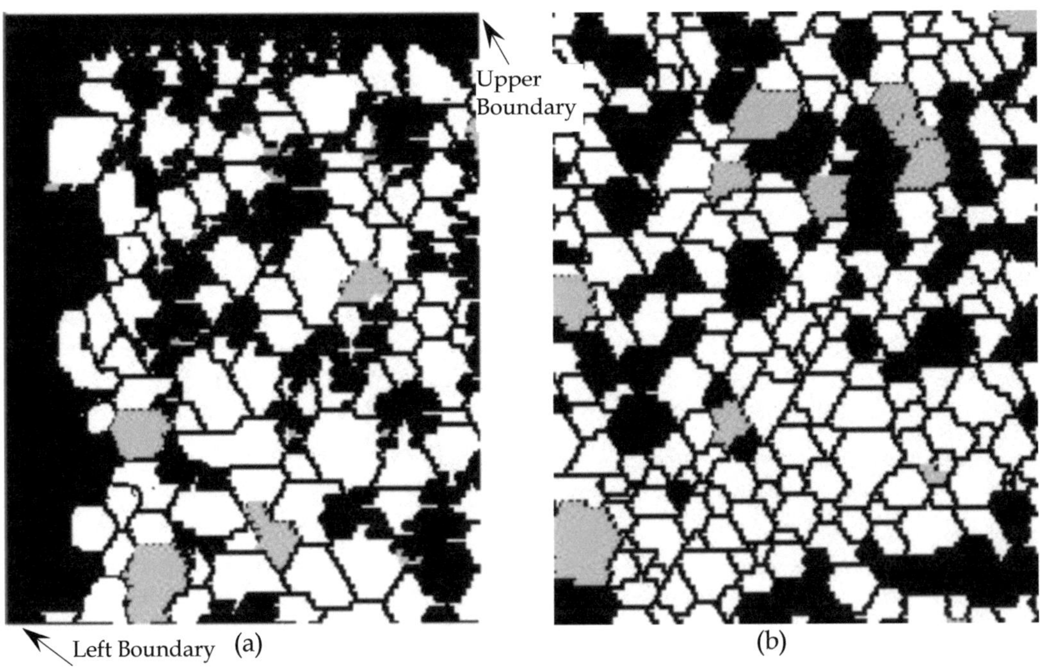

(a) (b)

(White region denotes composite phase, black region denotes pore and grey region denotes second phase)

Fig. 7. Two-phase Microstructure of Simulation Domain

slower than that of single-phase grain because the pores continuously migrate and some second-phase particles play a role in pinning effect. The inhibition of pore for composite grain growth is weaker than that of second-phase inhibition for composite grain growth, so that the size of grain around pore is commonly bigger than that of grain around second-phase particle. Just like the simulation results of single-phase system, the pore diffusion along the grain boundary is also found in Fig.7.

The pores in the simulated microstructure gradually disperse and become smaller, and the mass center migrates towards the pores because of pore diffusion along the grain boundary and pore annihilation in both single-phase and two-phase simulation results. And most pores exist at the triple junction point of grain boundary.

4.3 Densification and grain growth

4.3.1 Densification

The simulated microstructure and SEM fracture surface of Al_2O_3 ceramic tool materials are shown in Fig.8. It can be seen from Fig.8 (a) and (b) that some vacancy are surrounded by grains and become voids (or air cavities). As shown in Fig.8 (c), some voids is also found in the Al_2O_3 ceramic tool materials. Voids not only lead to the lower densification, but also lead to the lower mechanical properties because of stress concentration. Some pores are separated from the grain boundary in Fig.8 (a) and (b). The vacancies at the grain boundary, which separate from the grain boundary and become the part of grain, form voids in the process of simulation or sintering, companied by grain growth.

Comparing Fig.8 (a) and (b), it is seen that there are more pores separated from grain boundary in Fig.8 (a) because the grains of single-phase material grow faster than that of two-phase material. Therefore, the perfect content of second phase is useful for the grain refinement as well as elimination or decrement of vacancies, which leads to the improvement of mechanical properties.

The densification of microstructure is calculated by Eq.(15) at simulation time t.

$$\rho = \frac{A_t}{A_0} = \frac{N_g}{N_g + N_p(t)} \tag{15}$$

where, ρ is the densification, A_t is the area of the simulation region at simulation time t, A_0 is the area of the simulation region at the beginning, N_g is solid-phase site number, $N_p(t)$ is the pore-phase site number which is not annihilated at simulation time t.

The relationship between the densification and the simulation time (MCS) is shown in Fig.9. The change trends of the densification curve for single-phase and two-phase material are similar. At some stage of simulation, the densification of two-phase material is slightly lower than that of single-phase material. This is the reason that the growth rate of two-phase grain is slower than that of single-phase grain. At the beginning of simulation, the rate of densification is slower because most pores don't migrate along the grain boundary to form vacancy. The densification is up to 90%, and the ration of pores drastically decreases. Then the rate of densification is gradually slow.

Fig. 10 is the calculated densification curves at final stage sintering of an alumina powder compact (Kang & Jung, 2004) . Comparing Fig.9 and Fig.10, it is found that the densification curves of simulation are very similar to those of calculated results. This means that the simulation result is correct and accurate.

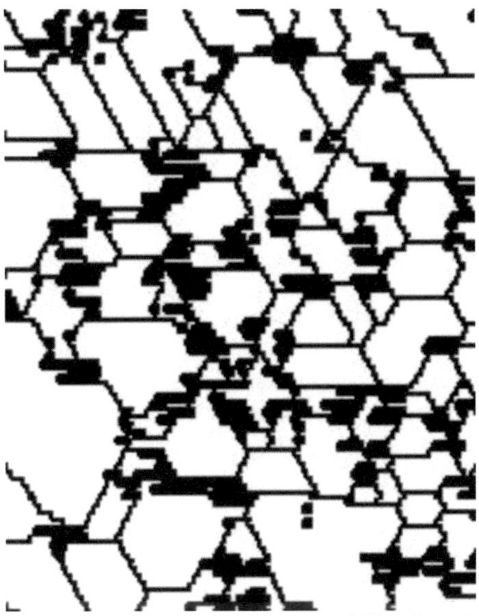

(a) Simulated Microstructure of Single-phase

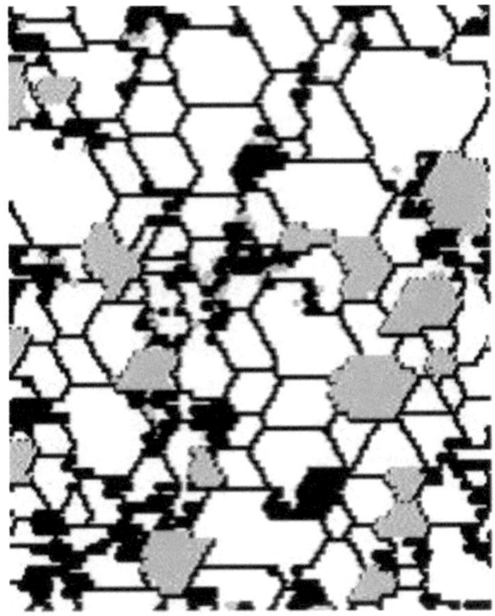

(b) Simulated Microstructure of Two-phase

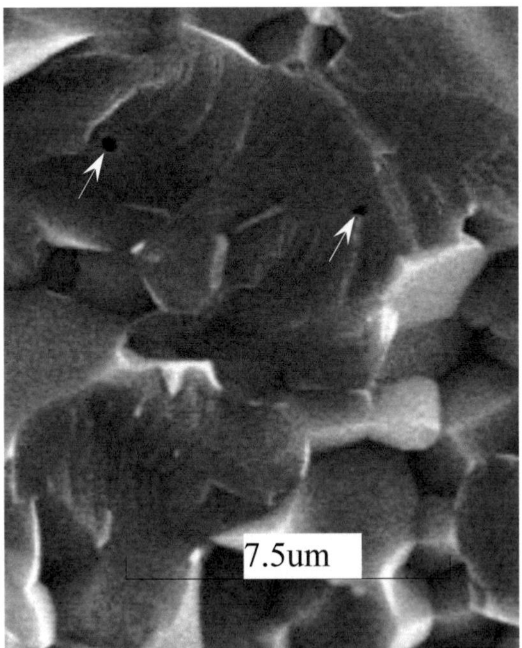

(c) SEM Fracture Surface of Al_2O_3 Ceramic Tool Materials

Fig. 8. Simulation Microstructure at 300MCS and Fracture Surface of Al_2O_3

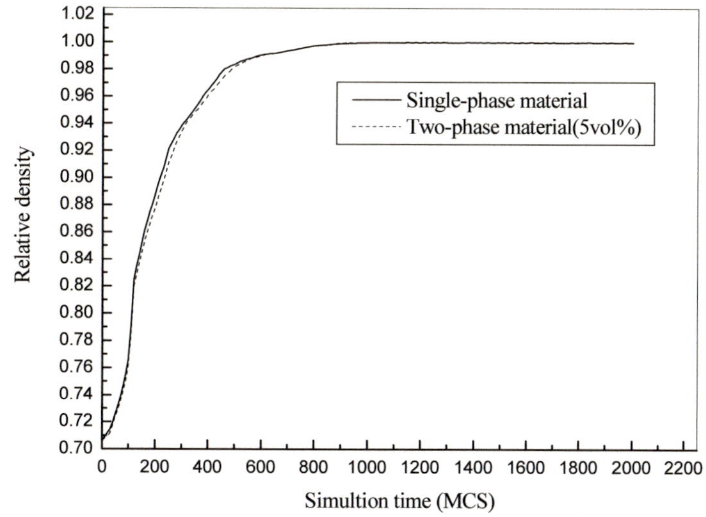

Fig. 9. Relative Density vs. Simulation Time

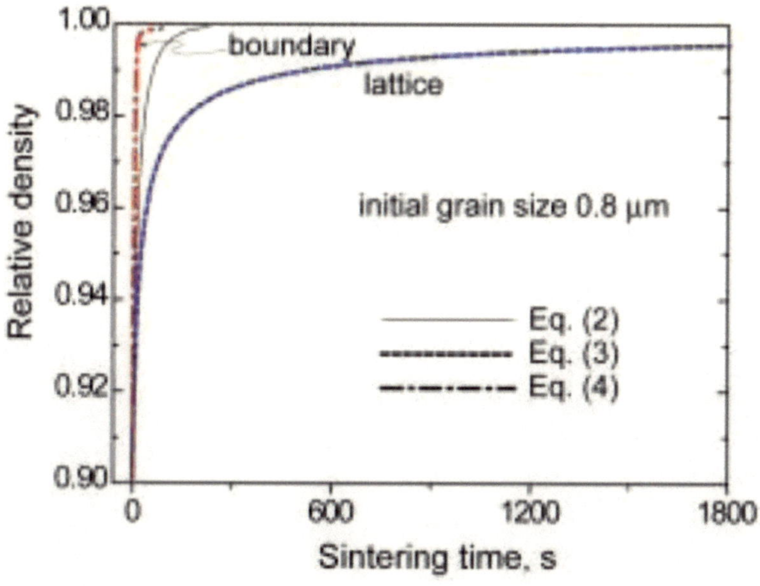

Fig. 10. Calculated Relative Density vs. Sintering Time of an Alumina Powder Compact (Kang & Jung, 2004)

4.3.2 Grain growth

Fig.11 is the relationship between average grain size $\overline{R}$ and simulation time t (MCS). The average grain size of single- and two-phase material grows slowly from 0 to 400 MCS. The average grain size of single-and two-phase material grows fast when the simulation time is more than 400 MCS. However, the average grain size of single-phase material is bigger than that of two-phase material during the whole process of simulation. The reason can be found from Fig. 9. The densification is lower, that is to say, the content of pore is more, when the simulation time is less 400 MCS. The strong pinned effect of pores on the grain growth leads to a low rate of grain growth. However, the pinned effect of pores on the grain growth which is gradually weak with a decrement of pores leads to a high rate of grain growth, especially after 400 MCS. The rate of grain growth is slow in the whole process because the grain growth is inhibited by not only pores but also second-phase particles and the inhibition of second-phase particles become stronger with the grain growth of composite.

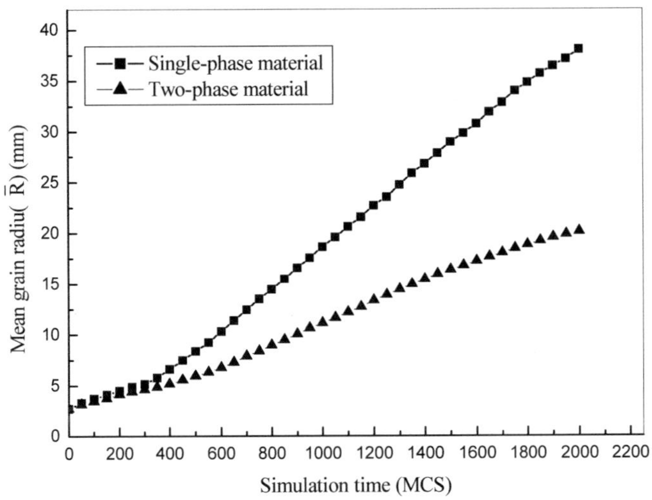

Fig. 11. Average Grain Size ($\bar{R}$) vs. Simulation Time (MCS)

The simulation results have been validated by the hot-press fabrication of Al_2O_3 ceramic tool materials. The fracture surface of the pure Al_2O_3 material and the Al_2O_3 /SiC composite material is shown in Fig. 12 (a) and (b), which are fabricated by the hot-press at the same

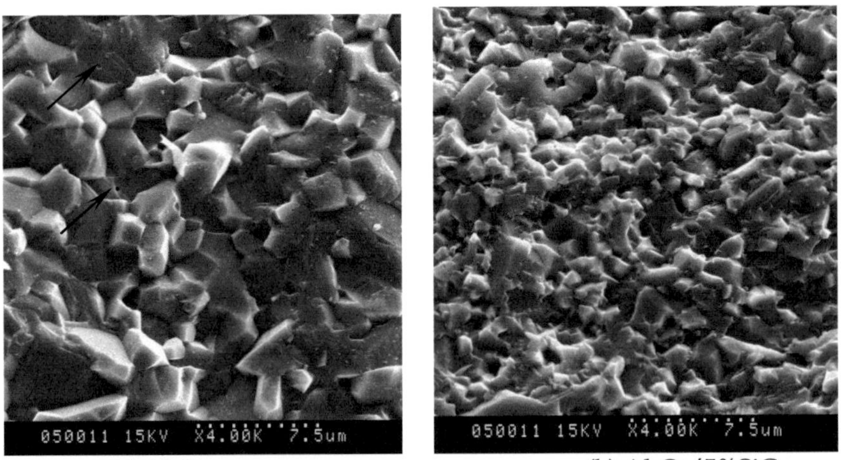

(a) Al_2O_3 (b) Al_2O_3/5%SiC

Fig. 12. SEM Fracture Surface of Al_2O_3 Ceramic Tool Materials fabrication technologies. By the comparison of Fig. 12 (a) and (b), it can be found that the grain size of Fig. 12 (b) is significantly smaller than that of Fig. 12 (a) because the growth of matrix, Al_2O_3, are inhibited by the second phase, SiC. As the arrows are shown in Fig.12 (a), there are voids inside the grains because the grains grow fast and vacancy is separated from the grain boundary, which leads to vacancy trapped by the grain.

According to the theory of sintering, at the final sintering stage the grains grow fast. It is considered that the simulation of microstructure evolution is the final sintering stage of ceramic tool material because the rate of grain growth becomes high and the content of pores drastically decrease after 400MCS.

5. Conclusions

A computer simulation for the sintering processes of single- and two-phase ceramics has been performed using a two-dimensional hexagon lattice model mapped from the realistic microstructure. Different types of boundary energy, the initial grain size, the composition content, the initial pore size and the content of pore are considered in the new simulation model. Because of the different boundary energy, the phase-two and the pore can pin the composite of phase-one and decrease the growth rate of phase-one during the microstructural evolution, which has been invalidated experimentally. The second-phase particle is useful for the refinement of composite grain, which can significantly improve the mechanical properties of ceramic tool material, especially the flexural strength and fracture toughness. The high rate of grain growth is disadvantage of the exhalation of pores according to the result of simulation and experiment. A preliminary investigation shows that the simulation results can accurately prove the relevant sintering kinetics theories and is consistent with the experiment results.

6. Acknowledgement

This project is supported by National Outstanding Young Scholar Science Foundation of NSFC (50625517), National Science Foundation of China (51075248) and Outstanding Young Scholar Science Foundation of Shandong Province (JQ201014).

7. References

Anderson, M. P.; Srolovitz, D. J.; Grest, G.S. (1984). Computer simulation of grain growth-I Kinetcs. Acta Metall., Vol. 32, No. 5, pp. 783-791

Belmonte, M.; Nieto, M. I.; Osendi, M. I.; Miranzo, P. (2006). Influence of the SiC grain size on the wear behavior of Al_2O_3/SiC composites. J. of the European Ceramic Society, Vol. 26, pp. 1273-1279

Bordère, S. (2002). Original Monte Carlo methodology devoted to the study of sintering processes. J. Am. Ceram. Soc., Vol. 85, No. 7, pp. 1845-1852

Braginsky, M.; Tikare, V.; Olevsky, E. (2005). Numerical simulation of solid state sintering. International Journal of Solids and Structures, Vol. 42, pp. 621–636

Chen, I.W.; Hassold, G. N.; Srolovitz, D. J. (1990). Computer simulation of final-stage sintering: II, influence of initial pore size. J. Am. Ceram. Soc., Vol. 73, No. 10, pp. 2865-2872

DeHoff, R. T. (1989). Stereological Theory of Sintering. In: Uskokovic, D.P. et al. (Eds.), Science of Sintering. Plenum Press, New York, pp. 55-71

German, R. M.; Olevsky, E. A. (1998). Modeling grain growth dependence on the liquid content in liquid-phase-sintered materials. Metall. and Mater. Trans., Vol. 29A, No. 12, pp. 3057-3067

Guo, J. K. (1998). The exploration on new approach of strengthening and toughening of ceramic materials. Journal of Inorganic Materials, Vol. 13, No. 1, pp. 23-26

Guo Shiju, (1998). The theory of powder sintering. Metall. Industry Press, Beijing. pp. 30-40

Handwerker, C. A.; Dynys, J. M.; Cannon, R. M. and Coble, R. L. (1990). Dihedral angles in magnesia and alumina: Distributions from surface thermal grooves. J. Am. Ceram. Soc., Vol. 73, No. 5, pp. 1371-1377

Hassold, G. N.; Chen, I-Wei; Srolovtz, D. J. (1990). Computer Simulation of Final-stage: I, Model, Kinetics, and Microstructure. J. Am. Ceram. Soc, Vol. 73, No. 10, pp. 2857-2864

Jagota, A.; Dawson, P. R. (1990). Simulation of the viscous sintering of two particles. J. Am. Ceram. Soc., Vol. 73, No.1, pp. 173-177

Jiao, S.; Jenkins,M. L.; Davidge, R. W. (1997). Interfacial Fracture Energy-mechanical Behaviour Relationship in Al_2O_3/SiC and Al_2O_3/TiN Nanocomposites. Acta mater., Vol. 45, No. 1, pp. 149-156

Jin, Z. H.; Gao, J. Q.; Qiao G. J. (2002). Ceramic Material for Engineering[M], Xi'an: Xi'an Jiaotong University Press, P. R. China.

Kang, S. L.; Jung, Y. (2004). Sintering kinetics at final stage sintering: model calculation and map construction. Acta Mater., Vol. 52, No. 15, pp. 4573-4578

Kingery, W. D; Francois, B. (1965). Grain Growth in Porous Compacts. J Am Ceram Soc., Vol. 48, pp. 546-547

Lang, F. F. (1982). Transformation toughening part 1 size effects associated with the thermodynamics of the constrained transformation. J. Mater Sci., Vol. 17, pp. 225-234

Myers, N.; Meuller, Tim; German, R. (1989). Production of Porous Refractory Metals with Controlled Pore Size. In Particle Packing Characteristics, Metal Powder Industries Federation, Edited by Randall German, pp. 298-300

Srolovitz, D. J.; Anderson, M. P. Sahni P. S. (1984). Computer simulation of grain growth-II Grain Size Distribution, Topology, and Local Dynamics. Acta Metall., Vol. 32, No. 5, pp. 793-802

Mori, K. (2006). Finite element simulation of powder forming and sintering. Comput. Methods Appl. Mech. Engrg., Vol. 195, pp. 6737-6749

Mukhopadhyay, A.; Basu, B.; Das Bakshi, S. (2007). Pressureless sintering of ZrO_2-ZrB_2 composites: Microstructure and properties. International Journal of Refractory Metals & Hard Materials, Vol. 25, pp. 179-188

Tanaka, K.; Kohyama, M. (2003). Atomic Structure Analysis of Σ=3, 9 and 27 Boundary, and Multiple Junctions in β-SiC. JEOL, Vol. 28, No. 2, pp. 8-10

Tikare, V.; Braginsky, M,;, Olevsky, E. A. (2003). Numerical simulation of solid-state sintering I, sintering of three particles. J. Am. Ceram. Soc., Vol. 86, No. 1, pp. 49-53

Tikare; V., Holm; E. A. (1998). Simulation of grain growth and pore migration in a thermal gradient. J. Am. Ceram. Soc., Vol. 81, No 3, pp. 480-484

Wakai, F.; Shinoda,Y.; Akastu,T. (2004). Methods to calculate sintering stress of porous materials in equilibrium. Acta Materialia, Vol. 52, pp. 5621-5631

Wakai, F; Yoshida, M.; Shinoda, Y.; Akastu, T. (2005). Coarsening and grain growth in singering of two particles of different sizes. Acta Mater. Vol. 53, pp. 1361-1371

Part 5

Ceramic Membranes

9

Synthesis and Characterization of a Novel Hydrophobic Membrane: Application for Seawater Desalination with Air Gap Membrane Distillation Process

Sabeur Khemakhem and Raja Ben Amar
Université de Sfax, Laboratoire des Sciences de Matériaux et Environnement, Faculté desSciences de Sfax, Sfax, Tunisie

1. Introduction

Ceramic membranes are usually prepared from metal oxides like alumina, zirconia, titania. The development of membrane processes is generally limited because the price of the commercial membranes is too expensive, which is particularly true for the inorganic membranes. One of the challenges for future development of the inorganic membranes consists to prepare membranes of low cost made with natural non-expensive material (Khemakhem, & al., 2009; Saffaj & al., 2006). Rapid development and innovation have already been realized in this area (Cot, L.; 1998). Clay minerals are a well-known class of natural inorganic materials, with well-known structural adsorption, rheological and thermal properties (Brigatti & al., 2006; Wang & al., 2007). Research on clay as a membrane material has concentrated mainly on pillared clays (Tomul & al., 2009). Studies of membranes prepared entirely from natural clays have just started (Mao & al., 1999). These materials originally have a hydrophilic character due to the presence of the surface hydroxyl (-OH) groups, which can link very easily water molecules (Larbot& al., 2004; Picard & al., 2004).

Modification of membrane material and/or membrane surfaces has considerable influence on separation characteristics. Modification of the membrane materials surface is done to increase its hydrophilicity. The technique of modification normally involves the use of hydrophilic polymers or copolymers (Zhao & al., 2010; Alias & al., 2008), blending of hydrophilic or charged polymers with hydrophobic polymers (Mohd Norddin & al., 2008; Bolong & al., 2009), grafting of polymers (Nishizawa & al., 2005; Zdyrko & al., 2006) and the surface modification of the membrane itself (Hu & al., 2008; Yu & al., 2008). Grafting process, leading to the increase of the hydrophobic properties, can be performed by reaction between -OH surface groups of the membrane and ethoxy groups (O-Et) presented in organosilane compounds (Krajewski & al., 2004 Faibish & Cohen 2001). The grafting process leads to a monomolecular layer of organosilane compound on the membrane surface (Yoshida & Cohen 2003; Schondelmaier & al., 2002).

This chapter describes the development of Novel mesoporous hydrophobic membrane obtained by grafting of fluoroalkylsilane onto microfiltration layer surface based on Tunisian clay. Ceramic membranes can be described by an asymmetric porous material formed by a macroporous support with successive thin layers deposited on it. Paste from Tunisian silty marls referred M11, is extruded to elaborate a porous tubular configuration used as supports. The support heated at 1190°C, shows an average pore diameter and porosity of about 9.20µm and 49%, respectively. The elaboration of the microfiltration layer based on Tunisian clay referred JM18, is performed by slip-casting method. The heating treatment at 900°C leads to an average pore size of 0.18µm. The obtained membrane was surface modified to change their hydrophilic character into hydrophobic one by the grafting of the triethoxy-1H,1H,2H,2H-perflluorodecylsilane $C_8F_{17}(CH_2)_2Si(OC_2H_5)_3$. The following grafting parameters were studied: the concentration of fluoroalkylsilane (C8) compound in the grafting solution and the time of grafting. The resulting products were investigated using Fourier transform infrared spectroscopy (FTIR), thermogravimetric analysis (TGA) and ^{29}Si CP/MAS NMR spectra. The new surfaces were examined by water contact-angle measurements and the water flux through the grafted membrane was also measured. The new hydrophobic membrane seems to be promising in the field of membrane distillation. High salt rejection rates higher than 99% were obtained for modified MF ceramic clay membrane. Air gap membrane distillation (AGMD) is a low cost process very efficient to produce fresh water from seawater.

2. Membrane preparation

2.1 Support shaping and characterisation

For this study, the supports were prepared from Tunisian silty marls (M11). The chemical composition of these materials is shown in Table 1.

Elements (%)	M_{11}
SiO_2	31.61
Al_2O_3	10.38
Fe_2O_3	6.53
CaO	24.17
MgO	19.92
Na_2O	2.21
K_2O	3.43
TiO_2	1.55

Table 1. Composition of silty marls (M11).

The chemical analysis reveals that this kind of silty marls is essentially formed with a great amount of silica and calcium oxide. Fig. 1 presents the XRD patterns of raw silty marls, it shows that quartz (Q) is the main crystalline mineral present in this powder.

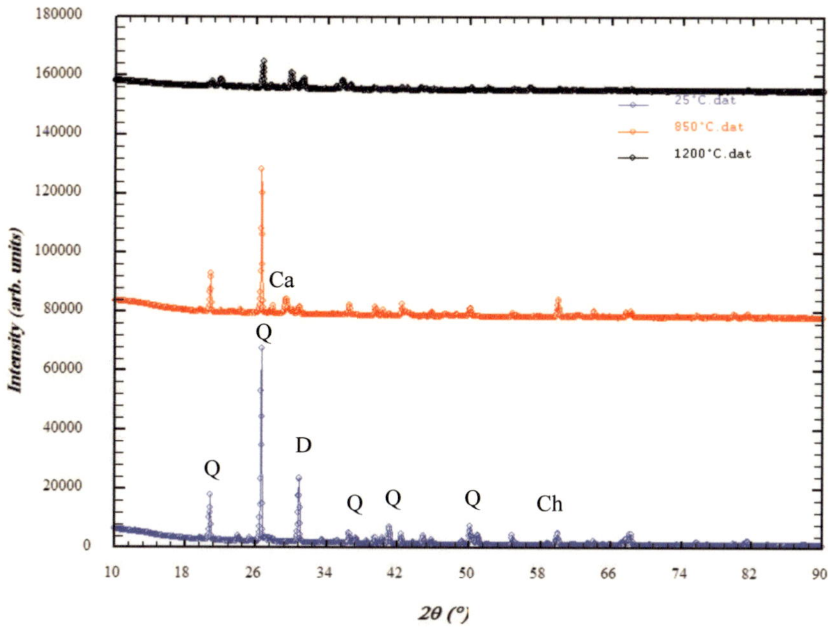

Fig. 1. X-ray diffractograms of the silty marls sample at different temperatures (Q = quartz, Ca = calcite, Ch = chlorite, D = dolomite).

The particle size analysis of the powder after crushing for 2 h with the assistance of a planetary crusher at a rate of 250 revolutions/min and calibrated with 100 mm was determined using a Particle Sizing Systems (Inc. Santa Barbara, California, USA Model 770 AccuSizer). The particle diameters range varied from 0.5 to 54 mm (Fig. 2).

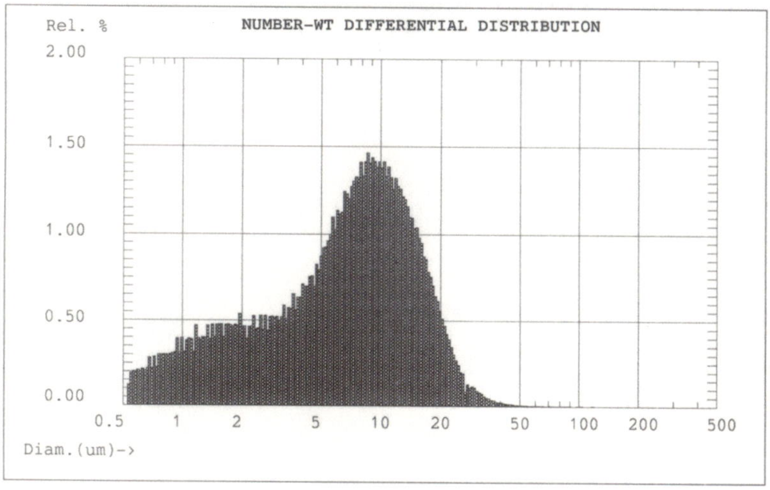

Fig. 2. Silty marls (M11) particle size distribution.

Plastic pastes are prepared from ceramic powder of silty marls mixed with organic additives:

- 4% (w/w) of Amijel: pregelated starch, as plasticizer (Cplus 12072, cerestar).
- 4% (w/w) of Methocel: cellulose derivative, as binder (The Dow Chemical Company).
- 8% (w/w) of starch of corn as porosity agent (RG 03408, Cerestar).
- 25% (w/w) of water.

The rheological properties must be studied to obtain a paste allowing shaping by extrusion process [18]. Fig. 3 shows the different configurations of tubes extruded in our laboratory (two monochannel of different diameter and one multichannel tube).

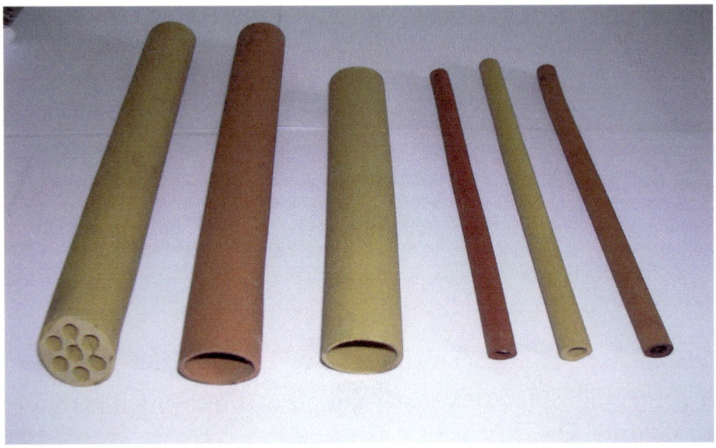

Fig. 3. A photograph of variety of configurations of porous ceramic supports.

Thermogravimetric analysis (TGA) and differential scanning calorimetry (DSC) were performed with simultaneous DSC-TGA 2960 TA instrument. The sample was heated at room temperature to 1250 8C at a rate of 5 8C/min under static atmospheric conditions. Two endothermic peaks were detected (Fig. 4).

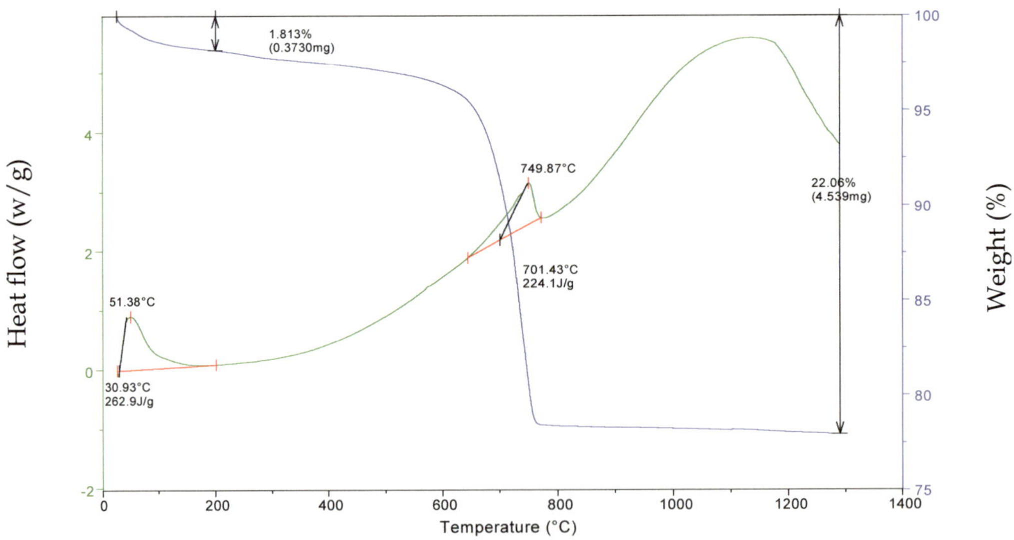

Fig. 4. Thermal analysis curve: DSC and TGA for silty marls powder.

First peak appears at 51.38 8C, due to a weight loss of 1.81% of the initial weight. It corresponds to the departure of water (moisture or adsorption) due to attraction on the surface of the sample and zeolitic water inserted between the layers or in the cavities of the crystalline structure. Second peak which maximum appears toward 749.87 8C corresponds to the dehydroxylation.

Sintering experiments of the support were carried out in air. Two steps has been realised: the first for the elimination of organic additives at 250 °C, and the second for the sintering at 1190 °C. The temperature–time schedule not only affects the pore diameters and porous volume of the final product but also determines the final morphology and mechanical strength. By controlling the sintering temperature of the ceramic, it is possible to increase the pore size and to obtain a higher mechanical strength. We have also observed that the obtained silty marls support presents the highest mean pore diameter for the highest mechanical strength: the support fired at 1190 °C and characterised by mercury porosimetry showed mean pore diameters and porosity of about 9.2 mm and 49%, respectively (Fig. 5).

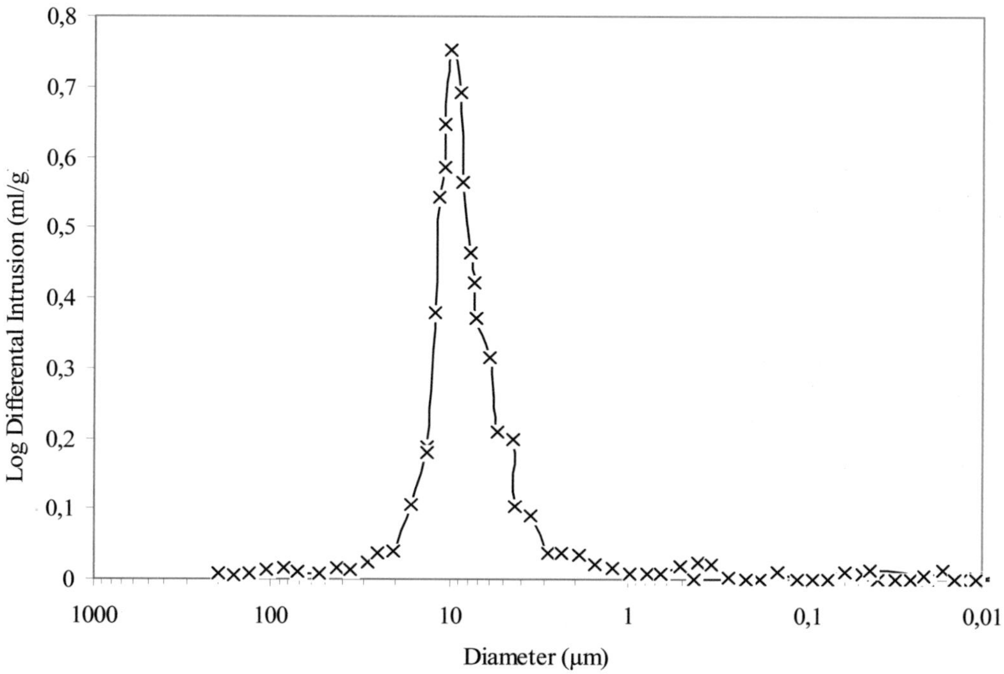

Fig. 5. Pore diameters of the silty marls supports.

It can also be observed that the porosity and the pore size parameters are strongly dependent on the sintering temperature and particle size of the powders (Table 2).

2.2 Microfiltration layer shaping and characterization

The material used for the membrane preparation is a Tunisian clay powder (JM18) taken from the area of Sidi Bouzid(Central Tunisia). This powder is crushed for 4 h with a planetary crusher at 250 revolutions/min and calibrated with 50 mm. The obtained particle diameters range from about 0.5 to 23 mm (Fig. 6).

Powders	Temperature (°C)	Pore size (μm)	Porous Volume (%)
Crushed during one hour and calibrated with 125 μm	1160	9.6	58
	1170	10.9	56
	1180	12.8	53
	1190	14.3	52
	1200	16.5	51
Crushed during two hours and calibrated with 100 μm	1160	5.9	52
	1170	7.3	50
	1180	8.5	49
	1190	9.2	49
	1200	12.5	46

Table 2.Variation of pore size and porous volume according to the powder particle sizes for the silty marls (M11).

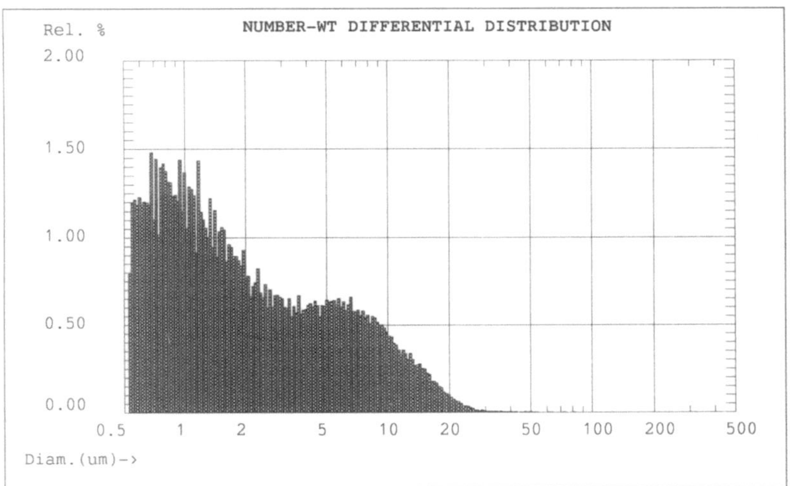

Fig. 6. Clay (JM18) particle size distribution.

The chemical composition of the clay (JM18) is shown in Table 3.

Elements (%)	JM18
SiO_2	62.64
Al_2O_3	17.09
FeO_3	8.5
MgO	0.07
Na_2O	0.32
K_2O	4.8
Mn_2O_3	0.02
SO_3	0.4
Loss on the ignition	6.16

Table 3.Composition of clay (JM18).

It reveals that this material is essentially formed with a large amount of silica 62.64%. For preparing a microfiltration layer with JM18, the suspended powder technique was used. A deflocculated slip was obtained by mixing 5% (w/w) of JM18, 30% (w/w) of polyvinyl alcohol (PVA) (12% w/w aqueous solution) as binder and water (65% w/w). The thickness of microfiltration layer can be controlled by the percentage of the clay powder added to the suspension and the deposition time. The viscosity of the slip elaborated according to the protocol described previously has been studied right before deposition. The used viscosimeter (LAMY, TVe-05) permits to use 5 speeds of rotation for the determination of the dynamic viscosity of the substance to characterize. Fig. 7 shows the rheogram of the slip used. It is done by the curve of shear stress (τ) versus speed of rotation (D). The slip has a plastic behavior of Bingham, controlled by the presence of PVA; the value of the limiting shear stress is 4 mPa. Such behavior permits the maintenance of particles in a stable suspension. The deposition of the slip on the M11 support was performed by slip-casting using a deposition time between 10 and 15 min.

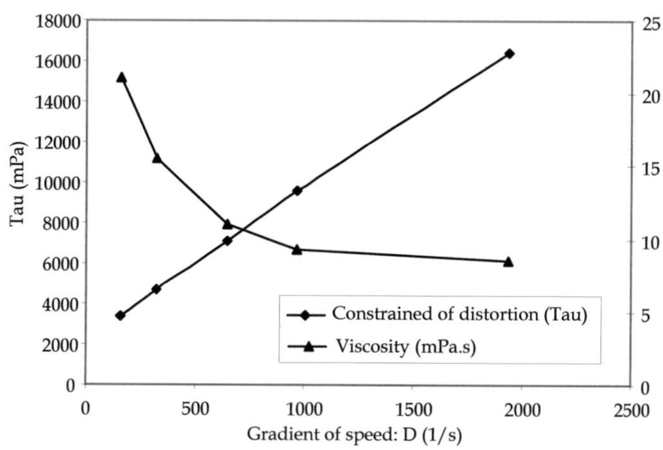

Fig. 7. Evolution of the stress (τ) and the viscosity (μ) vs. deformation of clay (JM18) slip.

After drying at room temperature for 24 h, the clay membrane was sintered at 900 °C for 2h, after debonding at 250 °C for 1 h. Total porous volume and pore size distribution are measured by mercury porosimetry. This technique relies on the penetration of mercury into a membrane's pores under pressure. The intrusion volume is recorded as a function of the applied pressure and then the pore size was determined. The pore diameters measured were centered near 0.18 mm (Fig. 8).

The pore size in the microfiltration layer can also be varied using powders with different particle size distributions. Different microfiltration membranes with different layers thickness (between 5 and 50 mm) were prepared. SEM (scanning electron microscopy) images of the resulting membranes are shown in Fig. 9.

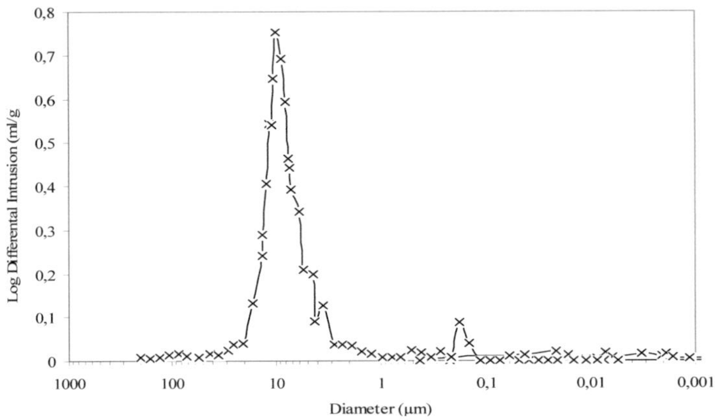

Fig. 8. Pore diameters of the clay (JM18) membrane.

Fig. 9. Scanning electron micrographs of clay (JM18) membrane (A: membrane thickness upper than 10 mm, B: membrane thickness less than 10 mm, 1: surface, 2: cross-section).

This figure gives information on the texture of the elaborated membrane surface. A defect free membrane was only obtained for membrane thickness less than 10 mm (in order to 7 mm). Crossflow microfiltration tests were performed using a home-made pilot plant (Fig. 10) at a temperature of 25 °C and transmembrane pressure (TMP) range between 1 and 4 bar.

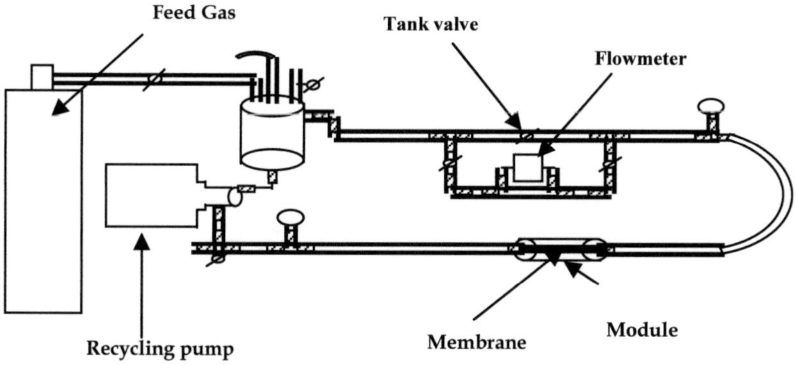

Fig. 10. Flow schema of experimental apparatus.

The flow rate was fixed at 2.5 m s^{-1}. Before the tests, the membrane was conditioned by immersion in pure deionised water for a minimum of 24 h. The working pressure was obtained using a nitrogen gas source. The membrane was initially characterized by the determination of water permeability which was 870 l $h^{-1}m^{-1}bar^{-1}$ (Fig. 11).

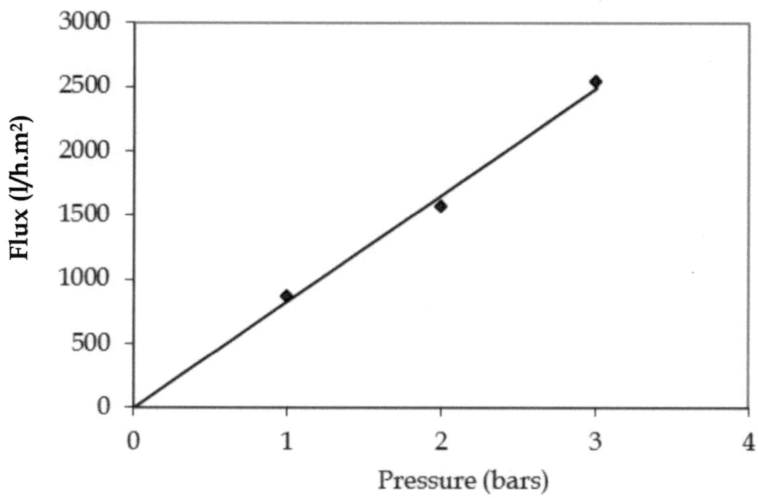

Fig. 11. Water fluxes vs. working pressure.

3. Modification of elaborated microfiltration membrane surface

3.1 Experimental

3.1.1 Materials

Tunisian clay membrane was prepared in our laboratory from the support to the finest layer and was grafted as described earlier (Khemakhem, & al., 2006; Khemakhem, & al., 2007). The triethoxy-1H,1H,2H,2H perfluorodecylsilane C8F17(CH2)2Si(OC2H5)3 (97%) from Sigma was used. Analytical grade ethanol (99%) was purchased from Riedel-de-Haën.

3.1.2 Grafting process

Grafting of membranes was performed by the use of fluorinated silanes: triethoxy-1H,1H,2H,2H-perflluorodecylsilane. Grafting it occurs with a succession of condensation reactions between the OH groups found in the surface membrane and the Si–O–alkyl groups of the silane. To realize grafting, solutions of the triethoxy-1H,1H,2H,2H-perflluorodecylsilane were prepared in ethanol at a concentration of 10^{-2} mol l^{-1}. Prior to the chemical modification, the membranes were cleaned in an ultrasonic bath in the presence of ethanol and acetone successively for 5 min and dried in an oven at 100°C. Samples, planar membranes as well as tubular membranes, were completely immersed in prepared solution for different times (15min, 30min and 60min) at room temperature. The grafted membrane was then rinsed in ethanol and acetone successively and placed in an oven at 100°C for 1 h.

3.1.3 Characterization

The FTIR spectra were obtained from KBr pellet using a Perkin-Elmer BXII spectrophotometer under transmission mode. The spectra were collected for each measurement in the spectral range 400–4000 cm^{-1} with a resolution of 4 cm^{-1}. Thermogravimetric analysis (TGA) was performed on a TGA2950 thermobalance. Samples were heated from 30 to 900°C at a heating rate of 10°C/min. The measurements of contact angles were performed at room temperature (20°C) using a OCA 15 from Dataphysics, equipped with a CCD camera, with a resolution of 752-582 square pixels, working at an acquisition rate of 4 images per second. Collected data were processed using OCA software. Distilled water was used for measurements and planar membranes realized with clay were grafted. The drop image was recorded by video camera and digitalized. Each contact angle is the average value of 20 measurements. Grafted and ungrafetd surfaces were also characterized by Scanning Electron Microscopy (Hitachi S-4500). ^{29}Si cross-polarization magic-angle-spinning nuclearmagnetic-resonance (CP/MAS NMR) spectra were gained on a Bruker DSX-300 spectrometer operating at 59.63 MHz. Water permeability was measured on grafted ceramic membranes by crossflow filtration experiments. Measurements were performed using home-made pilot plant.

3.2 Results and discussion

3.2.1 Contact angle measurements

The hydrophobic character of the resulting material was tested by measuring the contact angle of water drop. This method gives us information and a determination of the hydrophobicity of the grafted samples. Results obtained on the planar membrane are reported in Fig.12. The low contact angle of the ungrafted membrane, results from the high hydrophilic character of the membrane surface as a consequence of the high density of the hydroxyl group on the surface of the membrane. The studied grafting times were (15, 30 and 60min). We noticed that the value of contact angle increases weakly with the increase of grafting time.

A high efficiency of molecule triethoxy-1H,1H,2H,2H- perflluorodecylsilane and a high hydrophobicity of grafted surface is shown in Fig. 12. Values of the contact angle were in the

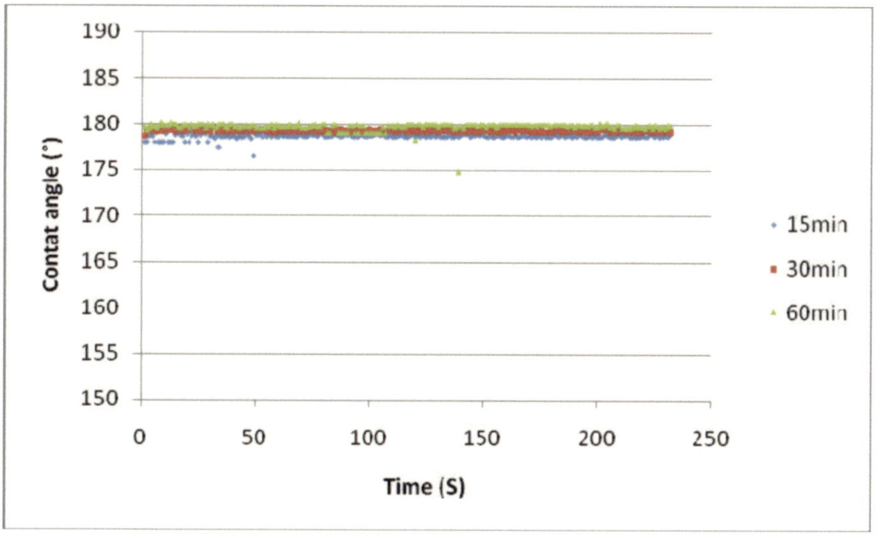

Fig. 12. Evolution of contact angle versus time of microfiltration membrane at different grafting times.

range of 177°-179° for all prepared membranes, what means that the tested membrane possessed the hydrophobic character. The high hydrophobicity of the grafted ceramic surface is illustrated in Fig.13 . We obtain non-wetting materials as we observed that there is no capillary suction of a drop of water deposited on the grafted planar membrane.

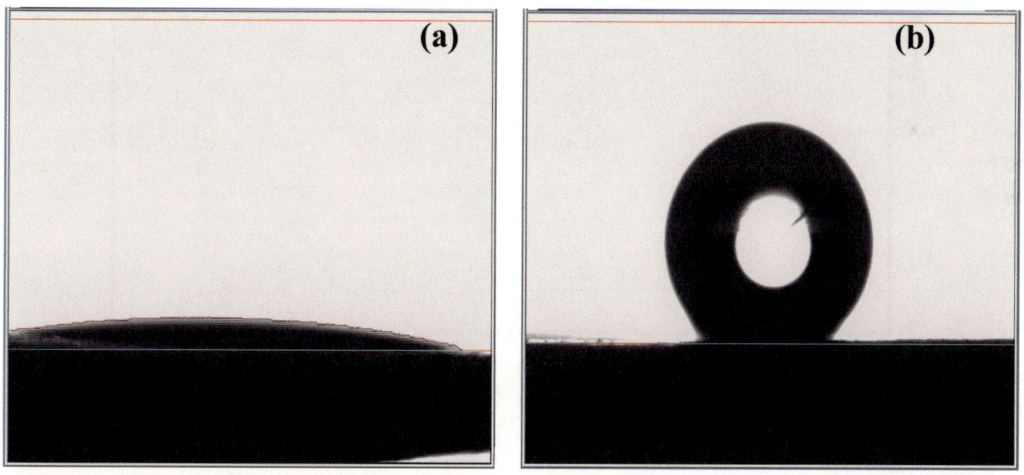

Fig. 13. Top view of a droplet of water, deposited on a membrane surface, (a) ungrafted membrane, (b) grafted microfiltration membrane.

3.2.2 Infrared absorption spectra

Infrared spectra on each grafted or ungrafted surface membrane were measured. The infrared spectra of the ungrafted clay membrane are shown in Fig. 14a. It could be confirmed that there were hydroxyl groups in the ungrafted surface clay membrane from the band at 3100–3500 cm^{-1}. The band at 3425 cm^{-1} corresponds to the -OH stretching vibration of the adsorbed water (Frost & al., 2000). A new band at 2978 cm^{-1} attributed to the anti-symmetric stretching of the $-CH_2$ group of the triethoxy-1H,1H,2H,2H-perflluorodecylsilane was showed in Fig. 14b, indicating the presence of silane in the grafted products. The bands at 1540 cm^{-1}, 1240 cm^{-1} and 1207 cm^{-1} corresponding respectively to the stretching vibration of the C-C, C_xF_{2x+1} and $Si-CH_2CH_2C_xF_{2x+1}$ groups. The pores size diameter and the size of the used particles for the elaborated membrane have an important influence on the amount of the grafted silanes. The grafting reaction conducted in the silane favors its entering into the clay layer.

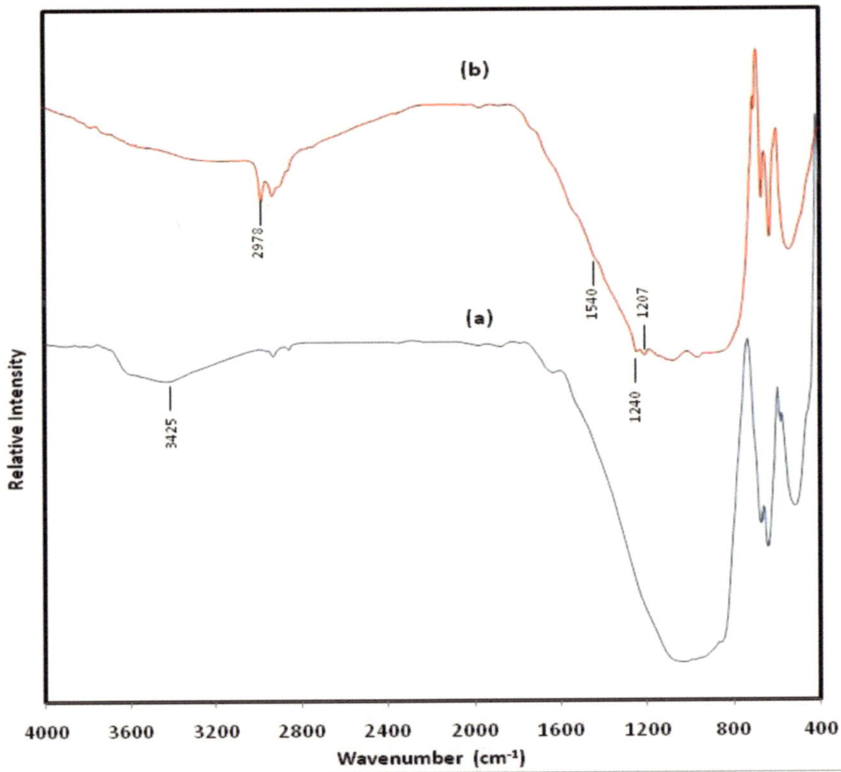

Fig. 14. Infrared spectra of (a) ungrafted membrane, (b) microfiltration grafted membrane.

3.2.3 TGA analysis

Thermogravimetric analysis (TGA) was provided as a simple method to measure the content of silane and adsorbed water. This method is based on the assumption that the dehydration and dehydroxylation reactions correspond to the two discrete mass loss steps in TG curves and they do not overlap each other. The TG curves of ungrafted membrane decreased rapidly when the decomposition temperature varied from 30 to 200°C, resulting from the loss of water molecules for the clay membrane contained (Fig. 15a). When increasing the temperature from 200 to 900°C, the curve was smooth with a mass loss of about 0.35%, suggesting that the membrane was very stable at high temperature. Three-stage decomposition procedure is shown in the TG curves of the grafting membrane (Fig. 15b). The initial mass loss took place between 30 and 200°C, which was attributed to the loss of water molecules attached to the membrane. The highest mass loss in the second stage from 300 to 650°C, results from decomposition of the grafted silane. A complete mass loss was not occurred even after heating the material up to 900°C.

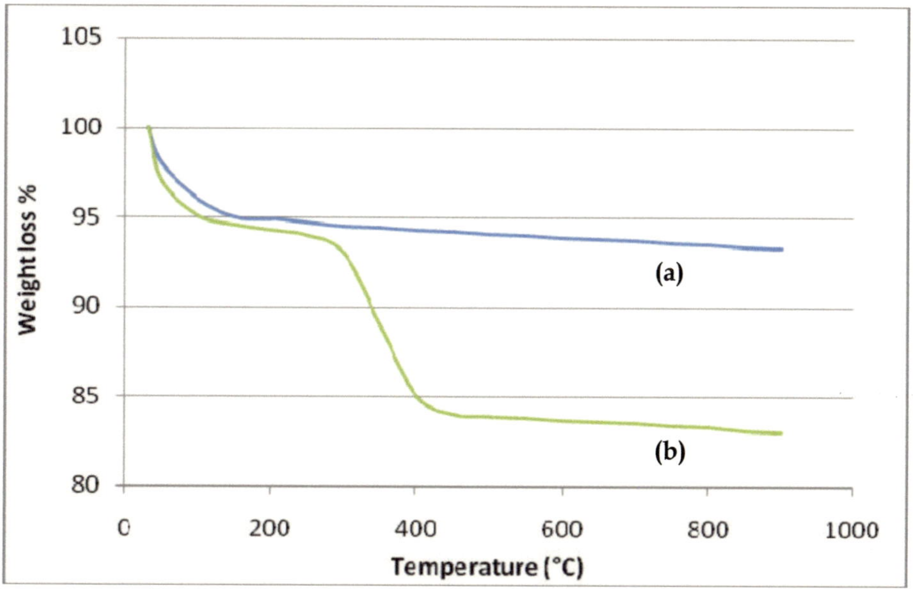

Fig. 15. TG curves of: (a) ungrafted membrane, (b) microfiltration grafted membrane.

3.2.4 CP/MAS ^{29}Si solid state NMR

The chemical shift of silicon is determined by the chemical nature of its neighbors, namely, the number of siloxane bridges attached to a silicon atom, and M, D, T, and Q structures are the commonly used notation corresponding to one, two, three, and four Si-O- bridges, respectively. According to the nomenclature recommended in the literature of M^i, D^i, T^i, and Qi (Glasser & Wilkes, 1989; Young & al., 2002; Gavarini, 2002), where i refers to the number of -O-Si groups bound to the silicon atom of interest, T^i corresponds to R-Si(-OSi)i-OR-(3−i). Four Q peaks can be present, namely Q^0 (−66 to−74 ppm), Q^1 (−78 to−83 ppm), Q^2 (−83 to −88 ppm), Q^3 (−90 to −100 ppm), and Q^4 (−107 to −110 ppm) and four T peaks can be present, namely T^0 (−37 to−39 ppm), T^1 (−46 to −48 ppm), T^2 (−53 to −57 ppm), and T^3 (−61 to −66 ppm) (Glasser & Wilkes, 1989).

Three distinct peaks observed in the ^{29}Si NMR spectra of the fluoroalkoxysilane indicating three distinct silicon chemical environments Fig. (16a). These peaks at chemical shifts of approximately −48.14 ppm, −54.52 ppm and −61.65 ppm are representatives respectively of T_1, T_2 and T_3 resonances. In Fig. (16b) we reported the ^{29}Si CP/MAS NMR spectrum of used

Tunisian clay. In this spectrum we observed tow ^{29}Si NMR peaks at −71.92ppm and −81.11ppm which correspond to the tow different Q^0 and Q^1 resonances of the different types of silicon atoms in the structure, respectively.

The ^{29}Si CP-MAS NMR spectra recorded for the modified MF membranes were given, on Fig. (16c). This spectra can be divided into two regions. The first region, from +50ppm to

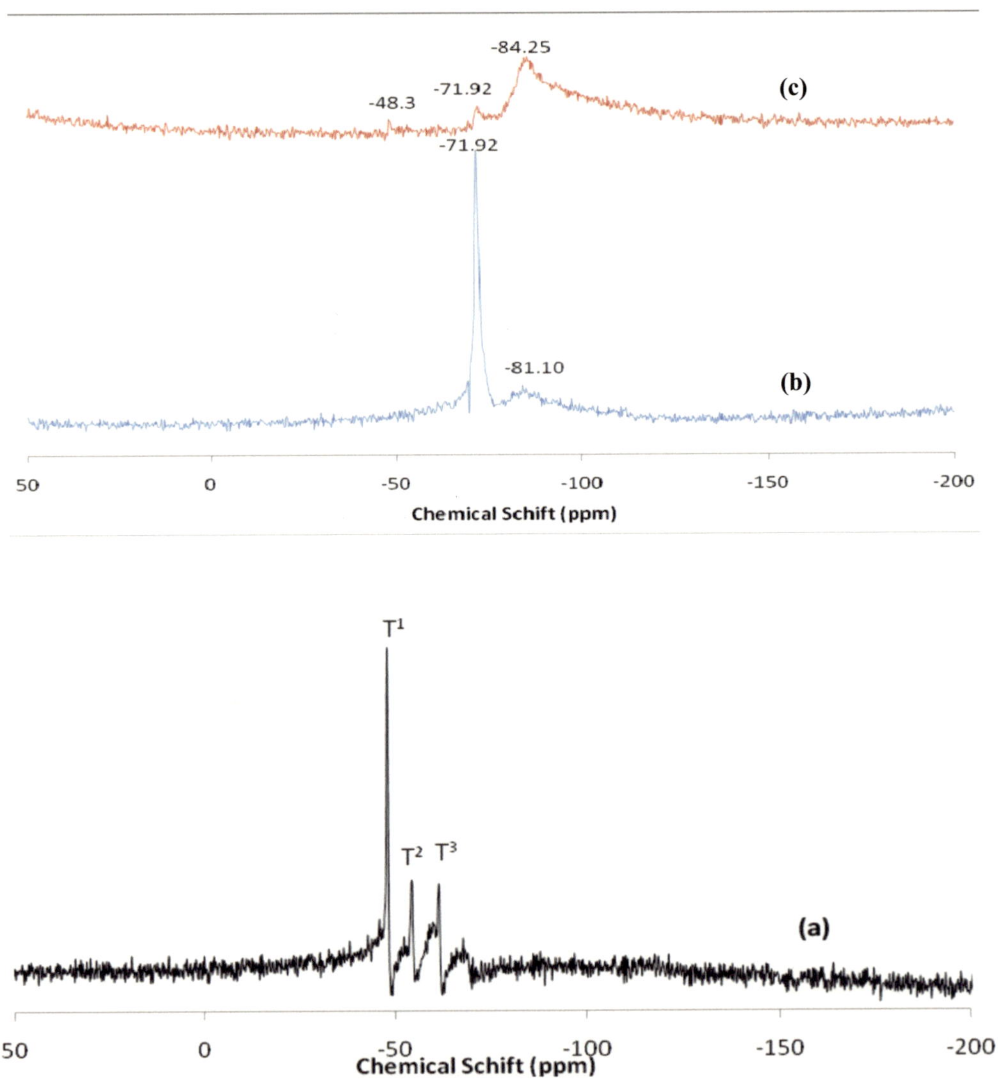

Fig. 16. ^{29}Si CPMAS NMR spectra of: (a) Fluoroalkylsilane molecule (b) ungrafted membrane, (c) microfiltration grafted membrane.

50ppm, contains the signal corresponding to the silicon atom of the fluoroalkoxysilane (T1). The second region range from −60ppm to −150ppm, corresponds to the signals of the various silicon in the clay structure. However the tow peaks observed in Fig. 16a at −71.92ppm and −81.11ppm correspond, to the Q_0 and Q1 silicons in the clay structure respectively. The presence of the T_1 peak indicates that the silane of fluoroalkylsilanes has been successfully grafted to the surface of microfiltration layer. In other words, the surface of microfiltration layer has been silylated.

3.2.5 Crossflow filtration experiments

Permeability measurements were used to test the hydrophobic character of the grafted membranes. Permeability of membranes was measured for grafted and ungrafted membranes. The ungrafted membrane exhibit a permeability of 867 l $hm^{-2}bar^{-1}$. After grafting, there is a high reduction of permeability. The permeability measured for grafted microfiltration membrane was 2.7 l h^{-1} m^{-2} bar^{-1} (Fig. 17). So, we can conclude a high effect of these molecules to decrease strongly the size of pores diameters and to reduce the permeability of the membrane.

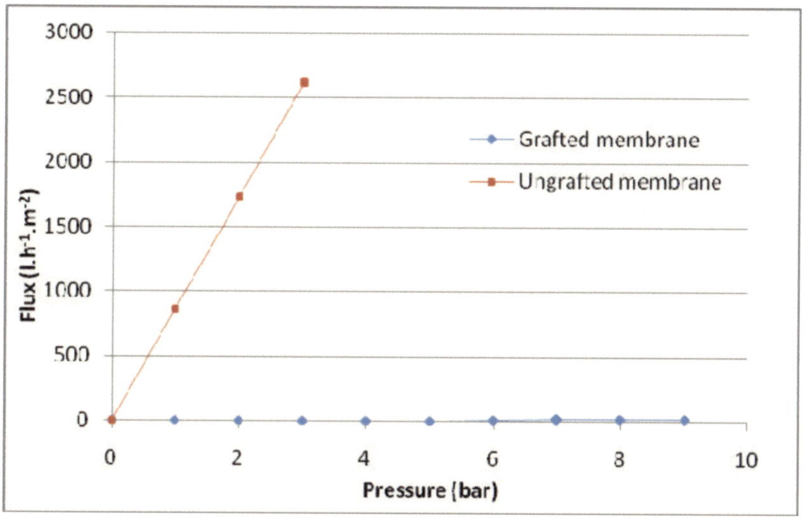

Fig. 17. Values of flux for microfiltration grafted and ungrafted membranes.

4. Application for seawater desalination

The experimental set-up presented in Fig. 18 was used for the performances of the triethoxy-1H,1H,2H,2H perfluorodecylsilane grafted membranes in the AGMD configuration process. The air-gap width was 10 mm. The used water was heated in a feed tank and it was circulating through the membrane module. During experiments, permeate vapors were condensed at a cooled stainless steel surface close to the membrane (Fig. 18). The permeate water flux through the membranes was determined by measuring permeate volume as a function of time.

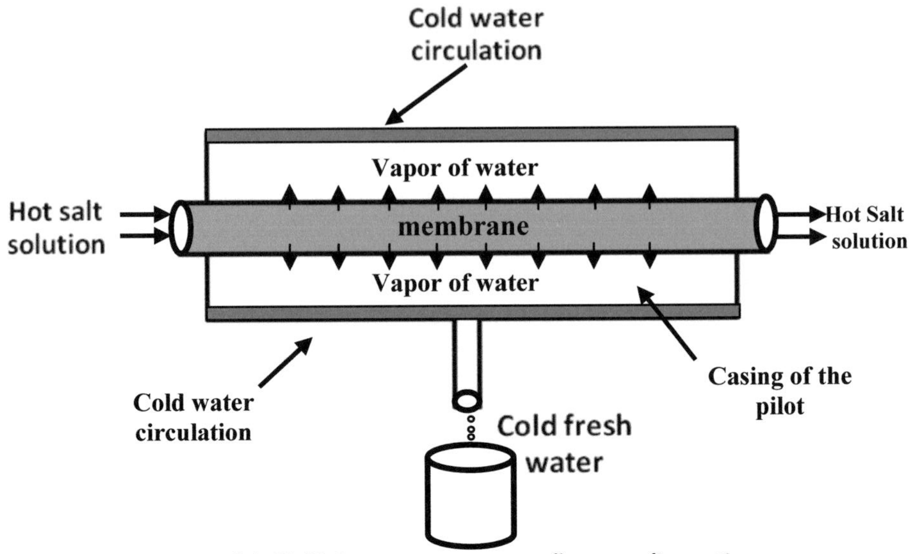

Fig. 18. Schematic presentation of AGMD in counter current flow configuration.

In each experiment, two liters of water were used and was circulated through the membrane module by using a variable flow peristaltic pump. The feed pressure was measured with a manometer at the inlet of the feed cell frame. The feed flow rate was measured with a digital flowmeter at the outlet of the feed cell frame, and it was kept constant during the experimental run by adjusting the pump speed. The feed and cooling plate temperatures were kept constant by using controlled heating and cooling thermostats, respectively.

The fine control of the feed temperature was achieved by using an auxiliary heat exchanger between the feed reservoir and the pump. The temperatures were measured with two thermometers located at the feed cell frame and at the cooling plates, respectively. The obtained distillate was collected directly in a calibrated graduated cylinder. The distillate flow through the membrane was determined from the temporal evolution of the liquid level in the cylinder. The temperature difference is defined as $T = T_2 - T_1$ where T_2 is the feed solution temperature and T_1 is the cooling water temperature. Salt concentration both in feed and permeate solutions were determined through conductivity measurement. The separation factor, R was calculated using the following expression:

$$R = \left(1 - \frac{C_P}{C_F}\right) \times 100$$

Where Cp and Cf are the NaCl concentration in the permeate and in the bulk feed solution, respectively.

Air Gap Membrane Distillation (AGMD) experiments were performed for seawater desalination using prepared hydrophobic membrane. Seawater desalination aims to obtain fresh water with free salt adequate for drinking. In our work seawater treated is collected from SIDI MANSOUR Sea, located at Sfax (Tunisia). Measurements of permeate flux and rejection rates were carried out by AGMD as a function of the temperature. The feed side temperature was thus varied from 75 °C to 95 °C, while keeping the cooling system temperature constant at 5 °C. It can be seen that the salt retention in AGMD process with grafted ceramic membranes is higher than 98% (Fig. 19).

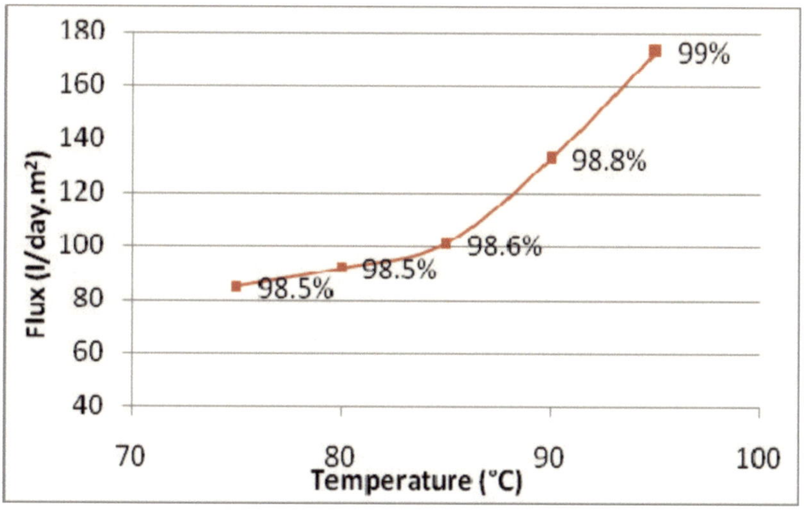

Fig. 19. Variation of the permeate flux as a function of the temperature. The values reported on the graph correspond to the rejection rates calculated for the different membranes after filtration of seawater.

These results proved that in the case of AGMD with aqueous solutions containing non-volatile compounds like NaCl, only water vapor is transported through the membrane. It can be seen that the salt retention in AGMD process with grafted ceramic membranes is close to 100%.

5. Conclusion

Surface modification of "silanized" MF Tunisian clay membrane via graft polymerization of perfluoroalkylsilane (C8) was carried out. Ceramic membranes surface grafted with perfluoroalkylsilane changed the hydrophilic character into a superhydrophobic one. The effect of silane treatment on the properties of the clay membrane surface studied was found to depend on the membrane pore structure. It was proven that the presented grafted silane constitutes an ideal method for producing improved ceramic NF membranes with a decreased pore size. The data obtained with different techniques indicated a polymer grafting on the treated ceramic grains on the surface and inside the pores to form a selective barrier whose permeation flux decreases. FTIR measurements determinate the nature of chemical bonds involved in the organic layer covering the clay surface. Contact-angle experiments provided important information on the wettability and the hydrophobic character of the modified surfaces. Modification of the clay surface with C8 led to an increase of the contact angle.

Membrane distillation is an emerging technology for desalination. Membrane distillation differs from other membrane technologies in that the driving force for desalination is the difference in vapour pressure of water across the membrane, rather than total pressure.

In our study, interesting results were obtained for membrane distillation experiments conducted using MF modified clay membrane, validating the ability of these modified ceramic membranes to act as membrane contactors for desalination. An important influence of the feed temperature and NaCl concentration on the permeate flux was observed. In the same time, high salt rejection rates were obtained in AGMD process with grafted clay ceramic membranes.

The membranes for MD are hydrophobic, which allows water vapour (not liquid water) to pass. The vapour pressure gradient is created by heating the source water, there by elevating its vapour pressure. The major energy requirement is for low-grade thermal energy. It is expected that the total costs for drinking water with membrane distillation, depending on the source of the thermal energy required for the evaporation of water through the membrane. Solar energy could very much help this process in our countries which are very sunny resulting in a reduction of energy costs. Thus, membrane distillation could become competitive relative to other processes.

6. References

Alias, J.; Silva, I.; Goñi, I & Gurruchaga, M. (2008). Hydrophobic amylose-based graft copolymers for controlled protein release, *Carbohydrate Polymers* , 74. 31-40.

Bolong, N.; Ismail, A.F.; Salim, M.R.; Rana, D & Matsuura, T. (2009). Development and characterization of novel charged surface modification macromolecule to polyethersulfone hollow fiber membrane with polyvinylpyrrolidone and water, *Journal of Membrane Science.*, 331. 40-49.

Brigatti, M.F.; Galan, E.; Theng, B.K.G. (2006). Chapter 2 Structures and Mineralogy of Clay Minerals, *Developments in clay Science.* 1, 19-86.

Cot, L.; (1998) Inorganic membranes: academic exercise or industrial reality, inorganic membranes, *Proc. Fifth Internal Conference on Inorganic Membranes,* Nagoya, June, 22–26.

Faibish, R.S & Cohen, Y. (2001) Fouling-resistant ceramic-supported polymer membranes for ultrafiltration of oil-in-water microemulsions, *J. Membr. Sci.* 185. 129–143.

Frost, R.L.; Kloprogge,J.T. (2000) *Spectrochim. Acta Part A.*, 56. 2177.

Gavarini, S. (2002) Durabilité chimique et comportement à l'irradiation des verres quaternaires LnYSiAlO (Ln = La ou Ce), matrice potentielle d'immobilisation d'actinides mineurs trivalents., *Thèse de Doctorat.*, Soutenue le 25 novembre.

Glasser, R.H. & Wilkes, G.L. (1989) Solid-state ^{29}Si NMR of Teos-Based Multifunctional Sol-Gel Materials, *Journal of Non-Crystalline Solids.*, 113. 73-87.

Hu, Y.; Wang, M.; Wang, D.; Gao, X & Gao, C. (2008). Feasibility study on surface modification of cation exchange membranes by quaternized chitosan for improving its selectivity, *Journal of Membrane Science,* 319. 5-9.

Khemakhem, S.; Ben Amar, R. & Larbot, A. (2006) Study of performances of ceramic microfiltration membrane from Tunisian clay applied to cuttlefish effluents treatment, *Desalination.*, 200. 307-309.

Khemakhem, S.; Ben Amar, R. & Larbot, A. (2007) Synthesis and characterization of a new inorganic ultrafiltration membrane composed entirely of Tunisian natural illite clay, *Desalination.*, 206. 210-214.

Khemakhem, S.; Larbot, A. & Ben Amar, R. (2009). New ceramic microfiltration membranes from Tunisian natural materials: Application for the cuttlefish effluents treatment, *Ceramics International*. 35. 55-61.

Krajewski, S.R.; Kujawski, W.; Dijoux, F.; Picard, C & Larbot, A. (2004). Grafting of ZrO2 powder and ZrO2 membrane by fluoroalkylsilanes, Colloids Surf., A: Physiochem. *Eng. Asp*. 243. 43–47.

Larbot, A.; Gazagnes, L.; Krajewski, S.; Bukowska, M & Kujawski, W. (2004). Water desalination using ceramic membrane distillation, *Desalination* 168. 367–372.

Mao, R. Le Van.; Rutinduka, E.; Detellier, C.; Gougay, P.; Hascoet, V.; Tavakoliyan, S.; Hoa, S.V & Matsuura, T. (1999). Mechanical and pore characteristics of zeolite composite membrane, *J. Mater. Chem.*, 9. 783–788.

Mohd Norddin, M.N.A.; Ismail, A.F.; Rana, D.; Matsuura, T.; Mustafa, A & Tabe-Mohammadi, A. (2008) Characterization and performance of proton exchange membranes for direct methanol fuel cell: Blending of sulfonated poly(ether ether ketone) with charged surface modifying macromolecule, *Journal of Membrane Science.*, 323. 404-413.

Nishizawa, N.; Nishimura, J.; Saitoh, H.; Fujiki, K & Tsubokawa, N. (2005) Graftig of branched polymers onto nano-sized silica surface: Postgrafting of polymers with pendant isocyanate groups of polymer chain grafted onto nano-sized silica surface, *Progress in Organic Coatings.*, 53. 306-311.

Picard, C.; Larbot, A.; Tronel-Peyroz, E.; & Berjoan, R. (2004) Characterisation of hydrophilic ceramic membranes modified by fluoroalkylsilanes into hydrophobic membranes, *Solid State Sci*. 6. 605–612.

Saffaj, N.; Persin, M.; Younsi, S Alami.; Albizane, A.; Cretin, M. & Larbot, A. (2006). Elaboration and characterization of microfiltration and ultrafiltration membranes deposited on raw support prepared from natural Moroccan clay: Application to filtration of solution containing dyes and salts, *Applied Clay Science*. 31. 110-119.

Schondelmaier, D.; Cramm, S.; Klingeler, R.; Morenzin, J.; Zilkens, Ch & Eberhardt, W. (2002). Orientation and self-assembly of hydrophobic fluoroalkylsilanes, *Langmuir* 18. 6242–6245.

Tomul, F & Balci, S. (2009). Characterization of Al, Cr-pillared clays and CO oxidation, *Applied Clay Science.*, 43. 13-20.

Wang, Guan-Hai. & Zhang, Li-Ming. (2007). Reinforcement in thermal and viscoelastic properties of polystyrene by in-situ incorporation of organophilic montmorillonite Original, Applied Clay Science. 38. 17-22.

Yoshida, W & Cohen, Y. (2003) Topological AFM characterization of graft polymerized silica membranes, *J. Membr. Sci*. 215. 249–264.

Young, S.K. ; Jarrett, W. L. ; Mauritz, K. A. (2002) Nafion®/ORMOSIL nanocomposites via polymer-in situ sol–gel reactions. 1. Probe of ORMOSIL phase nanostructures by ^{29}Si solid-state NMR spectroscopy, *Polymer.*, 43. 2311-2320.

Yu, H.Y.; Liu, L. Q.; Tang, Z. Q.; Yan, M. G.; Gu, J. S. & Wei, X. W. (2008) Mitigated membrane fouling in an SMBR by surface modification, *Journal of Membrane Science.*, 310. 409-417.

Zhao, Yong-Hong.; Wee, Kin-Ho & Bai, Renbi. (2010). Highly hydrophobic and low-protein-fouling polypropylene membrane prepared by surface modification with sulfobetaine-based zwitterionic polymer through a combined surface polymerization method, *Journal of Membrane Science.*, 362. 326-333.

Zdyrko, B.; Swaminatha Iyer, K & Luzinov, I. (2006). Macromolecular anchoring layers for polymer grafting: comparative study, *Polymer.*, 47. 272-279.

10

Fabrication, Structure and Properties of Nanostructured Ceramic Membranes

Ian W. M. Brown[1,2], Jeremy P. Wu[2] and Geoff Smith[2]
[1]*The MacDiarmid Institute for Advanced Materials and Nanotechnology*
[2]*Industrial Research Ltd (IRL), Lower Hutt*
New Zealand

1. Introduction

The fabrication of nanostructured ceramic membranes by anodising pure aluminium in various acidic electrolytes has attracted considerable attention in recent years because the anodisation process enables the preparation of films with well controlled, uniform pores from tens to several hundreds of nanometres in diameter. The basic cell structure containing cylindrical pores has long been known (O'Sullivan & Wood, 1970; Furneaux et al., 1989) and methods for preparing pore arrays in a hexagonal pattern with interpore distances of between 50 to 500 nm have been reported (Masuda & Fukuda, 1995; Masuda et al., 1997, 1998; Li et al., 1998). The preparation of regular pore arrays typically involves electrolytic polishing and multiple anodising steps or even mechanical pre-texturing (Asoh et al., 2001). As-prepared porous anodic alumina (PAA) membranes are amorphous to X-ray diffraction (XRD). Their chemical composition is not stoichiometric Al_2O_3 but incorporates a considerable quantity of anion impurities and hydroxyl groups incorporated from the anodising electrolyte into the alumina structure or bound to the alumina surface (Thompson, 1997).

PAA has proved to be valuable as a template for fabricating many materials, especially for the synthesis of metallic or semiconductor nanometre-scaled wires and particles (Choi et al., 2003; Lombardi et al., 2006; Sander & Tan, 2003). After chemical functionalising, PAA membranes can be utilised for catalytic (Cho et al., 2005) or optical (de Azevedo et al., 2004) purposes. Further, the porous film itself may be employed for filtration (Sano et al., 2003), gas separation (Lira & Patterson, 2002) or as a photonic crystal (Masuda et al., 2000). Because it is a ceramic oxide, PAA has considerable potential in high-temperature applications, although problems can occur when certain types of PAA membranes are heated. For example, phosphoric acid-derived membranes tend to buckle or crack when heated above 700°C, due to high mechanical tension arising from a phosphorus gradient along the pore direction (Brown et al., 2006).

This review details fabrication methodologies for PAA membranes and examines the ceramic chemistry of their physical and chemical structures formed under different anodising conditions. It describes how these structures change in response to thermal treatment, which is important in the context of their use in high temperature applications, such as for gas separation technologies. It also examine two specific applications of PAA membranes: their use as templates for the formation of palladium-based ultra-thin film hydrogen gas separation devices, and as host structures for the fabrication of metal nanowire arrays.

2. Fabrication methodology

The anodisation of aluminium can be undertaken using a range of acid electrolytes, most commonly oxalic, sulphuric and phosphoric acids, whose conduction and electrochemical characteristics lead to voltage control of the interpore distance, the pore diameter of the fabricated ceramic and the thickness of the barrier layer, a concept first introduced by O'Sullivan and Woods (1970) and later elaborated and refined by many others (eg. Li et al., 1998). These relationships broadly conform to the equations: interpore distance ~ 2.8 nm/V, pore diameter ~ 1.29 nm/V and barrier layer thickness ~ 1.04 nm/V although these relationships were developed for a limited set of conditions using H_3PO_4 with 80-120 V applied anodisation voltage. Subsequent work indicates that this is a useful guideline but there are plenty of exceptions to these relationship rules. The schematic in Figure 1 broadly summarises our own experience which indicates a general trend of ~ 2.5 nm/V for interpore distance.

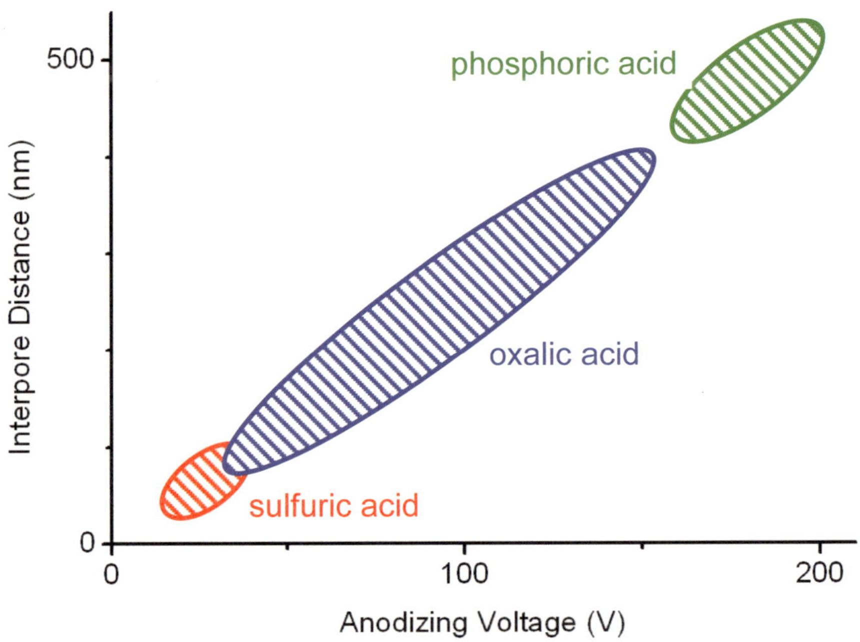

Fig. 1. Schematic relationship between acid electrolyte, voltage and pore structure.

Figure 2 shows a collation of Scanning Electron Microscope (SEM) images illustrating the electrolyte/voltage/pore structure relationship. In this case the four combinations of electrolyte and voltage illustrated show a clear relationship between pore diameter and anodising voltage although this does not conform to the O'Sullivan model, with pore diameter ~ 0.67 nm/V in this case.

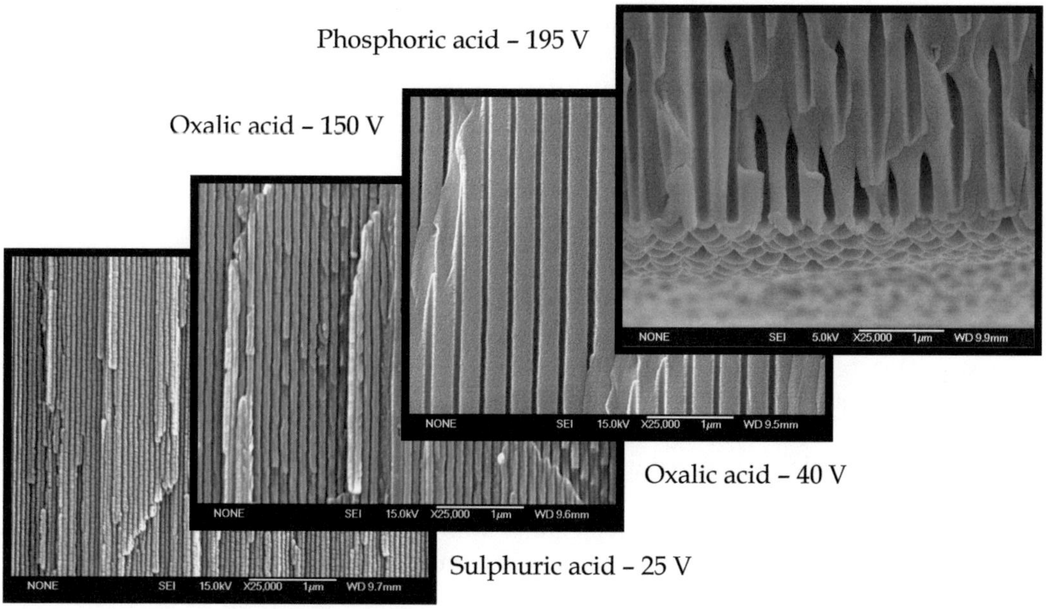

Fig. 2. SEM images showing the relationship between electrolyte, voltage and pore structure.

The selection of acid electrolyte and anodising voltage enables the fabrication of a very wide range of ceramic membrane pore sizes and interpore spacings, all based on the same hexagonal pore topology. This topology is a consequence of self ordering caused by mutual repulsive forces between oxide growth sites in the aluminium metal. Other topologies can be forced by techniques such as nano-indentation patterning of the aluminium foil substrate using tools such as Electron Beam Lithography (EBL) and Focussed Ion Beam (FIB) techniques (Lee et al., 2006a; Robinson et al., 2007). The degree of perfection of pore ordering is usually enhanced by a two step fabrication process. A preliminary anodisation step (either initiated from a nano-indentation site or by self initiation) is terminated early and the resultant film is stripped by acid dissolution, leaving residual etch pits in the metal substrate. These pits form sites which direct the anodic growth during a second anodisation cycle. The electrochemical mechanism of membrane formation has been discussed widely in the literature over many years (see eg. Sulka, 2008) and a simple précis will suffice in this case.

The anodisation process commences with the formation of an ultrathin aluminium oxide barrier layer on the aluminium surface, as electrochemically generated O^{2-} ions from the electrolyte solution ($H_2O = 2H^+ + O^{2-}$) interact with Al^{3+} ions released by the electrochemical reaction $Al = Al^{3+} + 3e^-$. The barrier layer rapidly reaches a limiting thickness (usually a function of the anodising voltage) at which point roughening or unevenness in the barrier layer develops. Few studies have been undertaken to answer the question as to why roughening occurs at the oxide/electrolyte interface, but it is well known that the pores are initiated from this roughening. The electric field is concentrated in the pits in the oxide/electrolyte interface and pores grow by either field assisted plastic flow of the oxide or by field assisted dissolution. There are two processes in balance during the growth of the characteristic collinear pore structure. The formation of new oxide is offset by the dissolution of (older) oxide at a distance from the zone of greatest electric field gradient. This means that the rate of growth slows during anodisation as the distance to the oxide/electrolyte interface increases. In practical terms this imposes a finite limit to the thickness of the PAA layer that can be generated before the dissolution rate balances or exceeds the deposition and growth rate (O'Sullivan & Wood, 1970).

2.1 Electrochemical fabrication

It is common to prepare PAA membranes by direct suspension of shaped aluminium foil pieces in an acid electrolyte. To ensure that anodisation occurs on one side only one Al face is protected with epoxy or nail polish, which is subsequently stripped following the anodisation chemistry. One such methodology is as follows (Kirchner, 2007a) : Annealed high purity (99.99%) aluminium foil of 0.25 mm thickness (Alfa Aesar) was used as a starting material. Pieces 25 mm x 10 mm were degreased in acetone, coated on one side with ADR246 epoxy (Adhesive Technologies) and cured for 2 h at 80 °C. In the case of a sulphuric acid electrolyte the specimens were partially immersed in 0.3 M H_2SO_4 and anodised for at least 12 h at a constant potential of 25.0 V with a platinum cathode. The electrolyte temperature was maintained at 7 °C by means of a stirred double jacketed vessel connected to a thermostat. After anodising, the epoxy layer was removed using dimethylformamide (DMF).

It has been shown to be practical to use a 'clamp on' cell where the PAA dimension is accurately controlled by an O-ring compression fitting such as shown schematically in Figure 3. In this case a typical methodology is as follows (Brown et al., 2010) : Aluminium foil, 0.25 mm thick, 99.995% pure (Alfa Aesar), was cut into 12 mm strips and cleaned by submersing in ethyl acetate and sonicating for 20 minutes. The cleaned foil was rinsed with distilled water and allowed to air dry. Once dry the strips of foil were cut to produce 12 x 12 mm squares. A schematic of the glass electrolytic cell used for anodisation is shown in Figure 3.

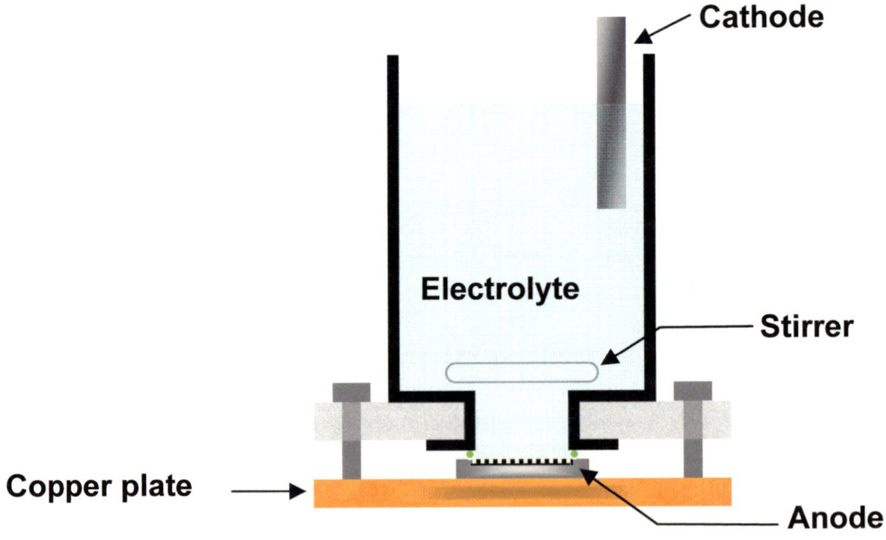

Fig. 3. Schematic of Anodisation Cell.

In this case the square of aluminium foil is placed directly on a chilled copper plate beneath the stirred electrolytic cell and is clamped in place with an O-ring which exposes an 8 mm diameter Al foil surface to the electrolyte. A platinum cathode is used and current is transferred to the aluminium anode via the copper plate. The copper plate effectively withdraws the heat dissipated by the electrochemical processing which could otherwise lead to uneven pore growth leaving the membrane unsuitable for further use. To prevent such damage cooling plate is maintained a constant temperature of 3 °C throughout the anodisation. The design is scalable to larger diameter specimens, conditional upon the need to dissipate process heat, which is both electrolyte and methodology dependent.

PAA membranes can be produced using both so-called 'mild' and 'hard' anodizing techniques (Kirchner et al., 2007a; 2008; Lee et al., 2006b; Brown et al., 2009). For mild anodisation (MA) the voltage is maintained at a minimal level compatible with obtaining alumina growth. For example, with an oxalic acid electrolyte, MA is typically carried out at 40 V. Under these conditions the membrane growth occurs very slowly and it may take a day to prepare a membrane of 100 µm thickness. The same thickness can be obtained using hard anodisation (HA) techniques in about two hours. HA uses a much higher voltage to obtain higher current density and an increased rate of alumina growth. HA using an oxalic acid electrolyte is carried out at 150 V. However if the anodisation is commenced at 150 V the aluminium oxide grows unevenly resulting in an unusable membrane. Successful HA requires formation of a stable barrier layer, which is normally prepared under MA conditions. Once this layer is formed (around 10 minutes) the voltage can be safely increased to that selected for the HA conditions. Figure 4 (left) shows a typical current-voltage-time profile for HA using 0.3 M oxalic acid solution as the electrolyte.

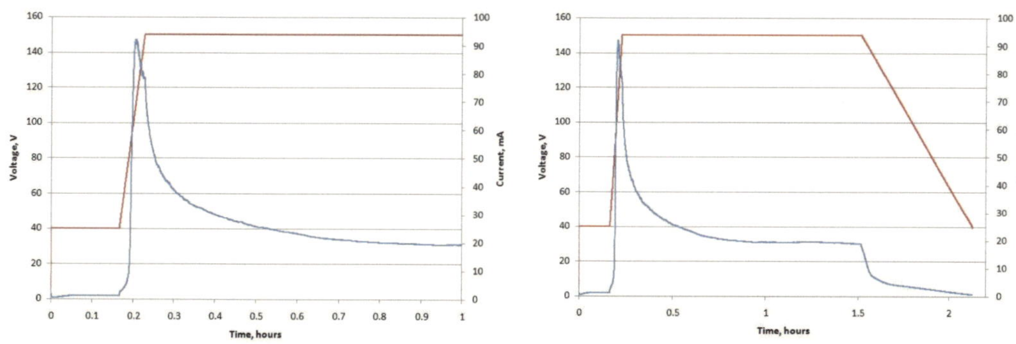

Fig. 4. Current-voltage-time relationships during anodisation, demonstrating progression from mild to hard anodisation conditions (left) and mild-hard-mild conditions (right).

The cell was equilibrated for 20 minutes at 3 °C prior to commencement of anodisation. The voltage was maintained at 40 V for 10 minutes, forming a thin barrier layer upon which the remainder of the membrane grows. The voltage was then increased at 0.5 V/s to 150 V. The thickness of the membrane is dependent on the amount of charge transferred. Higher voltage allows a greater rate of charge transfer; hence less time is required at a higher voltage to obtain the same thickness of PAA. In this case the voltage was held at 150 V until 120 C of charge had been transferred (typically 12 h) after which the voltage was decreased at a rate of 0.05 V/s to 40 V. The 120 C corresponds to a PAA thickness of 120 µm. During anodisation, the electrolyte was stirred at a constant rate of 250 rpm using a magnetic stirrer.

The cell design allowed the fabrication of 8 mm diam. circular PAA discs, which retained their optically transparency even after subsequent detachment, pore widening and thermal procedures (Figure 5).

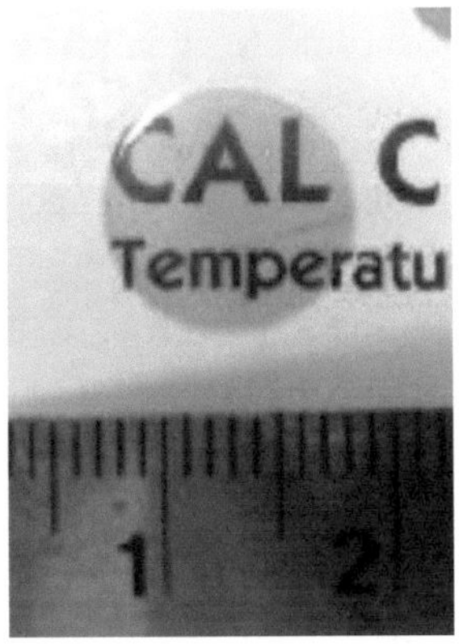

Fig. 5. Transparent PAA template.

Lower voltage (MA) conditions result in slow growth of the membrane, typically a few microns per hour, whereas higher voltage (HA) conditions can lead to growth rates of up to 100 microns per hour. A consequence of this is that conventional pore widening techniques, usually carried out on MA membranes by phosphoric acid etching, can be quite ineffective on the denser and more acid resistant HA substrates (see section 2.3). The development of smart anodisation cycles that incorporate both MA and HA process steps provides a means to mitigate this problem. Figure 4 (right) illustrates how these processes are combined to fabricate membranes in which the final phase of the growth cycle is undertaken using reduced voltage conditions, permitting easier pore opening and widening of the near surface ceramic pore structure.

2.2 Membrane detachment and pore opening

The as-anodized PAA membrane remains attached to the aluminium substrate. To obtain free-standing membranes numerous methods have been reported to dissolve or detach the aluminium. In terms of dissolution methods, the use of $HgCl_2$ solutions, although effective, has generally been superseded by dissolution of the remaining aluminium metal in a saturated iodine-methanol (or ethanol) solution at 50ºC (Kirchner, 2007a). The transparent PAA membranes are rinsed in methanol to clean off excess iodine. PAA fabrication using MA techniques results in relatively slow growth of the alumina template, typically requiring 12-24 h or more to achieve 100 μm thickness. Figure 6 shows MA membranes prepared using conditions of 0.3 M H_2SO_4, 25 V, 5 °C, 12 h. The membranes were detached using iodine-ethanol dissolution of the residual aluminium metal and in this case the iodine etching process took 12 h. Fig. 6 (left) shows the closed pores on the metal contact (basal) side of the film prior to opening by etching in 5% H_3PO_4 for 60 min at 30 °C shown in Fig. 6 (right). During this treatment the opposing face was protected with nail polish, which was subsequently removed with acetone. The hexagonal pore topology is clearly visible in both images.

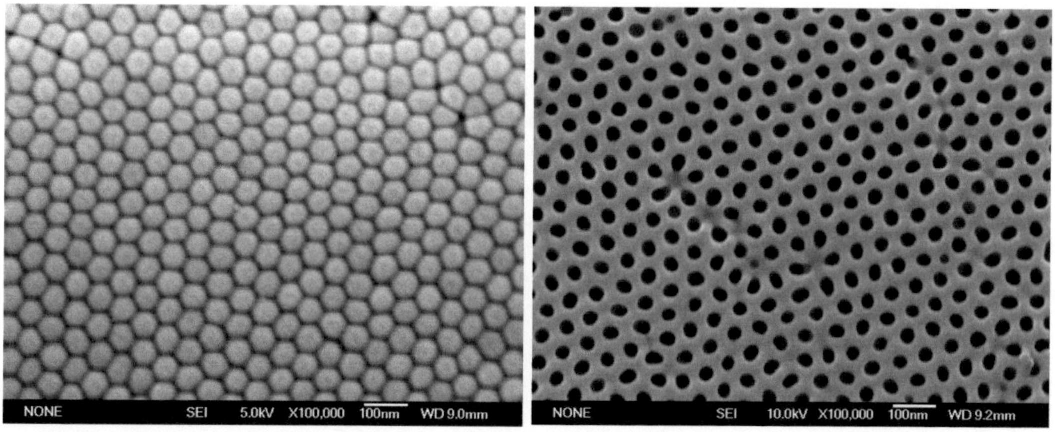

Fig. 6. MA membranes (0.3 M H_2SO_4, 25 V, 5 °C, 12 h). *Left*: basal face revealed by iodine etching. *Right*: Pores opened using 5% H_3PO_4 (60 min, 30 °C).

The attraction of employing HA techniques is the relative speed with which a PAA template can be formed, with growth rates typically 100 $\mu m.h^{-1}$. In this context the 12 h or so required for iodine etching is a significant issue. To address this, anodisation pulse techniques have been developed (Yuan et al., 2006; Chen et al., 2007; Brown et al., 2008) to accelerate removal of the membrane from the Al substrate. The SEM image in Figure 7 shows a HA specimen prepared in 0.3 M oxalic acid at 6°C (40 V 10 min, ramp 0.5 $V.s^{-1}$ to 150 V, hold for 35 min, 90 C charge transferred). Following anodisation the oxalic acid electrolyte was discarded and after rinsing the cell with distilled water, was replaced with a 1:1 $HClO_4$ (60%)/ethanol solution. The membrane was detached by applying a 175 V pulse for 15 s. This method has the advantages of being rapid and not requiring any heavy metals which may contaminate the membrane. After the single voltage pulse the perchloric acid-ethanol mixture was removed and the free-standing membrane was rinsed in distilled water. Membranes may be stored in distilled deionised water until further use.

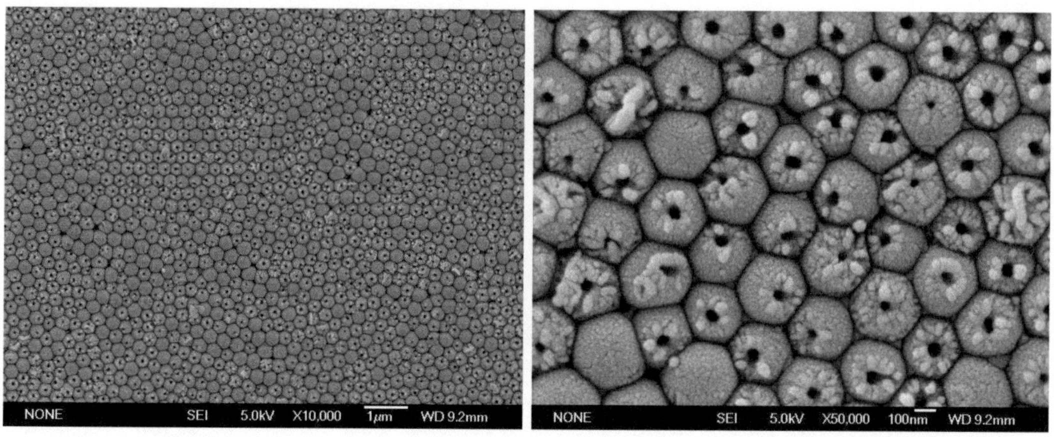

Fig. 7. HA membrane (0.3 M oxalic acid, 150 V) showing pore opening caused by 175 V detachment pulse.

The acid attack on the Al causes immediate detachment of the membrane. The detachment procedure commences the pore opening process, with a high proportion of the pores shown in Fig. 7 featuring small 'pin-hole' openings near the centre of the basal pore cap. The close-up image shows that this process is accompanied by deposition of considerable alumina debris around the pin-hole. Increasing the detachment voltage to 200 V appeared to give an increase in the proportion of opened cells. These HA materials are chemically durable and present a considerable challenge to create a uniform open pore cross section, such as required as a template for a hydrogen separation membrane. Subsequent etching of the pulse detached specimens required 30-60 min exposure to 5% H_3PO_4 at 45 °C. These conditions opened most of the pores but created considerable surface roughness, damage and debris on both the basal and electrolyte sides of the membranes. The iodine removal and acid etch procedure on the less chemically durable MA membranes was more controllable and predictable in its outcome. The result of applying these (much slower) processes to HA membranes was unexpected. Figure 8 shows an HA specimen (0.3 M oxalic acid, 5 °C, 20 mA/180 V), with Al removed in iodine solution, etched in 5wt% H_3PO_4 at 45 °C for 30 min. This shows preferential dissolution at cell junctions with no pore opening at cell centres. To successfully prepare controlled porosity templates using HA processing techniques, a different approach is required in which a low voltage step is introduced at the end of the HA processing (Fig 4 (right)). This effectively creates an MA zone in the vicinity of the barrier layer that is more readily pore-opened.

Fig. 8. HA membrane (0.3 M oxalic acid, 150 V), Al removed in iodine solution, etched in 5 wt% H_3PO_4, showing preferential dissolution at cell junctions.

Following detachment of the PAA from the aluminium substrate the pores of the PAA are closed at one end. Removal of this barrier layer opens the pores to form continuous collinear channels through the ceramic. In addition to removing the barrier layer the pores may also be widened. MA amorphous alumina membranes are susceptible to attack by acids and bases. Relatively mild conditions can be used to etch the MA membrane to remove the barrier layer and widen the pores.

For materials prepared using a 'mild-hard-mild' anodising sequence such as that illustrated in Fig 4 (right), the small 'mild' section of the membrane is more susceptible to attack by acids than the main body of the membrane prepared under 150 V 'hard' conditions. By initiating and closing the anodising process at 40 V (in this case) the barrier layer is thus more susceptible to acid attack than the remainder of the membrane so the pore caps can be removed with little impact on the body of the membrane. Membrane etching was undertaken using a 5 wt% solution of phosphoric acid, placed in a beaker in a water bath maintained at 30 °C. Both the temperature and time of immersion affect the final pore size and structure of the membrane. Immersing the membranes for 60 minutes at 30 °C was determined as the ideal conditions to remove the barrier layer and partially widen the pores without damaging the structure of the membrane. The PAA membranes were rinsed in distilled water and allowed to air-dry.

2.3 Hierarchical pore structures

By integrating various anodisation schemes, a hierarchical pore structure may be formed on a single aluminium substrate. It is possible to increase the pore diameter and spacing, and vice versa during anodisation, but the key is to control the anodisation voltage. The voltage, specific to each acidic electrolyte, determines the pore spacing and the thickness of the barrier layer (O'Sullivan & Wood, 1970). Hence, any abrupt change in anodisation voltage can result in either a barrier layer that is too thick for charge transfer, so the anodisation stops, or a barrier layer that is too thin to handle the large amount of current, so electric-field assisted dissolution processes dominate.

Figure 9 shows SEM images of the cross-section of PAA templates with a hierarchical pore structure (Wu et al., 2010a). Figure 9 (left) shows a smooth transition in pore structure fabricated using a transition from 0.3 M oxalic acid at 40 V-150 V to 0.1 M phosphoric acid at 160 V (forming the bottom part of the image - the barrier layer structure). Fig. 9 (right) illustrates a sudden step change from 195 V (phosphoric acid) to 80 V (oxalic acid) in pore diameter transition in strong contrast to the progressive change in pore structure as shown in Fig. 9 (left). The temperature and concentration of the acidic electrolyte are extremely important in attaining a smooth transition. An integrated process that allows gradual change in electrolyte type, concentration, and temperature while the anodisation is controlled by a computer-interface is ideal for the fabrication of hierarchical anodic alumina templates.

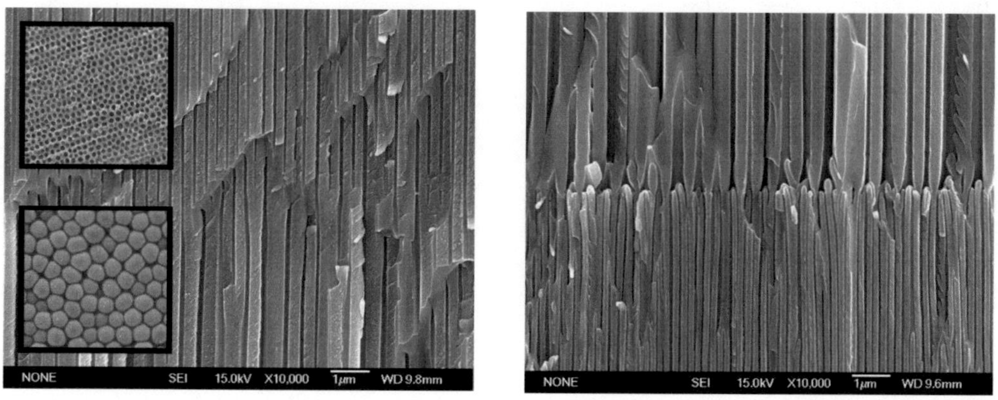

Fig. 9. SEM images of the cross-section of PAA templates with a hierarchical pore structure.

Left. Insets: (top) electrolyte facing side, 0.3 M oxalic at 40 V-150 V; (bottom) barrier layer, 0.1 M phosphoric at 160 V.
Right. Sudden step change from 195 V (phosphoric acid) to 80 V (oxalic acid).

3. Thermal and structural transformations of porous anodic alumina

Understanding the thermal transformation behaviour of PAA is essential if it is desired to use the PAA materials for high temperature applications such as templates for gas separation devices operating above 300 °C. The principal research tools used to characterise the materials are SEM, XRD, Thermal analysis (DTA/DSC/TGA), Evolved Gas Analysis by Mass Spectrometry and Solid State Nuclear Magnetic Resonance (NMR). XRD and thermal analysis techniques have been used to characterise crystalline phase transitions in PAA prepared in phosphoric acid (Wang et al., 2004), oxalic acid (Mardilovich et al., 1995) and sulphuric acid (Ozao et al., 2001) at temperatures up to 600 °C, 1200 °C, and 1250 °C respectively. Irrespective of the acid electrolyte used during their preparation, the as-formed PAA membranes are X-ray amorphous. During heat treatment, PAA crystallises initially as a metastable transition-alumina phase, finally transforming to the stable hcp α-phase (corundum) at higher temperature.

Magic-angle spinning nuclear magnetic resonance (MAS NMR) is a powerful tool to determine short-range structural changes in poorly crystallised or amorphous inorganic environments (MacKenzie & Smith, 2002). ^{27}Al MAS NMR has been used to study the Al-O coordination environment in as-prepared PAA (Farnan et al., 1989; Iijima et al., 2005; Brown et al., 2006). The as-formed materials show distinctive distributions of 4, 5 and 6 coordinated Al^{3+} in an X-ray amorphous structure which is stable to >700 °C. The ^{27}Al MAS NMR spectrum of an MA (H_2SO_4) membrane is shown in Figure 10, showing the contributions from 4, 5 and 6 coordinated species (59.8 ppm, 14%; 30.9 ppm, 57%; 3.2 ppm, 29%, respectively). AlO_5 is an extremely unusual coordination environment, and is very similar to that of amorphous ρ-Al_2O_3 derived from gibbsite ($Al(OH)_3$) heated in vacuo (MacKenzie & Smith, 2002). Typical instrument setup and operating conditions are described by Brown et al., (2006).

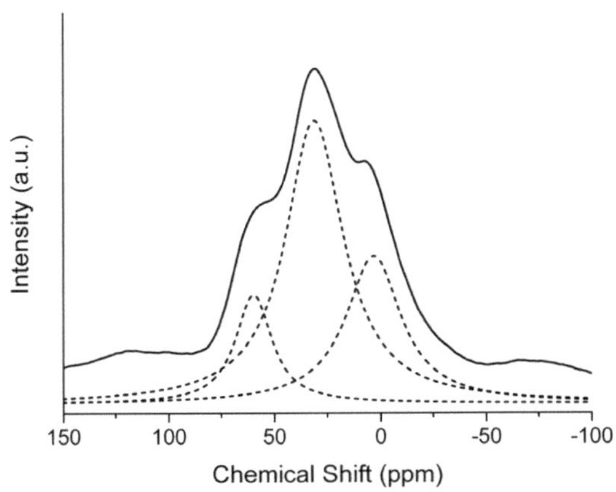

Fig. 10. ^{27}Al MAS NMR spectrum of a heated MA (H_2SO_4) membrane heated to 200 °C, curve resolved.

Previously reported thermal analysis data (Brown et al., 2006; Kirchner et al., 2007a, 2008) shows that further heating causes a sharp exothermic event due to formation of crystalline transition alumina phase(s). This event occurs at 850 °C for H_3PO_4 electrolyte, 900 °C for $H_2C_2O_4$ electrolyte and 970 °C for H_2SO_4 electrolyte, although the precise crystallisation temperature varies with sample heating rate. The thermal and structural characteristics of PAA prepared using these three acid electrolytes are markedly different and are explored in greater detail in the following sections.

3.1 Thermal and structural transformations of phosphoric acid membranes

Phosphoric acid-derived membranes are prepared under high voltage conditions, typically 150-200 V, leading to pore diameters in the 100-200 nm range. Here, we examine the thermal and structural transformations of commercially prepared Whatman Anodisc® alumina membrane filters, fabricated as discs 13 mm diameter x 60 μm thick. Such membranes, whether commercially or laboratory prepared, display a particular characteristic, which is that they buckle or curl into a tubular structure when heated above 850 °C (Brown et al., 2007), thereby limiting (or eliminating) their ability to be used for high temperature applications. In particular, such heating results in a sintering process that leads to closure of the surface pore structure on basal (anode) face of the membrane, shown in Figure 11, although the relics of the original 200nm diameter surface pore structure remain clearly visible.

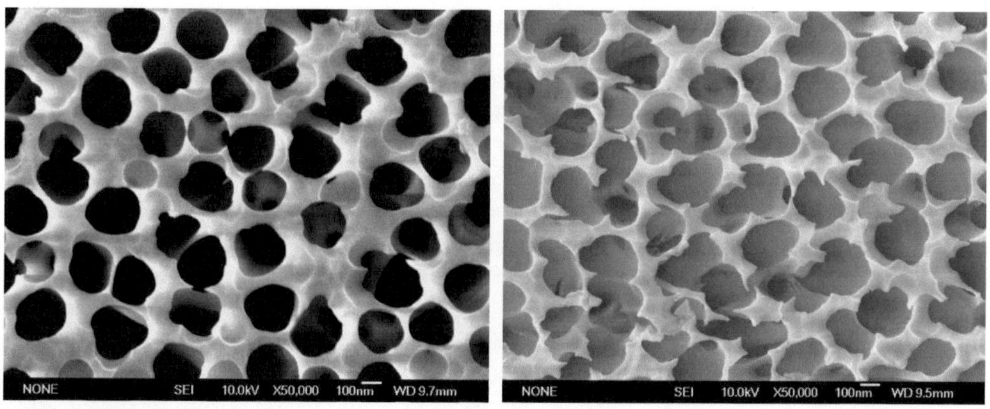

Fig. 11. SEM images of PAA prepared using phosphoric acid electrolyte. (Left) Unheated. (Right) Heated to 1000 °C.

The DSC-TGA data for PAA prepared using phosphoric acid electrolyte is shown in Figure 12. The characteristics are (i) effectively zero weight loss on heating to 1400 °C and (ii) three significant exothermic events at 850 °C, 1020 °C and 1340 °C, respectively. These events form the basis for heating schedules for XRD and NMR examination.

The XRD sequence is reproduced in Figure 13 and shows the progressive, but slow, development of crystalline phases with increasing temperature. The sequence should be viewed in the context of both the DSC data in Figure 12 and the sequence of ^{27}Al MAS NMR spectra shown in Figure 14. The XRD trace is completely amorphous until 790 °C, when the first weak reflections begin to appear. By 824 °C the key peaks of poorly crystallized θ-Al_2O_3 are observed. The short furnace hold time at 824 °C has been sufficient to push the crystallisation through the 850 °C exotherm in Fig.12. The second exotherm at 1020 °C is characterised in the XRD only by the appearance of a strong single peak at d = 4.11 Å, which improves in strength and crystallinity with continued heating. No other XRD peaks appear to be associated with this reflection, which is assigned to the cristobalite form of $AlPO_4$ (ICDD 31-0028).

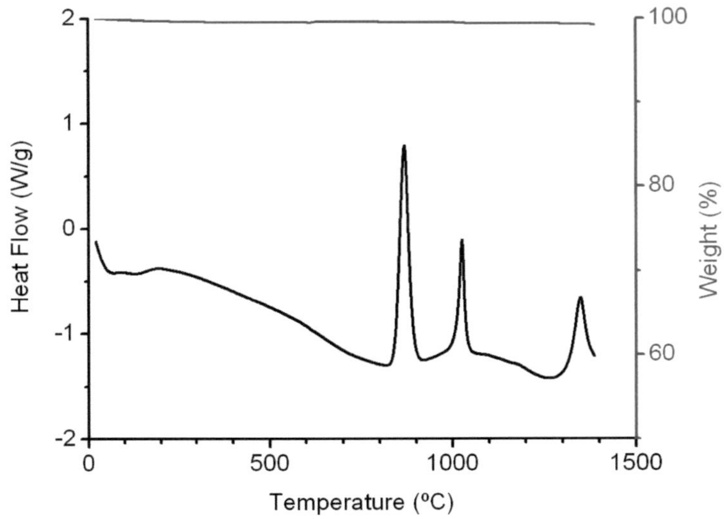

Fig. 12. DSC-TGA data for PAA prepared using phosphoric acid electrolyte.

With increasing heating temperature the first significant change to the NMR spectrum takes place at and above 790 °C with the development of a strong feature at 4–5 ppm, characteristic of 6-coordinated AlO_6 structural units. This is followed at and above 824 °C by a sharp feature at 38–40 ppm, characteristic of well-ordered AlO_4 or AlO_5 structural units. Both structural events occur ahead of or at the very commencement of the first exothermic DSC peak centred at 850 °C and underline the value of NMR in identification of the first signs of short range structural ordering in amorphous reaction systems. The vertical compression of Fig. 14 tends to de-emphasize the spectral features that are apparent in Fig. 10, although the progressive growth and collapse of specific features is evident. The broad peak at 30-32 ppm assigned to AlO_5 begins to decline in relative intensity beyond 925 °C and its elimination appears to be the only significant change in the nature or distribution of ^{27}Al NMR signals that coincide with the second DSC exotherm at 1020 °C. The broad strong peak centred at 63 ppm can be assigned to a range of AlO_4 environments that are bonded through neighbouring AlO_6 structural units (Müller et al., 1986). The interpretation is consistent with the presence of the transition alumina phases such as θ-Al_2O_3 and potentially δ-Al_2O_3 in which aluminium is distributed equally between octahedral and tetrahedral sites. Significantly, this peak grows in parallel with the strong AlO_6 peak centred at 4–5 ppm. Both 63 ppm and 5 ppm peaks are abruptly eliminated by heating above 1295 °C. The final 1340 °C NMR spectrum is dominated by a strong AlO_6 peak at 12 ppm, characteristic of α-Al_2O_3 (corundum) (MacKenzie & Smith, 2002).

The first NMR sign of this corundum peak is evident as a distinct shoulder in the penultimate spectrum at 1295 °C. The origin of the small sharp resonance at 38 ppm requires discussion. The peak develops at 824 °C and becomes increasingly sharp with increasing temperature. The origin of the peak lies in the fabrication methodology of the membrane itself. EDX analytical data (shown in Fig. 15 as a face-to-face phosphorus analysis profile) give an atomic ratio of 2.9 : 97.1 P : Al, confirming that the membrane was manufactured in phosphoric acid and that 'PO_4' structural units have been incorporated into the alumina ceramic.

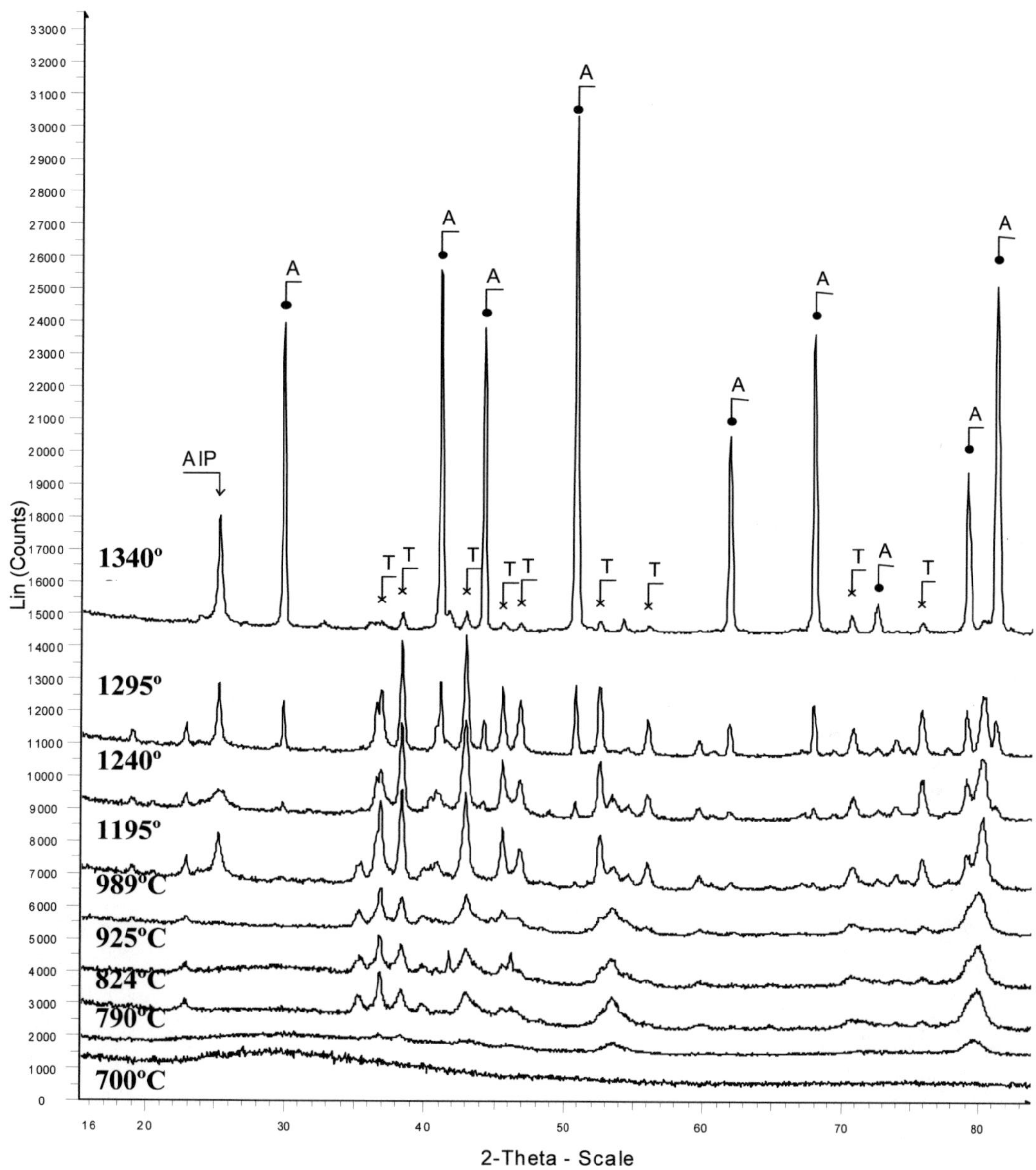

Fig. 13. XRD of heated phosphoric acid-derived PAA. (Key: AlP = $AlPO_4$; A = α-Al_2O_3; T = θ-Al_2O_3. Co Kα radiation).

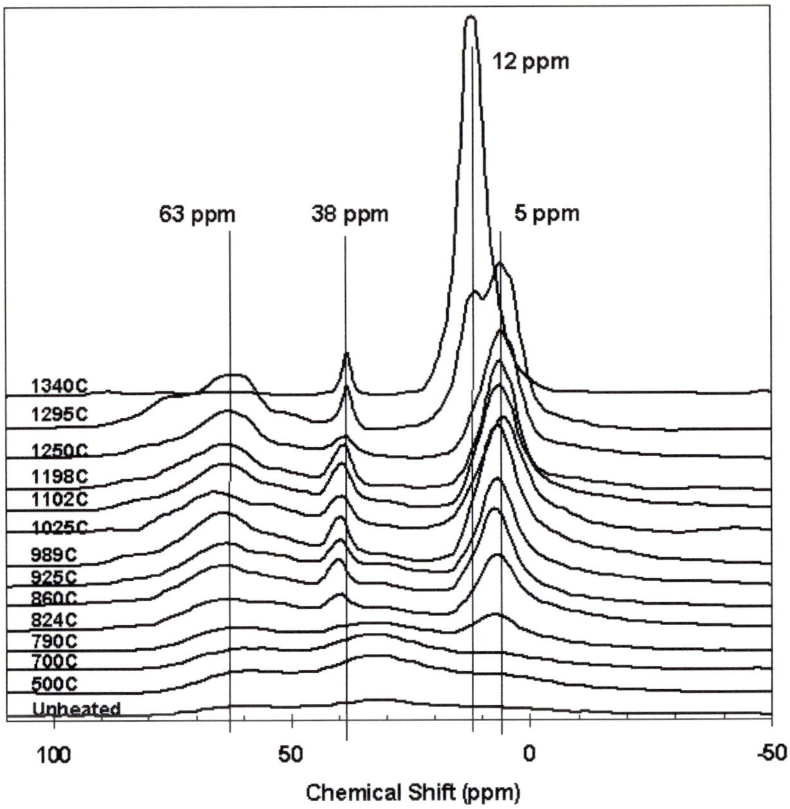

Fig. 14. ^{27}Al MAS NMR spectra of heated phosphoric acid-derived PAA.

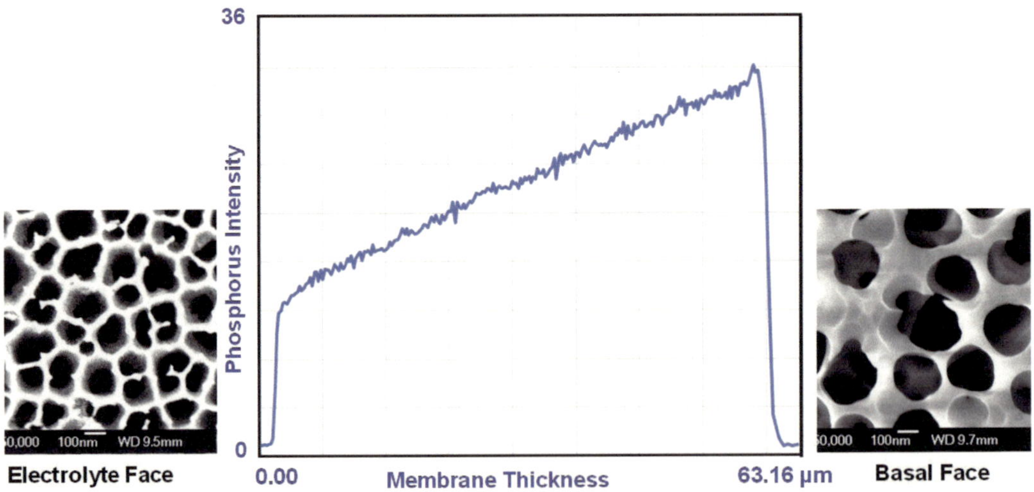

Fig. 15. EDX analysis profile for phosphorus across heated phosphoric acid-derived PAA.

Given this information and the XRD analytical data, the 38 ppm peak can be assigned to tetrahedral Al in a Q4 environment, indicating that all Al is connected through oxygen to P as a next neighbour (Müller et al., 1989). The analysis has been confirmed by ^{31}P NMR examination of both unheated and 1340 °C specimens. Brown et al., (2006) report that the 1340 °C ^{31}P spectrum displays a strong peak at 28.1 ppm, in agreement with previous $AlPO_4$ studies (Müller et al., 1989). The curve fitted area ratio of the AlO_4 ($AlPO_4$): AlO_6 (corundum) NMR peaks at 1340 °C is 2.7 : 97.3, in excellent agreement with the EDX analysis.

The onset of α-Al_2O_3 (corundum) formation is seen in the major XRD reflections by 1195 °C, which, unusually, is ahead of its visibility by NMR. However, the proximity of the strong AlO_6 NMR signal at 5 ppm arising from octahedral θ-Al_2O_3 is certain to have obscured any early observation of corundum crystallisation. By 1340 °C the corundum phase is dominant, accompanied by residual θ-Al_2O_3 and the minor recrystallised $AlPO_4$ phase.

PAA membranes prepared in phosphoric acid, such as commercially produced Whatman Anodiscs®, retain PO_4^{3-} entities in the ceramic structure, forming amorphous $AlPO_4$ which recrystallizes on heating to form the cristobalite structural phase of $AlPO_4$. Not only are the phosphate phases retained in the ceramic on heating to elevated temperatures (above 1300 °C) they are distributed asymmetrically across the thickness of the membrane (Fig. 15) resulting in severe loss of membrane planarity on heating above 700 °C.

3.2 Thermal and structural transformations of sulphuric acid membranes

Thermal analysis and Evolved Gas Analysis traces for sulphuric acid-derived PAA are shown in Figure 16. In marked contrast to phosphoric acid derived membranes, the heated sulphuric membranes display multi-stage weight loss behaviour and only two exothermic events are seen. Heated specimens generally retain their original planarity.

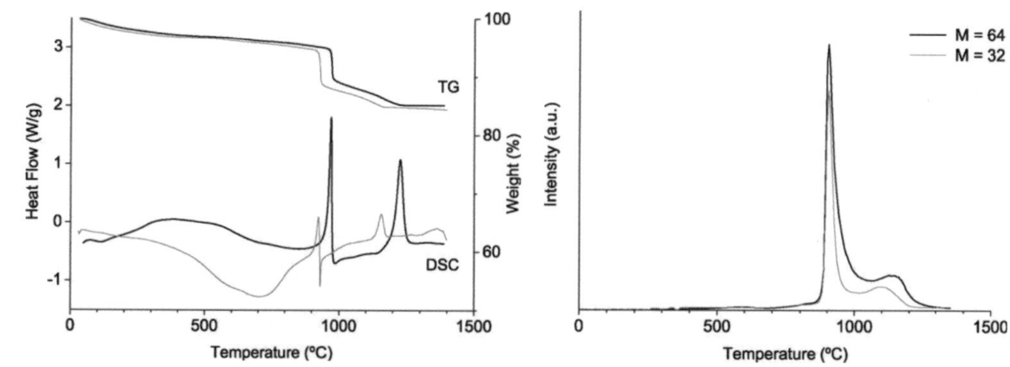

Fig. 16. (Left) DSC (weight-corrected heat flow) and TGA of PAA up to 1400 °C. Black line: scanning rate 20 °C/min, grey line: scanning rate 2 °C/min. The DSC signal of the latter is shown 2.5× for clarity. (Right) MS-Signals for mass numbers 64 (black, SO_2) and 32 (grey, mainly O_2).

The DSC curve determined at 20 °C/min shows a very sharp exothermic peak centred at 970 °C and another slightly broader exotherm centred at 1228 °C. The first exotherm is followed immediately by an endotherm, which is much more marked at low heating rates. These events are irreversible and do not reappear in consecutive heating cycles of the same sample. The thermal analysis data reported by Kirchner et al. (2007a) and Ozao et al. (2001)

are in good agreement. Decreasing the heating rate to 2 °C/min shifted the peak temperatures of the exotherms from 970 °C to 925 °C and from 1228 °C to 1157 °C, respectively, and resulted in a more pronounced endotherm at 930 °C. The heats of transition of the two exotherms were 119 J/g and 151 J/g, respectively. While integration of the high temperature peak is straight-forward, the low temperature result is less reliable due to the partial overlap with the endothermic process. The TGA curve consists of four sections, consisting of a gradual weight loss of 5.0% from ambient to 950 °C, followed by a very pronounced weight loss of 5.7% within the next 30 °C. The temperature of the maximum rate of weight loss was 970 °C, coincident with the first exothermic event. In the third section between 980 °C and 1230 °C the sample gradually lost 4.2 mass%. Above 1230 °C (the temperature of the second exothermic event), the sample weight remained constant.

Mass spectrometry (MS) showed that the strongest signals were from mass numbers 64, 32, and 48. The mass numbers 64, 48, and 66 had the same temperature profile and were found at a ratio of 100 : 55 : 5, consistent with SO_2 (Kirchner et al., 2007a). SO_2 is also expected to give a mass 32 fragment with approximately 10% of the intensity of the mass 64 signal. The mass 32 signal observed here was much larger than this, and follows a noticeably different temperature profile from the mass 64 signal (Fig. 16 (right)). The mass signal 32 was therefore interpreted as being composed predominantly of O_2. Neither gas species was detected at temperatures below 870 °C, but both reached a maximum at 903 °C, with secondary maxima at 1139 °C and 1100 °C for the SO_2 and O_2 signals, respectively. Furthermore, while SO_2 continued to be detected up to 1220 °C, the upper temperature limit for O_2 was significantly lower (1180 °C). No signal for mass number 80, corresponding to SO_3, was detected during the experiment.

The XRD data in Figure 17 show that no crystalline phases were detected in sulphuric acid-derived PAA samples heated to ≤800 °C, comparable with the phosphoric PAA data in Figure 13. Step-wise heating to 900 °C and 1000 °C resulted in the formation of ccp γ-alumina. At 1100 °C, the sample consisted mainly of δ-alumina, with a small amount of α-alumina (corundum). At 1200 °C only α-alumina was observed in the heated PAA.

The ^{27}Al MAS NMR spectra (Figure 18) indicates that in the as-prepared PAA, as well as in specimens heated to ≤800 °C, the aluminium exists in four-, five-, and six-fold coordination, corresponding to the resonances at 55 ppm, 30 ppm, and 7 ppm, respectively (MacKenzie and Smith, 2002). Again, this is comparable to the phosphoric PAA data in Figure 14.

From 900 °C upwards, the resonance corresponding to five-fold coordination disappeared and the peak corresponding to four-fold coordination shifted to 65 ppm. This is characteristic of AlO_4 environments bonded through AlO_6 structural units (Müller et al. 1986) as found in the ccp γ-alumina phase. No differences were found between the NMR spectra of the samples heated at 900 °C and 1000 °C (containing γ-alumina) and that of the 1100 °C sample, shown by XRD to contain principally δ-alumina. There is close structural similarity between the transition-alumina phases γ and δ, both of which contain Al in tetrahedral and octahedral coordination. At 1200 °C only the octahedral resonance at 11 ppm remains, as expected for hcp α-alumina which a published chemical shift of 12.5 ppm (Farnan et al., 1989; Mackenzie and Smith 2002). Direct comparison between XRD, NMR, DSC/TGA, and MS requires some caution since the dynamics of the heating regimes may impact upon the results. For example, the DSC/TGA results illustrate how the transformation temperatures can change with heating rate. The sample preparation for XRD and NMR characterisation experiments involves repetitive heating, with samples quenched once an equilibrium temperature was achieved. This may shift the phase transitions to lower temperatures.

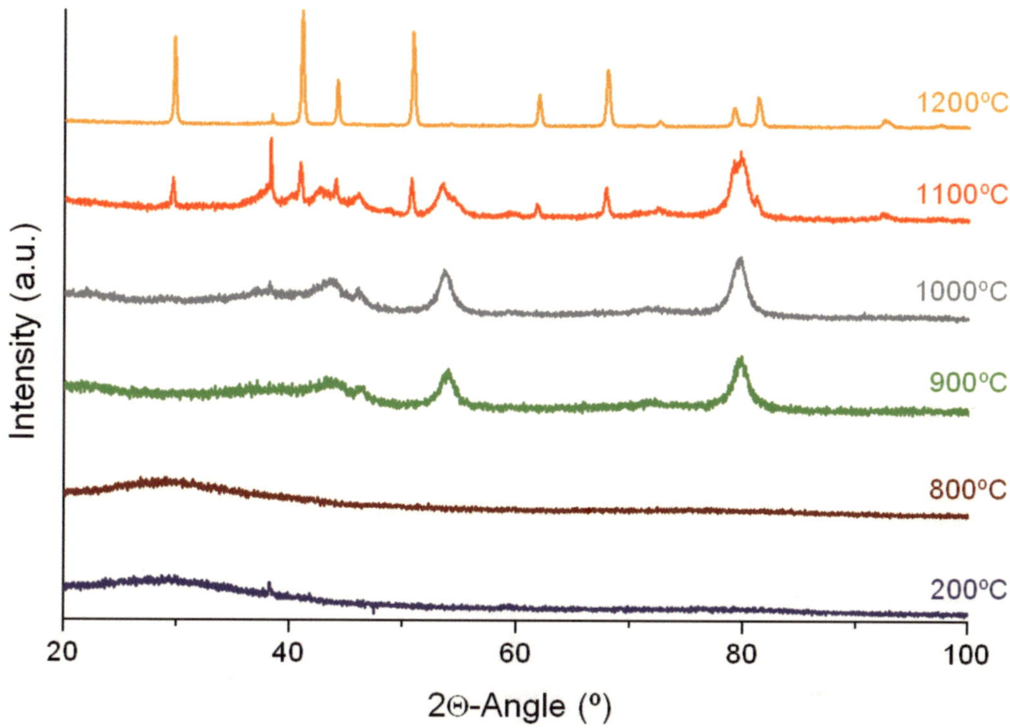

Fig. 17. XRD powder diffraction patterns of sulphuric acid-derived PAA heated to selected temperatures (CoKα radiation).

The complete thermal process can be summarised as follows for a heating rate of 20 °C/min: as-prepared PAA contains bound and structural hydroxyl species which are progressively lost with increasing temperature up to the first exotherm at 970 °C. At this point the amorphous alumina crystallises into γ-alumina. This is linked to a rapid exothermic weight loss caused by decomposition of the sulphate anions with the release of SO_2 and O_2. This is very similar to the thermal behaviour of $Al_2(SO_4)_3 \cdot 16H_2O$ which was measured for comparison (Kirchner et al., 2007a). This material shows an endothermic process associated with a weight loss occurring at 900 °C, with MS detection of both SO_2 and O_2. The release of SO_2 from PAA continues up to 1230 °C. The weight loss between 950 °C and 1230 °C (9.9% of the initial weight) may be attributed solely to the decomposition of SO_4^{2-}. No loss of sulphur species was observed by EDX below 800 °C, indicating an initial sulphur content of 4.3 mass% (allowing for the structural water). This corresponds to a weight loss of 10.7% (as SO_3), in agreement with the observed loss above 950 °C.

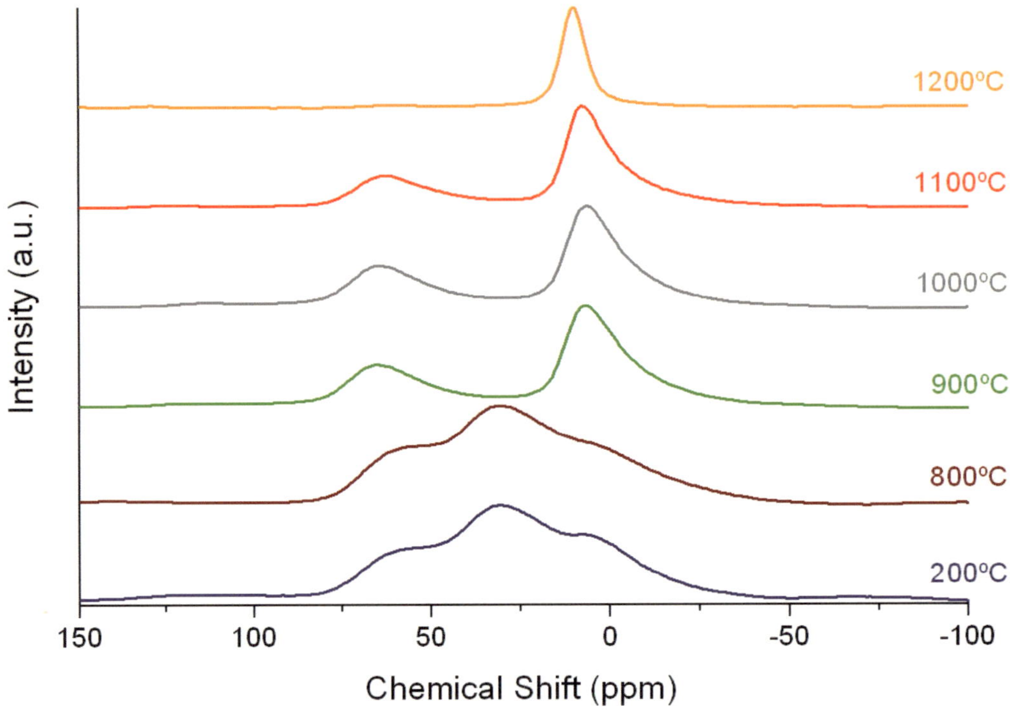

Fig. 18. 11.7 T ^{27}Al MAS NMR spectra of sulphuric acid-derived PAA heated to selected temperatures.

Although the weight loss at the higher temperature can be accounted for by the loss of SO_3, the difference in the temperatures at which O_2 and SO_2 are observed suggests a more complex decomposition mechanism. Since the evolution of O_2 clearly ceases before that of SO_2 the decomposition reaction is believed to be:

$$SO_4^{2-} \rightarrow SO_3^{2-} + \frac{1}{2} O_2$$

The sulphite species remains in the alumina matrix and decomposes subsequently according to the equation:

$$SO_3^{2-} \rightarrow O^{2-} + SO_2$$

At 1228 °C the alumina undergoes a phase transition to the stable corundum phase. The measured heat for this transition (151 J/g) is in reasonable agreement with the theoretical value for the $\delta \rightarrow \alpha$ transition (132 J/g (Chase, 1998)).

The pore architecture observed by SEM following heat treatment is shown in Fig. 19. On heating sulphuric acid-derived PAA membranes to 800 °C, the pore shape and size remain virtually unchanged. No overall shrinkage was observed at temperatures up to 1000 °C despite the weight loss and crystallisation processes. The basal face (Fig. 19, right) retains its open pore structure, in contrast to the pore closure observed in phosphoric acid membranes (Fig. 11, right).

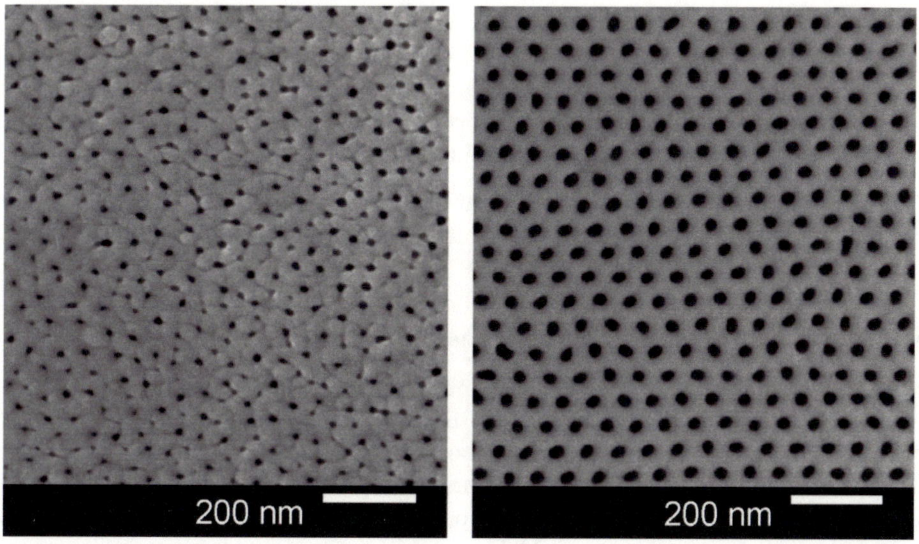

Fig. 19. SEM images of upper (left) and basal faces (right) of a sulphuric acid-derived PAA membrane heated to 800 °C.

3.3 Thermal and structural transformations of oxalic acid membranes

The thermal decomposition data for HA PAA oxalic membranes is summarised in Figure 20, which shows evolved gas measurement data superimposed onto the thermal analysis trace. Comparable with sulphuric acid-derived membranes, but in contrast to phosphoric acid-derived membranes, 6 wt% is lost in a sharp exothermic reaction commencing at about 900 °C. Mass Spectrometry (MS) data shows that this is coincident with the loss of CO_2 from the alumina lattice. A subsequent exothermic event centred at 1230 °C represents the

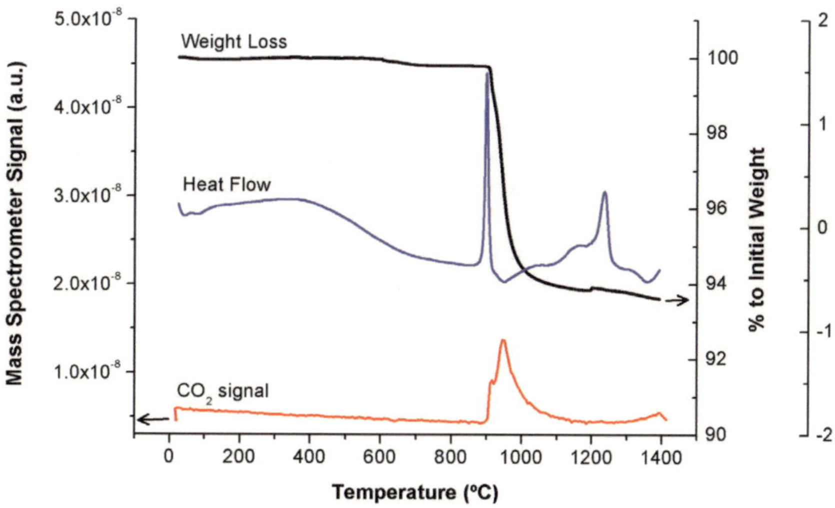

Fig. 20. DSC-TGA data for HA PAA prepared using oxalic acid electrolyte and showing evolved gas Mass Spectrometry data for CO_2.

crystallisation of α-Al_2O_3, showing a very similar event sequence to that observed in sulphuric PAA specimens (section 3.2). The MS trace for mass 44 (CO_2) shows the possibility of more than one crystallographic site for CO_2 decomposition but the absence of events due to mass 32 (O_2) suggest a simple, single step decomposition of carbonate ions to carbon dioxide plus retained oxide ions in the alumina matrix, namely $CO_3^{2-} \rightarrow CO_2 + O^{2-}$.

The XRD and ^{27}Al MAS NMR data for heated oxalic acid derived PAA membranes display very similar characteristics to those of the heated sulphuric membranes shown in section 3.2 and are not reproduced here.

3.4 Summary of thermal and structural transformations of PAA membranes

It is important to understand the thermally induced structural changes undergone by PAA membranes if these materials are to be used in elevated temperature applications, such as for hydrogen separation devices operating above 300 °C. Irrespective of the acid electrolyte used, all ceramic membranes are X-ray amorphous when first fabricated. They display a unique short range ordering, with a high proportion of 5 coordinated Al^{3+} in their structure, best revealed and monitored using ^{27}Al MAS NMR techniques. Further, relics of the acid electrolyte used during the anodisation process become chemically bound and remain stable within the alumina structure to temperatures in excess of 850 °C, when their release may be monitored using online Mass Spectrometric gas analysis techniques. The combination of NMR, XRD, MS and Thermal Analysis enables knowledge of a complete picture of the structure and thermal transformations of the membrane from totally amorphous through intermediate transitional alumina structures to a fully crystallized α-Al_2O_3 membrane above 1200 °C.

PAA prepared using phosphoric acid electrolyte shows no weight loss associated with any of the three exothermic events observed on heating through to 1400 °C but the material distorts badly due to a gradient of PO_4^{3-} species across the thickness of the membrane, which closes off the pores on the basal face. For oxalic acid electrolyte ~6.5 wt% CO_2 is evolved simultaneously with the 900 °C exotherm, but a planar body and optical transparency are maintained. For H_2SO_4 electrolyte the structure takes up high levels of bound SO_4^{2-} and OH^- structural units, with 4.5 wt% S incorporated uniformly across the thickness of the membrane. OH^- species are lost from the matrix continuously with heating up to the 970 °C exotherm at which point rapid weight loss occurs with the release of SO_2 and O_2 and the formation of γ-alumina. Similar to oxalic acid, a planar body and optical transparency are maintained above 1000 °C. SO_2 release continues up to the 1230 °C exotherm that marks the formation of corundum. The 9.9% weight loss observed between 950 °C and 1230 °C is attributed solely to the decomposition of bound SO_4^{2-}.

4. PAA Ceramics as templates for hydrogen separation filters

The uptake of sustainable hydrogen energy technologies is dependent on many technical, commercial and political factors, but a dominant technology issue is the provision of suitable quantities of high purity hydrogen. Low temperature fuel cells in particular are very sensitive to trace levels of CO, SO_2, H_2S and other impurities in the gas supply, which cumulatively damage and deactivate the electrode catalysts and reduce fuel cell efficiency (Baschuk & Li, 2003; Uribe et al., 2004). The nanometre scaled collinear pores in these PAA ceramic templates provide a means to differentiate gas species in their own right by means of Knudsen controlled diffusion processes (Roy et al., 2003). Expressed simply, if the nanopore diameter is smaller than the mean free path of the gas molecules at the temperature of interest then gas diffusion is controlled by the collisions between the gas molecules and the pore wall rather than by collisions between the gas molecules themselves. Knudsen's relationship states that the gas flux through the pore is inversely proportional to square root of the molecular weight of the gas. Hence, light gases such as H_2 and He will have a much higher flux than N_2, Ar, CO_2 etc. A gas separation system using this concept alone is possible but would behave more like a nanoscaled gas chromatograph than a gas filtration and purification device.

However, membrane filters do provide a means to achieve very high purity hydrogen gas, especially through the use of palladium or palladium alloy materials that provide hydrogen-specific conduction paths (Flanagan & Sakamoto, 1993; Paglieri, 2006). Hydrogen diffusion through these membrane filters is a multistep process: dihydrogen molecules adsorb onto the surface of the Pd membrane and dissociate into hydrogen atoms, which then diffuse through the interstitial octahedral sites in the fcc Pd lattice and reform as gaseous diatomic hydrogen on the exit face of the palladium membrane. Palladium has a high level of hydrogen permeability while remaining impermeable to other gases, which makes it ideal for the separation of hydrogen from gas mixtures. Difficulties arise because palladium hydride exists in two different phases, α and β, both fcc structured as in the parent metal. When hydrogen is absorbed into Pd below 300 °C the high hydrogen capacity β hydride $PdH_{0.67}$ predominates. Formation of this phase results in a 10% volume expansion causing stress on the Pd metal lattice and resulting in embrittlement and delamination of the material on repeated adsorption-desorption cycling (Goods & Guthrie, 1992; Bhat et al., 2009). However at temperatures above 300 °C the lower hydrogen capacity α hydride phase $PdH_{0.02}$ predominates and the volume expansion upon adsorption of hydrogen is very small. Consequently, Pd films used in the presence of hydrogen are useful membrane materials only at temperatures above 300 °C.

The introduction of other metals to form palladium based alloys has had promising results. In particular doping of the palladium with silver has been shown to improve the stability of the film and increase the solubility of hydrogen. Further, the temperature above which the α palladium hydride occurred was lowered with increasing silver content (Uemiya et al.,1991; Kikuchi & Uemiya, 1991). The hydrogen permeability was optimized when the silver content of the alloy was around 23 wt%. Silver occupies interstitial sites in the palladium lattice and so moderates the lattice expansion and contraction due to hydrogen absorption/desorption.

The fabrication of ultrathin Pd and Pd-Ag films supported by robust and thermally stable nanoscaled ceramic membranes offer specific advantages, including fast hydrogen transport via reduced diffusion path lengths through the metal and significantly reduced costs through reduction in the quantity of precious metal used. This section describes the fabrication of nanoscaled membranes by deposition of ultrathin Pd and Pd-Ag films on durable ceramic templates fabricated using hard anodisation (HA) techniques (Lee et al., 2006; Kirchner et al., 2007b; Brown et al., 2008). Section 4.2 describes methods to control plating to obtain a Pd : Ag ratio of approximately 75 : 25; sections 4.3 and 4.4 describe alloying and thermal cycling of the composite membrane and section 4.5 and demonstrates efficient hydrogen separation using these membranes.

4.1 Heat treatment

To ensure that the membranes could withstand exposure to highly basic plating solutions they were heat treated after pore opening to convert the amorphous alumina to γ-alumina. This was achieved by increasing the temperature at 10 °C / min to 890 °C and holding for 5 minutes before cooling over four hours. γ-alumina is able to withstand the high pH plating solutions without damage to the chemical or physical structure of the membrane. Heat treatment to 890 °C also mitigates the risk of membrane cracking due to thermal expansion mismatch when curing glass seals used for gas test samples at 860 °C. As noted above, most test membranes prepared for use as high temperature gas filters were fabricated using HA techniques to enhance resistance to high pH.

4.2 Metal deposition on PAA membranes

Pure palladium films undergo hydrogen embrittlement when maintained in a hydrogen containing atmosphere at temperatures below ~350 °C. The introduction of alloying metals such as silver has been shown to decrease the degree of hydrogen embrittlement and increase the hydrogen permeability of the coating. Palladium and silver containing films can be fabricated either by co-deposition or by individual deposition. In both cases the resulting film must be heat treated to alloy the silver and palladium.

A number of different methods can be used to plate PAA with metals. The electroless methodology of Brown et al., (2010) described here is easy to apply and more economic than the more typical electroplating methods. Electroless plating involves using a reducing agent to reduce metal ions in solution to the elemental metal. For electroless plating to be efficient and successful the surface of the membrane must be seeded with palladium crystals so that the reduced metal has a growth site (Mardilovich et al., 1998). If this activation is not undertaken the reduced metal will form on any available surface within the container. Activation also ensures rapid growth of the metal coating, decreasing the time required for plating. This method of activation is commonly used to prepare surfaces for electroless plating (Collins & Way, 1993; Mardilovich et al., 1998). The three solutions used for activation are a sensitizing solution consisting of $SnCl_2 \bullet 2H_2O$ (1 g L^{-1}) and HCl (33 %) (1 mL L^{-1}); an activation solution consisting of $PdCl_2$ (0.1 g L^{-1}) and HCl (33 %) (1 mL L^{-1}); and a solution of 0.01M HCl. The sensitizing solution should be freshly made before use, since insoluble Sn(OH)Cl forms over time via the reaction scheme in equation (1). Hydrochloric acid is added to the solution to extend the amount of time for the precipitation to occur by increasing the concentration of the products.

$$SnCl_2 + H_2O \rightarrow Sn(OH)Cl + HCl \quad (1)$$

The membranes are attached to plastic holders using a small amount of double-sided tape, then immersed in solution one for five minutes, then rinsed with distilled water. This sensitization step deposits tin (II) hydroxide particles onto the membrane surface. The sensitized membranes are then immersed in the activation solution for five minutes, during which the tin (II) hydroxide particles reduce the Pd (II) to Pd metal via equation (2). The membrane is then briefly immersed in a dilute acid solution to prevent the hydrolysis of the freshly deposited palladium. The entire sensitization and activation process is repeated six times to obtain a sufficient coating of Pd to seed the electroless plating process.

$$Pd^{2+} + Sn^{2+} \rightarrow Pd^0 + Sn^{4+} \quad (2)$$

An electroless plating method is used for all co-deposition and individual depositions. Electroless plating involves use of a plating solution containing a metal salt and stabilizing agents. A reducing agent is added just prior to plating which causes the metal salt to reduce to the elemental metal. 25 mL of the plating solution is placed in a plastic cup in a water bath heated at 60 °C. Once equilibrated, hydrazine is added to the solution which is stirred briefly before submerging the membrane in the solution. The samples are left for varying times depending on the desired thickness of the coating. The time was varied in the case of metal individual deposition to control the palladium to silver ratio of the coating. After plating has been completed the samples are rinsed in distilled water and allowed to dry. Palladium and silver can be co-deposited onto the PAA membranes using electroless plating solutions containing different ratios of Pd and Ag. Co-deposition has the advantage of interspersing the Pd and Ag particles resulting in increased intermetallic contact and more effective alloying (Cheng & Yeung, 1999). A range of plating solutions were used composed of different Pd:Ag ratios. The metal salts used were $(NH_3)_4Pd(NO_3)_2$ and $AgNO_3$ and the amount used of each was varied to give solutions with different ratios of Pd and Ag whilst maintaining a metal concentration of 10 mM. The overall composition of the solutions and the detailed methodology is not reproduced here (see Brown et al., 2010; Wu et al., 2010b).

Individual deposition of the Pd and Ag was also undertaken using an electroless method. By plating each metal individually the process can be controlled to obtain a range of different metal ratios. Again, the compositions of the plating solutions and the deposition methodology have been detailed elsewhere (Brown et al., 2010; Wu et al., 2010b). The components of the solutions were the same as the co-deposition solutions except for palladium where a chloride salt was used in place of the nitrate. Membranes that were coated with Pd only were used to compare with the bimetallic sample. Samples were also plated with Pd then sequentially with Ag as comparisons for the co-deposited samples.

Activation is used to seed the surface of the PAA membrane with palladium grains to form sites for growth of the palladium coating during electroless plating. The sensitization step deposits tin hydroxide particles on the surface of the PAA. During immersion in the activation solution the tin hydroxide particles reduce the palladium to palladium metal which then deposits on the surface of the membrane. No tin hydroxide residue remains on the surface of the PAA after activation. The surface of an activated PAA membrane is shown in Figure 21.

Co-deposition of Pd and Ag is desirable as it creates a layer of interspersed Pd and Ag grains, which aids alloying. A ratio of Pd:Ag 75:25 was sought based on previous work (Kikuchi & Uemiya, 1991) which reported that an alloy content of 23% Ag had the most selective hydrogen permeability, and the lowest critical temperature. Experiments were carried out using a number of plating solutions with different Pd:Ag ratios. All solutions gave an even coverage of Pd and Ag. EDX analyses carried out on samples plated for 30 minutes in solutions containing Pd:Ag 75:25 and Pd:Ag 50:50 showed that the films had a higher Ag content than that of their parent solution; eg. for the 75:25 solution the composition of the coating after 60 minutes was 63:37. This phenomenon has been previously reported (Cheng & Yeung, 1999) and indicates that Ag is preferentially plated, resulting in Ag enrichment. A Pd : Ag 90 : 10 solution examined using EDX analysis showed that a 90 : 10 film co-deposited for 60 minutes resulted in a Pd : Ag ratio of 73 : 27. This inconsistency prompted comprehensive EDX mapping analyses (Brown et al., 2010) which confirmed that Ag is deposited first in nodes on the PAA surface followed by a slower build-up of Pd in the area around the nodes, indicating that modifying the duration of plating will affect the ratio of Pd to Ag in the film. It was also observed that fresh solutions gave more reproducible results than aged solutions, so all gas test samples were prepared using freshly made solutions. SEM analysis and gas leak testing (in a cell similar to that described in section 4.5) indicated that a minimum of 30 mins. deposition time was required to ensure a uniform and even coat with no pinholes.

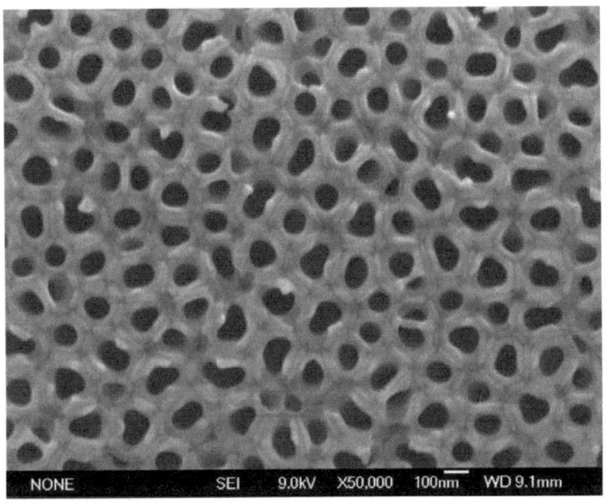

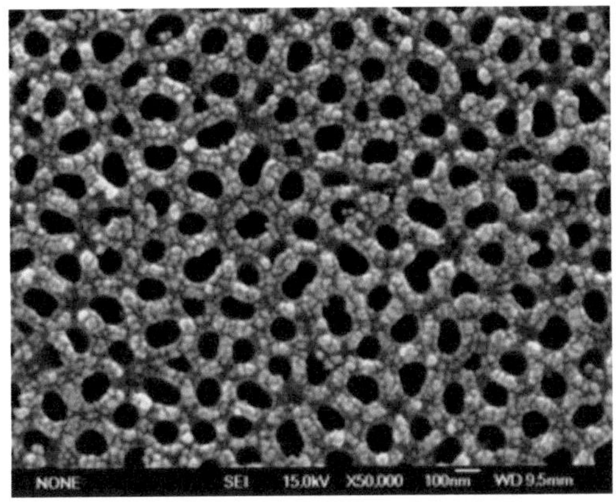

Fig. 21. SEM image of PAA membrane before (left) and after (right) activation. Scale bar is 100 nm.

4.3 Alloying

Freshly plated samples containing both Pd and Ag phases were alloyed to assess the influence of the addition of silver on the gas separation and thermal stability behaviour of the films. The optimised alloying conditions required heat treatment to be undertaken in forming gas (5% H_2 in 95% N_2) at 500 °C for 12 h. The success of the heat treatment was determined by XRD as shown in Fig. 22. The 'as-plated' scan of a pre-alloyed sample (Fig. 22 (left)) shows two discrete phases, with major peaks at approx. 44.6° 2θ and 47° 2θ due to Ag and Pd respectively. The 'post heat treatment' scan shows successful alloying revealed by a single peak at approx. 45.5° 2θ. The position of the alloyed peak can be used as a tool to estimate the relative proportions of Pd and Ag in the alloy, as demonstrated in the XRD patterns in Fig. 22 (right).

The surfaces of two different samples after alloying are shown in the SEM images in Fig. 23. These samples were heat treated under Ar to 550 °C for 12 h. The surface of the membranes show islands of alloyed metal surrounding areas where the metal is no longer present. This indicates that while the metal covered the PAA surface completely after plating, this amount of metal is insufficient to cover the surface after alloying. As a result the gas test samples were plated for multiple and longer times to ensure complete coverage after alloying. The desired coverage was achieved by lengthening the plating time to 90 min. as can be seen in the Pd : Ag 90 : 10 sample (Fig. 23, right).

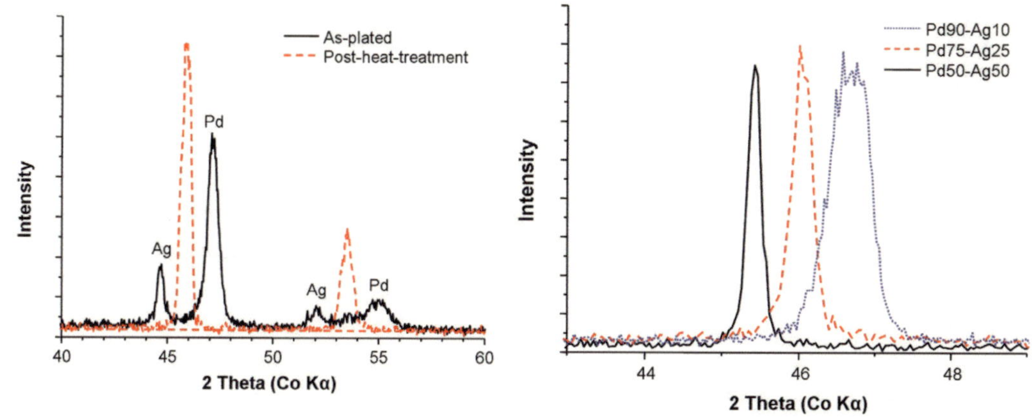

Fig. 22. XRD scans. Left: before and after alloying co-deposited Pd : Ag 50 : 50; Right: Alloyed co-deposited Pd : Ag 50 : 50 (45.4°), 75 : 25 (46.0°) and 90 : 10 (46.7° 2θ).

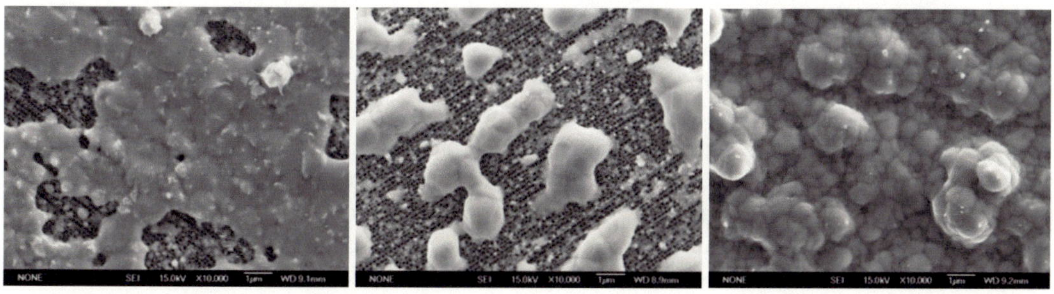

Fig. 23. SEM images of membrane surfaces after alloying: *Left*: Pd : Ag 50 : 50, 30 min; *Centre*: 75 : 25, 30 min; *Right*: 90 : 10, 60 min.

4.4 Thermal cycling

Thermal cycling tests were carried out to determine the relative stability of pure Pd film and Pd : Ag 75 : 25 co-deposited films. The PAA membranes were plated for 1 h, one with Pd only and one co-deposited with Pd : Ag 75 : 25 co-deposition solution. Following deposition, the bimetallic membrane was alloyed by heating to 500 °C for 12 h in forming gas. The Pd only and alloyed membranes were tested simultaneously. The membranes were placed in a tube furnace under 5% H_2 in N_2 gas. After gas equilibration in the chamber the furnace was ramped to 400 °C at 10 °C/min, the sample was held was 400 °C for 10 min. before cooling to 100 °C at 50 °C/h. The program was repeated to give a total of six cycles. The SEM results shown in Fig. 24 indicate a substantial difference between the membranes. The Pd-only membrane developed large cracks across the surface, whilst the bimetallic sample has no obvious defects. Before cyclic testing, the Pd : Ag, 75 : 25 co-deposited membrane had large nodes of silver on the surface (Fig 24, centre). The cyclic test has caused these to become less crystalline and more rounded in appearance. While it is possible that over time this change may cause defects in the membrane, the addition of silver to the membrane has increased the thermal stability and as a result the life of the membrane.

Fig. 24. SEM images of membrane surfaces after thermal cycling. *Left*: Pd only, after cycling; *Centre*: Pd : Ag 75 : 25, before cycling; *Right*: Pd : Ag 75 : 25, after cycling.

4.5 Gas separation devices

The samples used for gas separation require additional preparation. Following heat treatment at 900 °C the PAA membrane was mounted onto a ceramic ring, with a 5 mm central hole to allow the passage of gases. The PAA membrane was mounted over this ring using a glass seal, which was cured by heat treatment at 860 °C via ramping at 10 °C/min, holding at 860 °C for 10 min then cooling over 4 h. The heat treatment to 900 °C was carried out prior to mounting the membrane onto the disc. During this treatment the ceramic was converted from amorphous alumina to γ-alumina. This thermal pre-treatment was necessary to eliminate any ceramic distortion which might otherwise result in cracking at the glass-ceramic interface due to thermal expansion mismatch during treatment at 860 °C. The glass sealant can be used up its softening temperature of ~860 °C. After the membrane was mounted onto the ceramic disc the activation and plating steps were carried out as normal. The ceramic disc was covered in tape during the activation and plating to limit the metal coating to the PAA substrate. Gas separation tests were carried out using a rig shown schematically in Fig. 25.

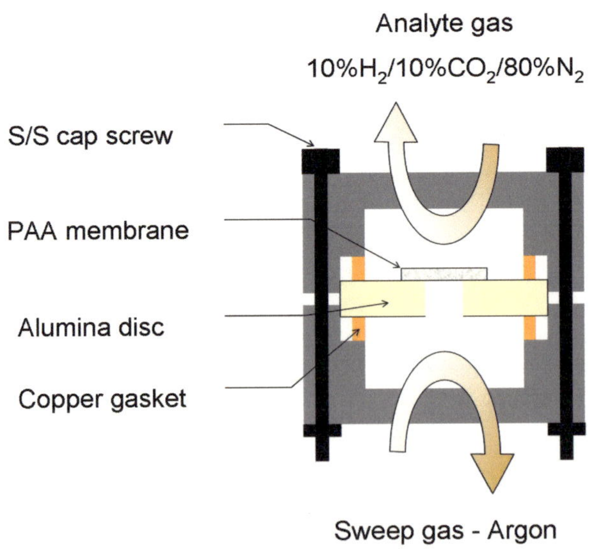

Fig. 25. Schematic of Gas Separation Cell. PAA membrane is 8mm dia. for scale.

An alumina ceramic support disc with the PAA mounted on it was placed between two copper rings inside a stainless steel rig. Copper rings were used because the high malleability of copper allowed the apparatus to be gas tight. Prior to the gas tests the copper rings were softened by heating to 900 °C and were also polished to ensure a good contact surface. The two halves of the stainless steel rig were clamped together using six stainless steel cap screws. The copper rings and stainless steel bolts were replaced before each test. The entire rig was placed inside a tube furnace to provide accurate temperature control.

The analyte gas was a hydrogen-containing mixture consisting of 10% H_2, 10% CO_2 and 80% N_2. The pressure was increased on the analyte side of the membrane using a bubbler. The increased pressure assists the diffusion of hydrogen through the membrane and encourages diffusion in one direction only. An argon sweep gas was used on the exit side of the membrane. After flowing through the cell the sweep gas was analyzed online by mass spectrometry. Prior to the gas separation test the rig was leak tested to ensure integrity of the coating and sealant. Leaks through the membrane and glass seal do occur at a minimal level as indicated by the occasional observation of signals for mass 28 (N_2) and mass 44 (CO_2) in the mass spectrometry results. Gas tests were carried out on co-deposited Pd:Ag 90:10, 80:20 and 75:25, a sequentially deposited sample and a Pd only sample summarised in Table 1.

Plating Method	Ratio Pd:Ag	Plating Time
Co-deposition	90 : 10	2 x 90 min.
Co-deposition	80 : 20	75 min. + 85 min.
Co-deposition	75 : 25	2 x 75 min.
Individual	Pd only	2 x 60 min.
Individual	78 : 22	Pd 75 min. Ag 2 x 20 min.

Table 1. Gas Test Samples.

The gas test samples were put through a thermal cycling procedure. Once the mass spectrometer signals for the various gases had reached equilibrium at room temperature the temperature was increased to 600°C over 24 h. The furnace was then cooled to 50°C over 12 h before heating to 600 °C over 24 h and cooling to 50 °C again. The final cycle was from 50 °C to 700°C. The Pd:Ag 90:10 sample was heated in 50 °C intervals allowing 30 min. at each temperature for the mass spectrometer signals to stabilize before taking a reading. This process was used to follow a similar heating schedule to 500 °C then 550 °C, returning to 50 °C by cooling between the heating cycles. The final ramp from 200 - 700 °C was undertaken using a steady heating rate for 36 h. The temperature schedule for the Pd only membrane was an exception to this program. As noted in the introduction to section 4, Pd undergoes hydrogen embrittlement if heated in a hydrogen containing atmosphere below ~ 300 °C. Hence, the schedule for Pd involved heating to 350 °C under a nitrogen or argon atmosphere then switching to the $10\%H_2/10\%CO_2/N_2$ analyte mixture. The mass spectrometer signals were allowed to settle overnight and the following day the sample was heated between 350 and 550°C over 8 h. The sample was then cooled to 350 °C over 4 h and heated back to 550 °C. The sample was finally cooled to 100 °C still under the hydrogen containing atmosphere.

Gas separation tests involve heating the samples at a constant rate while exposed to a hydrogen-containing gas mix (10% H_2/10% CO_2/80% N_2) and analyzing the composition of the Ar sweep gas using mass spectrometry. H_2 or other gases flowing through the membrane are picked up by the sweep gas and seen by the mass spectrometer. The rig used was not ideal and some analyte gas was able to enter the sweep gas by flowing past the ceramic support disc. As such a slight baseline level for H_2, CO_2 and N_2 is present in the data. The Pd-only results shown in Fig. 26 (left) have no data between room temperature and 350 °C as the membrane was heated under a H_2 free atmosphere in this temperature window to prevent embrittlement. Above 350 °C the H_2 flux (permeance) increases significantly while the N_2 and CO_2 fluxes remain at a minimal level.

The Pd:Ag 90:10 co-deposited specimen was sealed effectively with minimal leaks. Though there is some variability in the data (Fig. 26 (right)), overall the H_2 flux increases with increasing temperature, while the flux of both N_2 and CO_2 remain stable at a minimal level, essentially at the detection limit of the mass spectrometer. Up to ~ 350 °C the β hydride phase dominates and these results show that above 350 °C there is a more rapid increase in H_2 flux through the membrane with accelerated transport through the α phase. This increase in H_2 flux above 350 °C is consistent with the results for the Pd only membrane (Fig. 26 (left)) indicating that increasing the Ag content to ~ 30% (by EDX) has no detrimental effect on the H_2 selectivity.

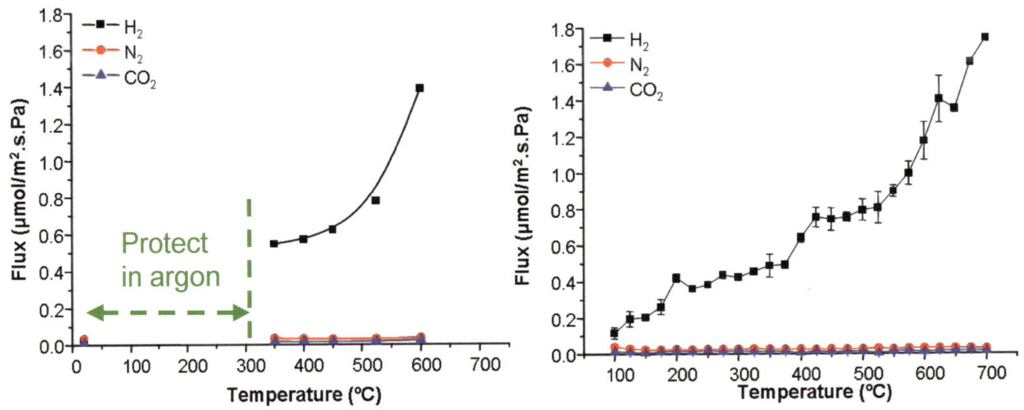

Fig. 26. Mass Spectrometer gas separation results for Pd membrane (left) and Pd : Ag 90 : 10 membrane (right).

Data for the Pd : Ag 80 : 20 and 75 : 25 co-deposited samples (not reproduced here) show that the change in H_2 permeance is much less dramatic than for the 80 : 20 and 75 : 25 samples however the measured permeance does increase steadily with temperature to ~1.0 μmol/m².s.Pa at 700 °C, compared to with ~1.8 μmol/m².s.Pa for the 90 : 10 membrane (Fig. 26 (right)). This is most likely a result of the increased Ag content in the Pd : Ag 80 : 20 and 75 : 25 materials. An ideal Ag content of ~23wt% has been shown to display the highest H_2 selectivity and thermal stability (Kikuchi & Uemiya, 1991). The actual Ag content in both of the 80 : 20 and 75 : 25 films is higher than this 23 wt% ideal and as such the films will be more thermally stable than those with higher Pd content. However, these higher Ag content alloys should also display reduced H_2 permeance, as shown in this data. EDX analysis indicated that the composition of the 90:10 film was actually close to the desired 23wt% Ag content so the H_2 permeance might be expected to increase compared to that of the Pd-only membrane, although the present data at 600 °C (Fig. 26) shows little distinction in flux.

The method used for these gas separation tests involved several cycles of heating and cooling, as previously noted. Each of these membranes shows a degree of thermal stability over the three cycles as no increase in N_2 or CO_2 was observed indicating no damage to the film. The results described here (Wu et al., 2010b) are those for the third (final) cycle of each of the tests and it is important to note that the flux increased considerably between cycles, indicating that greater benefit may well be obtained following more extended thermal cycling.

5. PAA Ceramics as templates for the growth of metal nanowires

Electrodeposition of conductive materials into PAA templates to form nanowire structures is readily undertaken under potentiostatic conditions, where the deposition rate remains constant and the deposited nanowires grow linearly with time. Based on the total measured amount of charge transferred and the approximate total volume of the nanowires, the electroplating process is 100% efficient when compared to the theoretical values. By measuring the charge transferred over a specific deposition time, the theoretical volume of material reduced in a 100% efficient process can be calculated. Thus, the height of the nanowires can be controlled and adjusted precisely. Figure 27 shows nickel metal nanowires electrodeposited in a PAA template.

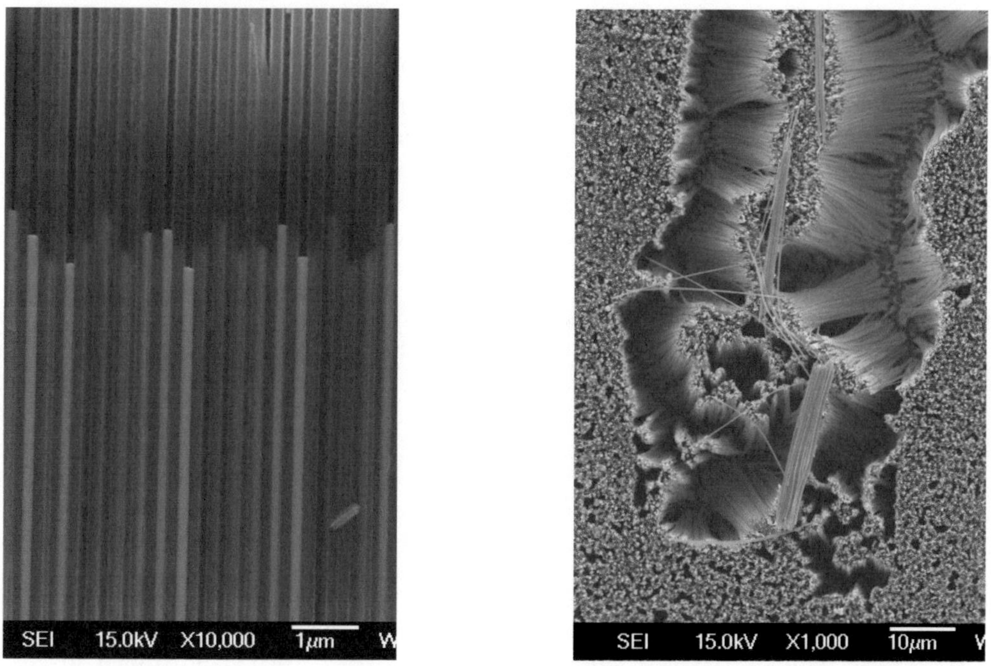

Fig. 27. Left: Nickel metal nanowires supported in their host alumina ceramic matrix, cross section view. Right: Plan view of free standing nickel metal nanowires whose PAA support matrix has been removed by etching.

5.1 Electrodeposition techniques for nanowire growth

Planar templates 120~150 μm thick and 10 mm in diameter were fabricated in 0.3 M oxalic acid (Brown et al., 2009). As described in section 2.2, the freestanding template was obtained by either dissolving the aluminium substrate in 1 M iodine-methanol solution at 50 °C or by voltage pulse detachment in 1:1 perchloric acid / ethanol mix at room temperature (Yuan et al., 2006). The as-anodised template was chemically treated with 5 wt% phosphoric acid at 30 °C for 50 minutes, during which the barrier layer oxide was etched away leaving the template with open-ended pore channels. Electrodeposition was carried out under potentiostatic and galvanostatic conditions using a 3-electrode system. The process was driven by NOVA software provided by Eco Chemie, and the deposition time was controlled precisely to the nearest second. For the working electrode, a 50 nm thick conductive layer (Ag or Au) was sputter-coated on one face of the template. An Ag/AgCl in 3 M KCl reference electrode was used, and the counter electrode was of the same material as the target species in the electroplating bath.

For nickel plating, an aqueous solution containing 300 g/L $NiSO_4.6H_2O$, 45 g/L H_3BO_3, and 45 g/L $NiCl_2.6H_2O$ was used. The deposition was carried out at -0.9 V potentiostatically at ambient temperature. For silver plating, the bath contained 17 g/L $AgNO_3$, 5.85 g/L NaCl, and 39.5 g/L $Na_2S_2O_3$. The deposition was carried out at ambient temperature, galvanostatically at 2 mA / cm^2. For copper plating, a solution containing 200 g/L $CuSO_4.5H_2O$ and 25 mL/L conc. H_2SO_4 was used at -0.1 V. The samples were rinsed in distilled water before and after deposition. 5 wt% phosphoric acid was used as the etching solution to remove the anodic alumina template.

5.2 Fabrication of free standing nanowires

Electrodeposition of conductive materials into PAA templates to form nanowire materials is a simple and versatile method. Fig. 28 demonstrates that under potentiostatic conditions, the deposition rate remained constant and the deposited nanowires had a linear increase in length after an extra 4 hours of deposition. Based on the total measured amount of charge transferred and the approximate total volume of these nickel nanowires, the electroplating process was 100% efficient when compared to the theoretical values.

Each template was kept wet prior to electrodeposition to ensure there were no trapped air bubbles to block the pore channels. The charge transferred over a specific deposition time can be used to calculate the theoretical volume of material reduced in a 100% efficient process, enabling control of the nanowire length. The alternative option is to closely monitor the current density (current per area) during deposition. Under potentiostatic conditions, any change in the surface area exposed to deposition will change the current density. After deposition and the removal of the template, the micrograph (Fig. 29) demonstrates how the ordered hexagonal array is translated from the PAA to the nanowires. In this case XRD analysis shows only the peak of the (220) plane for Ni, indicating a preferred growth orientation in the [110] direction.

The concept of template-assisted fabrication of nanowires is further illustrated in Fig. 30 where the pore branching created by hierarchical fabrication techniques (section 2.3) is mirrored by the branched pore structure developed in the metal nanowires, in this case copper nanowires. Provided there is clear passage for the plating solution, it is simple process to design hierarchical nanowire structures using an anodic alumina template (Wu et al, 2010a).

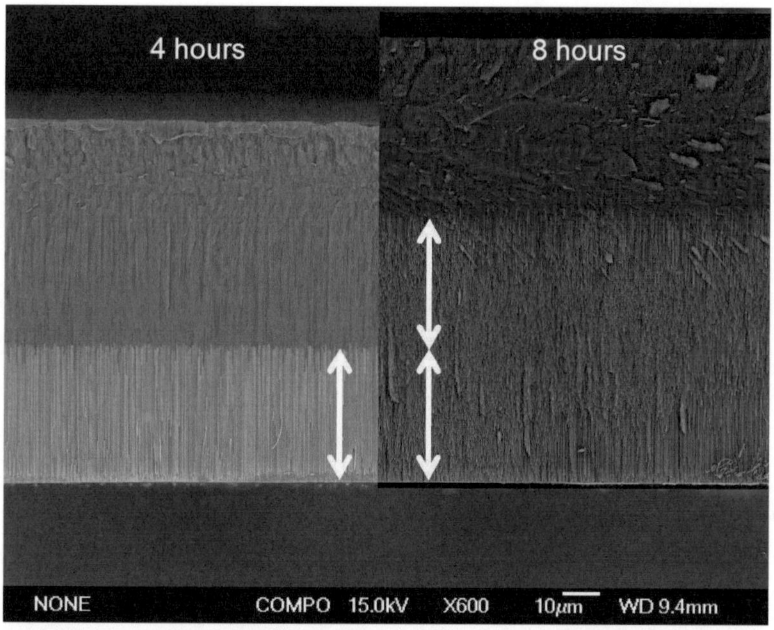

Fig. 28. Electrodeposited nickel in nano-channels of anodic alumina templates after 4 and 8 hours.

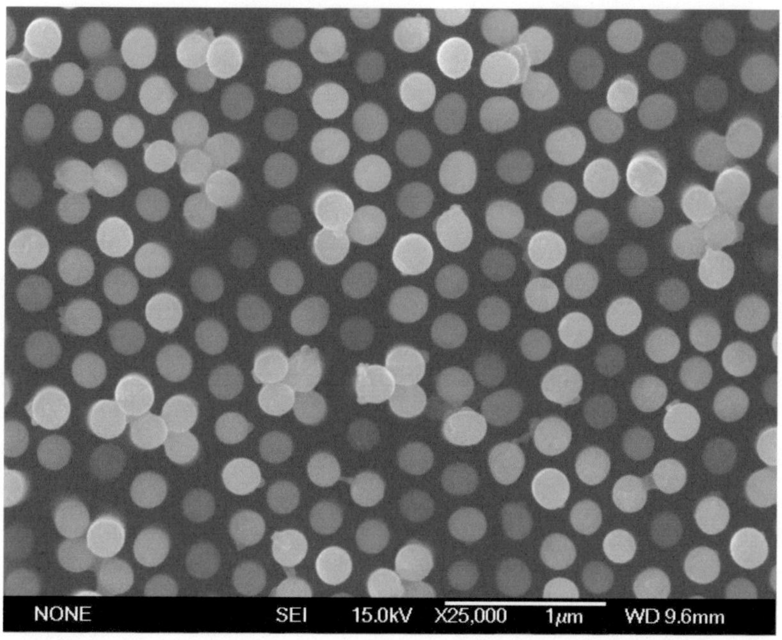

Fig. 29. Plan view of freestanding nickel nanowires (25 μm long, ~100 nm in diameter) after removal of PAA.

Fig. 30. Branched copper nanowires. Inset: branched nano-channels of an etched anodic alumina template, showing copper deposits formed *in situ*.

Segmented nanowires containing a combination of nickel, copper and silver have also been successfully prepared (Wu et al., 2010a) (Fig. 31). The backscattered electron image shows the contrast created by the atomic mass difference between the metals. However, on close examination, the silver deposition appeared to be the most problematic, providing poor coherence and less dense deposition. This was caused by the relatively high current density used during deposition when compared to nickel and copper, indicating that milder electrodeposition conditions are required for electroplating of silver species in these mixed metal systems.

Considerable research is active internationally in metal nanowire growth in PAA templates (Choi et al., 2003; Lombardi et al., 2006; Sander & Tan, 2003) and is well summarised in a recent comprehensive review by Kartopu and Yalçın (2010). As well as metal deposition techniques, other species such as ZnO piezoelectric materials (Wang et al., 2005) and conductive polymers are currently being investigated to make nanowires and nanowire arrays for sensors, actuators, capacitors, and ultrasonic devices by utilising the high aspect ratio, high surface area, wave-guiding ability of these collinear structures (Harris et al., 2007).

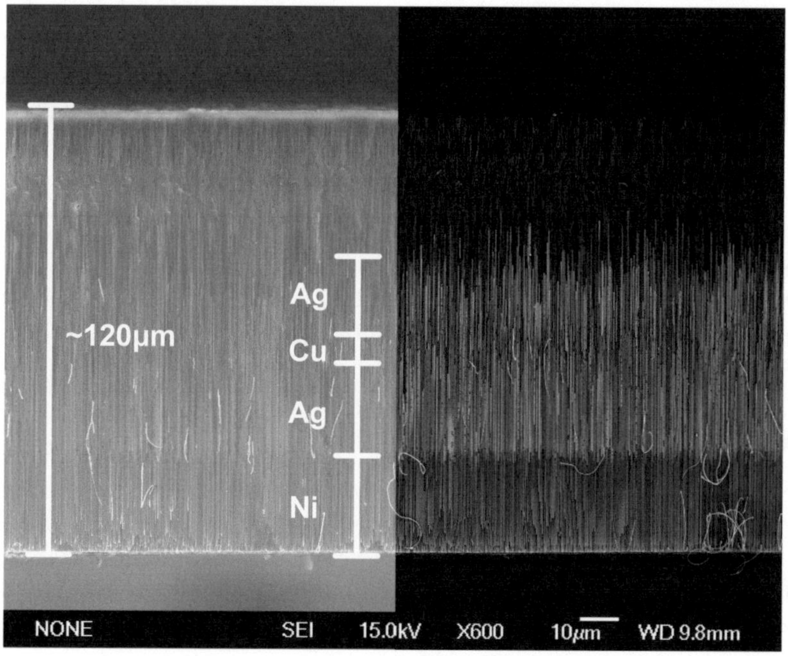

Fig. 31. Ni-Ag-Cu-Ag nanowires electroplated inside PAA membrane, 120 µm thick.

6. Conclusion

Fabrication methodologies for porous anodic alumina (PAA) membranes using phosphoric, sulphuric and oxalic acid electrolytes have been presented. Mild anodisation (MA) techniques employ the minimum voltage conditions compatible with oxide growth with a penalty of long membrane growth times, typically 12-24 h. Hard anodisation (HA) techniques commence with the deposition of an oxide barrier layer formed under MA conditions, prior to ramping up the anodising voltage to the highest sustainable voltage compatible with a stable oxide growth regime and with heat dissipation in the anodisation rig. The growth rates achieved using HA conditions may be up to 100 $nm.h^{-1}$. Both MA and HA membranes may be detached from their host metal substrates by dissolution of the Al metal in Iodine-EtOH solutions (slow) or by application of a brief high voltage pulse, typically 10-20 V higher than the maximum anodisation voltage (fast). The high voltage pulse punctures the closed basal pore caps and the acid environment causes immediate detachment of the membrane from the Al metal. The penalty for HA use is the greatly increased resistance of the ceramic to conventional acidic pore opening treatments, necessitating careful process design to establish MA type (etchable) zones at the basal face of the membrane.

Irrespective of the acid electrolyte used during their formation all PAA structures are X-ray amorphous. Examination of the structural and thermal properties of these membranes shows that they display unique short range ordering, having a mix of 4, 5 and 6 coordinated Al^{3+}, with the approximate proportions AlO_4 14%, AlO_5 57% and AlO_6 29%, respectively. This structure is stable until >700 °C at which point a series of thermal events occurs whose detailed outcome is strongly directly by ionic species (PO_4^{3-}, CO_3^{2}, SO_4^{2-}) incorporated into the alumina matrix from the acid electrolyte. A combination of ^{27}Al SS NMR, XRD, MS and Thermal Analysis enables knowledge of a complete picture of the structure and thermal transformations of these membranes from totally amorphous through intermediate transitional alumina structures to a fully crystallized α-Al_2O_3 membrane above 1200 °C.

Knowledge of these structural changes in response to thermal treatment is an important condition for use of these materials in high temperature applications. In this context the use of PAA membranes as templates for the formation of palladium-based ultra-thin film hydrogen gas separation devices has been described. Finally, a brief introduction is given to the use of PAA host structures for the fabrication of metal nanowires, a technique with the potential to be a key methodology for future electronic device development.

7. Acknowledgments

The authors gratefully acknowledge the key contributions of principal researchers Mark Bowden, Alexander Kirchner, Ken MacKenzie and Tim Kemmitt and the valued contributions of numerous student researchers, including Elly Jay, Melanie Nelson, Naser Al-Mufachi and Jules Carvalho. This research has been supported by the New Zealand Ministry of Science and Innovation and the MacDiarmid Institute for Advanced Materials and Nanotechnology.

8. References

Asoh, H., Nishio, K., Nakao, M., Tamamura, T. & Masuda, H. (2001). Conditions for fabrication of ideally ordered anodic porous alumina using pretextured Al, *J. Electrochem. Soc.* Vol. 148, B152-156.

de Azevedo, W.M., de Carvalho, D.D., de Vasconcelos, E.A. & da Silva Jr., E.F. (2004). Photoluminescence characteristics of rare earth-doped nanoporous aluminium oxide, *Appl. Surf. Sci.* Vol. 234, p.457-461.

Baschuk, J.J. & Li, X. (2003). Modelling CO Poisoning and O_2 Bleeding in a PEM Fuel Cell Anode, *Int. J. Energy Res.* Vol.27, pp1095–1116.

Bhat, V.V., Contescu, C.I. & Gallego, N.C. (2009). The role of destabilization of palladium hydride in the hydrogen uptake of Pd-containing activated carbons, *Nanotechnology* Vol. 20, pp.204011-20 (doi:10.1088/0957-4484/20/20/204011).

Brown, I.W.M., Bowden, M.E., Kemmitt, T. & MacKenzie, K.J.D (2006). Structural and Thermal Characterisation of Nanostructured Alumina Templates *Current Applied Physics* Vol. 6 (3), pp.557-561.

Brown, I.W.M., Bowden, M.E., Kemmitt, T., Kirchner, A. & MacKenzie, K.J.D. (2007). New Ceramic Membrane Materials for Gas Purification. *Advanced Materials Research* Vol. 29-30, pp.15-20.

Brown, I.W.M., Bowden, M.E., Jay, E., Kemmitt, T., Kirchner, A., MacKenzie K.J.D. & Smith, G., (2008). Nanostructured Alumina Ceramic Membranes for Hydrogen Separation, in *Global Roadmap for Ceramics – Proceedings of ICC2*, (ISTEC-CNR,

Institute of Science and Technology for Ceramics - National Research Council, pp.319-328.

Brown, I., Bowden, M., Kemmitt, T., Wu, J & Carvalho, J.(2009) Nanostructured Alumina Ceramic Membranes for Gas Separation. *International Journal of Modern Physics B.* Vol. 23, Nos. 6 & 7. pp.1015-1020.

Brown, I., Wu, J., Nelson, M., Bowden, M. & Kemmitt, T. (2010) Hydrogen Separation Membranes from Nanostructured Alumina Ceramics Pp 1-16 in *"Nanostructured Materials and Systems": Ceramic Transactions of the American Ceramic Society,* Volume 214, Eds Sanjay Mathur & Hao Shen, ISBN: 978-0-470-88128-6, 168 pages, July 2010.

Chase, M. (1998). *NIST-JANAF Thermochemical Tables, 4th ed., J. Phys. Chem. Ref. Data,* Monograph 9.

Chen, W., Wu, J.S., Yuan, J.H., Xia, X.H. & Lin, X.H. (2007). An environment-friendly electrochemical detachment method for porous anodic alumina, *Journal of Electroanalytical Chemistry*, Vol 600, pp.257-264.

Cheng, Y.S. & Yeung, K.L. (1999). Palladium-silver composite membranes by electroless plating technique. *Journal of Membrane Science* Vol.158, pp.127-141.

Cho, S.H., Walther, N.D., Nguyen, S.T. & Hupp, J.T. (2005). Anodic aluminium oxide catalytic membranes for asymmetric epoxidation, *Chem. Commun.* pp.5331-3.

Choi, J., Sauer, G., Nielsch, K., Wehrspohn, R.B. & Gösele, U. (2003). Hexagonally arranged monodisperse silver nanowires with adjustable diameter and high aspect ratio, *Chem. Mater.* Vol.15, pp.776-779.

Collins, J.P. & Way, J.D. (1993). Preparation and Characterization of Palladium-Ceramic Composite Membranes, *Ind. Eng. Chem.*, Vol.32, pp.3006-13.

Farnan, I., Dupree, R., Forty, A.J., Jeong, Y.S., Thompson, G.E. & Wood, G.C. (1989). Structural information about amorphous anodic alumina from 27Al MAS NMR, *Philos. Mag. Lett.* Vol.59, pp189-195.

Flanagan T.B. & Sakamoto, Y. (1993). Hydrogen in Disordered and Ordered Palladium Alloys, *Platinum Metals Rev.*, Vol.37(l), pp.26-37.

Furneaux, R.C., Rigby, W.R. & Davidson, A.P. (1989). The formation of controlled porosity membranes from anodically oxidized aluminium, *Nature* Vol.337 pp.147-149.

Goods, S.H. & Guthrie, S.E. (1992). Mechanical properties of palladium and palladium hydride, *Scripta Metallurgica* Vol.26(4), pp. 561-566.

Harris, P., Dawson, A., Young, R. & Lecarpentier, F. (2007). High frequency propagation in structured solids, in *Proc. 2007 IEEE International Ultrasonics Symposium,* New York, pp. 690-693.

Iijima, T., Kato, S., Ikeda, R., Ohki, S., Kido, G., Tansho, M. & Shimizu, T. (2005). Structure of duplex oxide layer in porous alumina studied by 27Al MAS and MQMAS NMR, *Chem. Lett.* Vol.34, pp1286-1287.

Kartopu, G. & Yalçın, O. (2010). Fabrication and Applications of Metal Nanowire Arrays Electrodeposited in Ordered Porous Templates. pp 113-140 in *Electrodeposited Nanowires and Their Applications,* Book edited by: Nicoleta Lupu. ISBN 978-953-7619-88-6, pp. 228, February 2010, INTECH, Croatia.

Kikuchi, E. & Uemiya, S. (1991). Preparation of supported thin palladium-silver alloy membranes and their characteristics for hydrogen separation, *Gas Separation & Purification*, Vol.5, pp.261-266.

Kirchner, A., MacKenzie, K.J.D., Brown, I.W.M., Kemmitt, T. & Bowden, M.E. (2007a). Structural characterisation of heat-treated anodic alumina membranes prepared using a simplified fabrication process, *Journal of Membrane Science* Vol. 287(2), pp.264-270.

Kirchner, A., Brown, I.W.M., Bowden, M.E., & Kemmitt, T. (2007b). Hydrogen Purification using Ultra-thin Palladium Films supported on Porous Anodic Alumina Membranes, in *Functional Nanoscale Ceramics for Energy Systems*, ed. E. Ivers-Tiffee and S. Barnett (Mater. Res. Soc. Symp. Proc. 1023E, Warrendale, PA, 2007), Paper 1023-JJ09-02.

Kirchner, A., Brown, I.W.M., Bowden, M.E., Kemmitt, T. & Smith, G. (2008). Preparation and high-temperature characterisation of nanostructured alumina ceramic membranes for high value gas purification, *Current Applied Physics* Vol. 8, pp.451-454.

Lee, W., Ji, R., Ross, C.A., Gösele, U., &. Nielsch, K. (2006a). Wafer-scale Ni imprint stamps for porous alumina membranes based on interference lithography. *Small*, Vol. 2, pp.978-982.

Lee, W., Ji, R., Gösele, U. & Nielsch, K. (2006b). Fast fabrication of long-range ordered porous alumina membranes by hard anodization. *Nature Materials* Vol. 5 (9) , pp 741-747.

Li, A.P., Müller, F., Birner, A. & Gösele, U. (1998). Hexagonal pore arrays with a 50–420 nm interpore distance formed by self-organization in anodic alumina, *J. Appl. Phys.* Vol. 84, pp.6023-6026.

Lira, H. de L. & Paterson, R. (2002). New and modified anodic alumina membranes. Part III. Preparation and characterisation by gas diffusion of 5 nm pore size anodic alumina membranes, *J. Membrane Sci.*, Vol. 206, pp.375-387.

Lombardi, I., Magagnin, L., Cavalotti, P.L., Carraro, C. & Maboudian, R. (2006). Electrochemical fabrication of supported Ni nanostructures through transferred porous anodic alumina mask, *Electrochem. Solid State Lett.* Vol. 9, pp.D13-D16.

MacKenzie, K.J.D. & Smith, M.E. (2002). *Multinuclear Solid-State NMR of Inorganic Materials*, Pergamon, Amsterdam. ISBN 0-08-043787-7.

Mardilovich, P.P., Govyadinov, A.N., Mukhurov, N.I., Rzhevskii, A.M. & Paterson, R. (1995). New and modified anodic alumina membranes. Part I. Thermotreatment of anodic alumina membranes, *J. Mem. Sci.* Vol .98, pp.131-142.

Mardilovich, P.P., She, Y. & Ma, Y.H. (1998). Defect-free palladium membranes on porous stainless steel support, *AIChE Journal*, Vol.44(2), pp.310-322.

Masuda, H., & Fukuda, K. (1995) Ordered metal nanohole arrays made by a two-step replication of honeycomb structures of anodic alumina, *Science* Vol. 268, pp.1466-1468.

Masuda, H., Hasegawa, F. & Ono, S. (1997). Self-ordering of cell arrangement of anodic porous alumina formed in sulphuric acid solution, J. Electrochem. Soc. Vol. 144, pp.L127-L129.

Masuda, H., Yada, K., & Osaka, A. (1998). Self-ordering of cell configuration of anodic porous alumina with large-size pores in phosphoric acid solutions, *Jpn. J.Appl. Phys.* Vol. 37, pp.L1340-L1342.

Masuda, H., Ohya, M., Nishio, K., Asoh, H., Nakao, M., Nohtomi, M., Yokoo, A. and Tamamura, T. (2000). Photonic band gap in anodic porous alumina with extremely high aspect ratio formed in phosphoric acid solution, *Jpn. J.Appl. Phys.* Vol.39, pp.L1039-L1041.

Müller, D., Gessner, W., Samosan, A., Lippmaa, E. & Scheler, G. (1986). *J.Chem. Soc. Dalton Trans.* Vol.1986, pp1277-1281.

Müller, D., Jahn, E., Ladwig, G. & Haubenreisser, U. (1984). *Chem. Phys. Lett.* Vol.109, pp.332-336.

O'Sullivan, J.P. & Wood, G.C. (1970). The morphology and mechanism of formation of porous anodic films on aluminium, *Proc. Roy. Soc. London A* Vol. 317 pp.511-543.

Ozao, R., Ochiai, M., Yoshida, H., Ichimura, Y. & Inada, T. (2001). Preparation of gamma alumina membranes from sulphuric electrolyte anodic alumina and its transition to alpha alumina, *J. Thermal Anal. Calorim.* Vol. 64, pp.923-932.

Paglieri, S.N. (2006). Palladium Membranes, in Nonporous Inorganic Membranes. Published Online: 3 Aug 2006, Eds. A.F. Sammells, M.V. Mundschau, Print ISBN: 9783527313426, Online ISBN: 9783527608799, pp77-105.

Robinson, A.P, Burnell, G., Hu, M. & MacManus-Driscoll, J.L. (2007). Controlled, perfect ordering in ultrathin anodic aluminum oxide templates on silicon. *Applied Physics Letters* Vol.91, pp.143123(3).

Roy, S., Raju, R., Chuang, H.F., Cruden, B.A. & and Meyyappan, M. (2003). Modeling gas flow through microchannels and nanopores, in *J. Appl. Phys.*, Vol. 93, pp.4870-4879.

Sander, M.S. & Tan, L.S. (2003). Nanoparticle arrays on surfaces fabricated using anodic alumina films as templates, *Adv. Funct. Mater.* Vol.13, pp.393-397.

Sano, T., Iguchi, N., Iida, K., Sakamoto, T., Baba, M. & Kawaura, H. (2003). Size-exclusion chromatography using self-organized nanopores in anodic porous alumina, *Appl. Phys. Lett.* Vol. 83, pp.4438-4440.

Sulka, G.D., (2008) Highly Ordered Anodic Porous Alumina Formation by Self-Organising Anodising. Chapter 1 in *Nanostructured Materials in Electrochemistry*, Ed A. Eftekhari, Publ. Wiley-VCH Verlag GmbH & Co., Weinhein, Germany. (2008). ISBN: 978-3-527-31876-6.

Thompson, G.E. (1997). Porous anodic alumina: fabrication, characterization and applications, *Thin Solid Films*, Vol. 297, pp.192-201.

Uemiya, S., Matsuda, T. & Kikuchi, E. (1991). Hydrogen permeable palladium-silver alloy membrane supported on porous ceramics, *Journal of Membrane Science*, Vol.56, pp.315-325.

Uribe, F.A., Valerio, J.A., Garzon, F.H. & Zawodzinski, T.A. (2004). PEMFC Reconfigured Anodes for Enhancing CO Tolerance with Air Bleed, *Electrochem. Solid-State Lett.* Vol.7, pp.A376-A379.

Wang, Q., Wang, G., Xu, B., Jie, J., Han, X., Li, G., Li, Q. & Hou, J.G. (2005). Non-aqueous cathodic electrodeposition of large-scale uniform ZnO nanowire arrays embedded in anodic alumina membrane. Materials Letters, Vol. 59 (11) pp.1378-1382.

Wang, Y.C., Leu, I.C. & Hon, M.H. (2004). Dielectric property and structure of anodic alumina template and their effects on the electrophoretic deposition characteristics of ZnO nanowire arrays, *J. Appl. Phys.* Vol. 95, pp1444-1449.

Wu, Jeremy P., Brown, Ian W.M., Kemmitt, Tim, & Bowden, Mark E. (2010a) Hierarchical Anodic Alumina Template-assisted Fabrication of Nanowires. *Proc. of International Conference on Nanoscience and Nanotechnology (ICONN), 2010.* Page(s): 29 - 32 DOI: 10.1109/ICONN.2010.6045163. Publ. IEEE Xplore.

Wu, Jeremy. P., Brown, Ian W.M., Bowden, Mark E. & Kemmitt, Timothy (2010b). Palladium coated porous anodic alumina membranes for gas reforming processes. *Solid State Sciences* Vol 12 pp1912-1916.

Yuan, J.H., Chen, W., Hui, R.J., Hu Y.L. & Xia, X.H. (2006). Mechanism of one-step voltage pulse detachment of porous anodic alumina membranes, *Electrochimica Acta,* Vol. 51, pp.4589–4595.